Artificial Intelligence in Healthcare

Artificial Intelligence in Healthcare

Edited by

ADAM BOHR
CEO and Co-Founder of Sonohaler, Copenhagen, Denmark

KAVEH MEMARZADEH
Commercial Field Application Scientist at ChemoMetec, Lillerød, Denmark

ACADEMIC PRESS

An imprint of Elsevier

Academic Press is an imprint of Elsevier
125 London Wall, London EC2Y 5AS, United Kingdom
525 B Street, Suite 1650, San Diego, CA 92101, United States
50 Hampshire Street, 5th Floor, Cambridge, MA 02139, United States
The Boulevard, Langford Lane, Kidlington, Oxford OX5 1GB, United Kingdom

Notices
Knowledge and best practice in this field are constantly changing. As new research and
experience broaden our understanding, changes in research methods, professional practices,
or medical treatment may become necessary.

Practitioners and researchers must always rely on their own experience and knowledge in
evaluating and using any information, methods, compounds, or experiments described
herein. In using such information or methods they should be mindful of their own safety
and the safety of others, including parties for whom they have a professional responsibility.

To the fullest extent of the law, neither the Publisher nor the authors, contributors, or
editors, assume any liability for any injury and/or damage to persons or property as a matter
of products liability, negligence or otherwise, or from any use or operation of any methods,
products, instructions, or ideas contained in the material herein.

British Library Cataloguing-in-Publication Data
A catalogue record for this book is available from the British Library

Library of Congress Cataloging-in-Publication Data
A catalog record for this book is available from the Library of Congress

ISBN: 978-0-12-818438-7

For Information on all Academic Press publications
visit our website at https://www.elsevier.com/books-and-journals

Publisher: Mara Conner
Editorial Project Manager: Isabella C. Silva
Production Project Manager: Kamesh Ramajogi
Cover Designer: Greg Harris
Front Cover design: Kaveh Memarzadeh

Typeset by MPS Limited, Chennai, India

Working together
to grow libraries in
developing countries

www.elsevier.com • www.bookaid.org

Endorsement

AI is increasingly improving sick care and enhancing healthcare globally. The unprecedented demand on and for medical services has meant that we need to harness new tools quickly, safely, and effectively. This text clearly and comprehensively covers where we are today and what is achievable in the near future.

Dr. Umang Patel
Director, NHS Services—Babylon Health

Contents

List of contributors

Arash Aframian
Trauma and Orthopaedics, Imperial College London, London, United Kingdom; NHS, London, United Kingdom

Adam Bohr
Sonohaler, Copenhagen, Denmark

Henrik Bohr
Chemical Engineering, Technical University of Denmark, Lyngby, Denmark

Johan Peter Bøtker
Department of Pharmacy, University of Copenhagen, Copenhagen, Denmark

Justin Cobb
Professor and the Chair of Orthopedics at Imperial College London, London, United Kingdom

Glenn Cohen
Harvard Law School, Cambridge, MA, United States

Stefano Colombo
Independent Scientist, Leon, Spain

Rajeev Dutt
AI Dynamics Inc., Redmond, WA, United States

Sara Gerke
The Petrie-Flom Center for Health Law Policy, Biotechnology, and Bioethics at Harvard Law School, The Project on Precision Medicine, Artificial Intelligence, and the Law (PMAIL), Harvard University, Cambridge, MA, United States

Jakub P. Hlávka
Health Policy and Management Department of the Price School of Public Policy and Schaeffer Center for Health Policy & Economics, University of Southern California, Los Angeles, CA, United States

Farhad Iranpour
Trauma and Orthopedic Surgeon, Imperial College London, London, United Kingdom

Zineb Jeddi
TICLab, International University of Rabat (UIR), Rabat, Morocco; National Institute of Statistics and Applied Economics (INSEA), Rabat, Morocco

Kaveh Memarzadeh
ChemoMetec, Lillerød, Denmark

Timo Minssen
Centre for Advanced Studies in Biomedical Innovation Law (CeBIL), University of Copenhagen, Copenhagen, Denmark

Reza Mirnezami
Department of Surgery and Cancer, Imperial College London, London, United Kingdom

Khanhvi Tran
Sonohaler, Copenhagen, Denmark

Thomas Ward
Department of Surgery, Massachusetts General Hospital/Harvard Medical School, Boston, MA, United States

Elan Witkowski
Department of Surgery, Massachusetts General Hospital/Harvard Medical School, Boston, MA, United States

About the editors

Adam Bohr, PhD, is the CEO and a cofounder of Sonohaler, an mhealth and medical device company focused on asthma management using acoustic signals and machine learning tools. He is also a cofounder of Zerion ApS, a pharmaceutical company aspiring to transform the pharmaceutical landscape for formulation of poorly soluble drugs. Previously, he was employed as an assistant professor at the Department of Pharmacy, University

of Copenhagen, where he was doing research on drug implants, nanomedicine, and microfluidic technology and teaching pharmaceutical technology subjects. He has published more than 45 peer-reviewed academic papers and book chapters and has a PhD in Biomedical Engineering from the University College London. He is a health futurist and a healthcare AI proponent with a passion for patient-centered healthcare technologies.

Kaveh Memarzadeh, PhD, is currently a commercial field application scientist at ChemoMetec, a biotechnology company that innovates in the field of automated cell cytometry. He oversaw research management and communications at Orthopaedic Research UK (ORUK), a UK-based medical char-

ity that funds projects into the betterment and improvement of human movement and augmentation. He has published numerous peer-reviewed academic papers and has a PhD in nanotechnology, biomaterials, and microbiology. He is also a visiting lecturer at the University College London, teaching on a range of topics from the future of prosthetics/bionics to utilization of nanotechnology for antimicrobial bone implants. In his spare time, he reads, paints, and builds his own gaming computers and utilizes the power of social media to share his passion for nature with hundreds of thousands of people.

Biographies

Farhad Iranpour is a Trauma and Orthopedic surgeon and a clinical lecturer at Imperial College London, London, United Kingdom. He specializes in the subjects of arthroplasty, knee surgery, and musculoskeletal biomechanics. He has a passion for cutting edge technology involving implants and medical devices.

Glenn Cohen is the James A. Attwood and Leslie Williams Professor of Law, Harvard Law School, Cambridge, MA, United States and the Faculty Director of Petrie-Flom Center for Health Law Policy, Biotechnology, and Bioethics. With more than 150 articles and 15 books, Cohen is one of the world's leading experts on the intersection bioethics (sometimes also called "medical ethics") and the law, as well as health law.

Jakub P. Hlávka is a Research Assistant Professor of Health Policy and Management and a Fellow at the Schaeffer Center for Health Policy and Economics, University of Southern California in Los Angeles, CA, United States. He works at the intersection of precision medicine, healthcare innovation, and aging.

Johan Peter Bøtker is an Assistant Professor at the Department of Pharmacy, University of Copenhagen, Copenhagen, Denmark, working within the field of pharmaceutical technology and engineering with a focus on computer vision set-ups. He uses imaging and machine learning tools in various pharmaceutical and manufacturing applications.

Justin Cobb is a Professor and the Chair of Orthopedics at Imperial College London, London, United Kingdom, where he runs the MSk Lab—a group of 35 surgeons, scientists, and engineers. He is advisor to the board of Stanmore Implants, a spin-out from UCL. He is a civilian advisor in orthopedics to the Royal Air Force and is Orthopedic Surgeon to Her Majesty the Queen.

Khanhvi Tran is the CTO of Sonohaler, a company focused on predictive asthma care. She holds a degree in pharmaceutical sciences and specializes in computational methods and software development. She is passionate about advancing healthcare technology.

Rajeev Dutt is a 20-year veteran of the high-tech industry, with prominent roles at household names like Hewlett Packard, Compaq, Microsoft, Intel, and the BBC. He has expertise in cloud computing, artificial intelligence, OS kernel design, media systems, and hardware and holds a degree in theoretical physics from Trinity College at The University of Toronto, Toronto, ON, Canada.

Reza Mirnezami is a Consultant colorectal surgeon at Royal Free Hospital NHS Foundation Trust. His work is primarily focused on colorectal cancer and he leads an active program of research aimed at applying precision medicine approaches to the management of patients with bowel cancer.

Sara Gerke is the Research Fellow in Medicine, Artificial Intelligence, and Law at the Petrie-Flom Center for Health Law Policy, Biotechnology, and Bioethics at Harvard Law School. She oversees the day-to-day work of the Center's Project on Precision Medicine, Artificial Intelligence, and the Law (PMAIL), including conducting law, policy, and ethics research; her research focuses on the ethical and legal challenges of artificial intelligence and big data in the United States and Europe as well as on other topics such as mitochondrial replacement techniques and stem cell research.

Stefano Colombo is an independent scientist with a focus on pharmaceutical formulation and Quality by Design development of biologics. He holds a degree in Bioinformatics from Universitá degli Studi di Milano, Milan, Italy and a PhD in drug delivery from the University of Copenhagen, Copenhagen, Denmark.

Timo Minssen is Professor of Law and the Founding Director of UCPH's Center for Advanced Studies in Biomedical Innovation Law (CeBIL) at the University of Copenhagen. His research and part-time advisory practice concentrates on Intellectual Property, Competition, and Regulatory Law with a special focus on emerging technologies in the health and life sciences, such as artificial intelligence and gene editing.

Thomas Ward is the Artificial Intelligence and Innovation Fellow at the Massachusetts General Hospital's Surgical Artificial Intelligence and Innovation Laboratory. His current research includes application of AI and Computer Vision for intraoperative video analysis.

Zineb Jeddi is a researcher in data science at the International University of Rabat (UIR). She holds a degree from the National Institute of Statistics and Applied Economics (INSEA) in Morocco and is an engineer in Operational Research and Decision Making.

Elan Witkowski is a minimally invasive surgeon at Massachusetts General Hospital and an instructor in surgery at the Harvard Medical School, Boston, MA, United States. His clinical interests include bariatric and metabolic surgery and he performs open, laparoscopic, and robotic surgery. His research interests include epidemiology, surgical outcomes, and artificial intelligence.

Arash Aframian is a trauma and orthopedic registrar at the Imperial College Healthcare NHS Trust and Chelsea and Westminster Hospital NHS Foundation Trust who loves his work. He applies his experience in IT to help develop cutting edge methods in healthcare and enjoys collaborating with colleagues to continue to push the envelope of digital health technology and big data.

Henrik Bohr is a physicist and currently a research scientist at the Institute for Chemical Engineering at the Technical University of Denmark, Lyngby, Denmark. His research is focused on biophysics, biomolecular structure calculation, and nuclear and molecular physics. He has published over 200 scientific articles including around 50 communications to scientific meetings in international journals in the area of theoretical physics, biophysics, and biotechnology.

Preface

About this book

Today, almost every healthcare discipline has accepted artificial intelligence as a viable tool that can provide an overall benefit to society. This tool is currently experiencing its latest spring (cycle of hype) but this time it has really made its way into every corner of research and development with numerous publications on big data, machine learning, and neural networks and a few on their applications for healthcare. These publications are often very technical and research focused. In this book, our aim is to cover the main applications of artificial intelligence in healthcare in an easy-to-understand, evidence-based perspective but also cover the essential socioeconomic topics. This book serves to cover topics surrounding the entire healthcare ecosystem from drug design, medical imaging, and surgery to data privacy, law, and ethics.

Intended audience

This book is written for a broad audience and specifically for those interested in the healthcare applications of artificial intelligence including clinicians, health and life science professionals, policy-makers, business leaders, university students, and patients.

How is this book organized

This book has two introductory chapters and ten topic-specific chapters surrounding artificial intelligence (AI) applications and considerations in healthcare. The first two chapters introduce the current landscape within healthcare and the rise of AI in this arena. Further, the different applications of AI within healthcare are introduced. The ten topic-specific chapters are written by specialists in each area, covering the whole healthcare ecosystem. Each of these chapters can be read as stand-alone chapters and cover the current AI developments in the area while providing specific case studies since not all technologies can be covered in this publication. First, the AI applications in drug design and drug development are presented followed by its applications in the field of cancer diagnostics, treatment, and medical imaging. Subsequently, the application of AI in medical devices and surgery, as well as remote patient monitoring, is covered. Finally, the book dives into the topics of security, privacy, information sharing, health insurances, and legal aspects of AI in healthcare.

Adam Bohr
Kaveh Memarzadeh

Introduction

We live in a mathematical world and our universe functions by the fundamental laws of physics. From this mathematical system and through millions of years of natural selection, living things have evolved on planet earth. How this intricate system with all its complexities came to be, is still a scientific and philosophical mystery. Until relatively recently, humans assumed that *Homo Sapiens* were the most intelligent and capable species on the planet, supposedly because we can carry out certain tasks exceedingly better than our co-earthlings. These core facets or abilities that presumably separated us from the rest of life include problem solving, learning quickly and from experience, reasoning, and understanding complex ideas. Of course, it goes without saying that while the question of consciousness is still a mystery to science, being intelligent is not unique to humans. In fact, if we narrowly focus independently on all the core facets covered above, it is possible that other species could outperform humans. However, humans are fantastic generalists and are rather imaginative. We ask fundamental questions about the nature of existence itself and can project thoughts into the future with an increased cognitive ability that is slightly different from the likes of Chimpanzees. This allows us to dream, contemplate, and shape our reality and world by constructing materials and technology that make life easier and more manageable. With this incredible ability to imagine things that never were, we as a species have transformed the world around us.

For millennia, modern humans have observed birds flying in the sky and have wondered about conquering the skies and after a multitude of attempts and failures, we now have safe flying machines that transfer hundreds of people from the United States to Central Asia in a matter of hours. The key inspiration here was the act of "flying." To make efficient devices that can carry hundreds of people in the air, it was not necessary to mimic the exact way of "flying" but to understand that it is possible to "fly," and perhaps perform it better and more efficiently than birds. In a conversation with Lex Friedman, the neuroscientist Jeff Hawkins suggested that for humans to understand what intelligence is, we need to understand how the brain (mostly the neocortex) functions and that this is already in place as a complex framework and all it takes is more time and effort. This is what the pioneers of artificial intelligence (AI) have worked

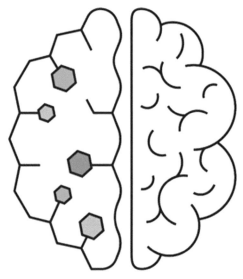

Figure 1 By utilizing the power of the brain, humans can understand how certain tasks (mental and physical) are performed and understanding the underlying mechanisms for these tasks can be the bases of future AI-based algorithms.

on for years. The goal would be to acquire inspiration from the phenomena of "intelligence" and not the development of a system that appear and functions precisely like a human brain (Fig. 1).

The promise of an intelligent machine

The moon landing was one of the greatest feats of humankind. This event occurred more than half a century ago and it goes without saying that the Apollo 11 computer (Apollo Guidance Computer) had the most cutting-edge technology at the time. However, when compared with the calculation and processing power of today's computers and even some of the cheapest smartphones, the Apollo Guidance Computer seems like ancient technology. This increase in the innovation rate, both on the hardware and software side, has made more room for creative pursuits, leading to advances in computer memory, processing power, and storage, which in turn has further led to the creation of vast amounts of data.

While humans are cognitively proficient at storing and memorizing information, we are also very good at forgetting and for the most part, extremely good at recording ideas, thoughts, and data on an external substrate and not in our brain. From ancient Egyptians to the Greeks, humanity has been recording various aspects of their lives and fantasies on inanimate materials for record keeping, culture, history, and science. Printed words and quantitative measures were memorialized with the advent of printing around 3500 BC by the ancient Persians and Mesopotamians that led directly to where we are today, in fact no less than 100 years ago, computers that did quantitative measures were once human, it is in the name—those who compute. However, it is important to note that the cost of mistakes in computing (by humans) for highly sensitive jobs and also the time wasted on correcting unnecessary errors seemed like a laborious and time-consuming exercise and yet this is a hallmark of human thinking, we are fallible.

The famous English mathematician, logician, and cryptanalyst, Alan Turing, understood this fallibility extremely well and in part of his famous article titled "Computing Machinery and Intelligence" highlights an imaginary scenario where a human and a machine are engaged in a Q&A session. At one point, the human asks the machine to add 34,957 to 70,764. Here, the machine pauses for about 30 seconds (apparently computing) and provides an answer of 105,621. The right answer would be 105,721, of course. In this scenario, the human does not know whether the opponent is a human or a machine and can only deduce this unknown based on their responses. It seems that with this example, Turing was trying to convey that humans often relate errors to other humans and expect a machine to be always right. The wrong answer then was a programming trick to fool the human into thinking that she/he was communicating with a human and not a machine. Turing would further comment on the production of a machine that somewhat gets its inspiration from the human body—he comments that "while this method is probably a sure way of producing a thinking machine, it seems to be altogether too slow or impracticable." In an ironic turn of events, he then turns his attention to the brain and proposes that perhaps we can provide and find "suitable branches of thoughts for the machine to exercise its powers in" with brain as an inspiration and propose several fields, including games, the learning and translation of languages, cryptography, and mathematics.

Turing turned out to be mostly accurate about this prediction. Historically, the field of machine intelligence and AI, as most call it, has some of its fundamental routes in the way that the nervous system functions at a basic level.

Implementations of all inspirations and findings from our nervous system and especially our brain's function can and will lead to various breakthroughs for the progress of humanity. It is also important to note that creating various algorithms will be an engineering-based problem with a certain goal and should not be an exact adoption of the biological function of the brain, of course there are researchers who will perhaps try and mimic the exact function of the brain; however, we find this approach rather unnecessary and encourage creative minds to find inspiration in the biological functions of the brain and use these as the means of finding novel ways to tackle the current healthcare challenges. We propose that because of the interlink and the circular influence of these two branches of study, AI alone will have a profound positive impact on human health in general and not just in the field of neuroscience.

Current applications and challenges in healthcare

Prevention of diseases should play a major role in healthcare. This can only happen if a society is provided with access to technology that can assist everyone with accurate information about the state of their health and well-being. While smart devices and wearables have multiple applications, they are now among various tools used by individuals to monitor their vitals. The analyzed data (often presented in an app or a web-based system) can be viewed by the individual or remotely monitored by healthcare professionals, allowing the necessary changes to be implemented by the patient before a life-threatening condition ensues.

Additionally, those patients who find themselves in clinical care will also have a higher chance of benefiting from an increasingly integrated system. The healthcare professionals who look after these patients can have immediate access to their reported symptoms and any previous and future imaging, biomarker tests, and also their current treatments. This integration of multiple data points can assist them with the workflow and the decision-making processes, but most importantly, these embedded procedures can also help patients benefit from other interventions that otherwise would have been overlooked or hard to implement. Furthermore, for tackling complicated diseases, multiple "treatment pathways" should be available. It is hypothesized that effective drug design and combinations that target specific molecular pathways could benefit from the field of machine learning. For instance, cancer patients could be individually analyzed and a novel "personalized" pathway can be devised to determine which drug/s and their relative combinations can be used for an optimal treatment outcome. These advancements in healthcare are mostly beneficial and will lead to longevity and prosperity for humanity.

Finally, as this book will concentrate on the function and applications of AI in healthcare, it is essential to remember that like most algorithms used today, the instances of applied AI in the following chapters are often referred to as narrow AI. This term refers to a system that can only perform and train for one or a small set of related tasks. From analyzing medical images to learning about human speech, effective implementation of narrow AI has dominated the healthcare system and should not be mistaken for human level intelligence or artificial general intelligence that can perform multiple complex tasks at the same time.

Despite this apparent limitation, operations running under narrow AI are likely to be highly competent and efficient at performing routine human tasks and in many cases can outcompete humans, allowing ample time for healthcare workers to engage in more productive tasks.

Adam Bohr
Kaveh Memarzadeh

CHAPTER 1

Current healthcare, big data, and machine learning

Adam Bohr[1] and Kaveh Memarzadeh[2]
[1]Sonohaler, Copenhagen, Denmark
[2]ChemoMetec, Lilleroed, Denmark

1.1 Current healthcare practice

The world is currently experiencing a major demographic transition where individuals aged 65 and older outnumber children aged 5 years and below. By 2050, the number of people aged 65 and above is estimated to reach 1.5 billion, equivalent to approximately 16% of the world population [1]. The global life expectancy has increased dramatically over the past few decades and continues to rise. People are becoming older in most major regions of the world, and the main health threats affecting people are changing toward conditions which reflect aging and changing lifestyle, especially in the developing world, where the rise in chronic noncommunicable diseases such as heart diseases, diabetes, and cancer now constitutes the greatest health issues [2]. The UN and WHO estimate that by 2025, 70% of all illnesses will be chronic, comorbid conditions. Such conditions often require continuous care over prolonged periods of time or increased attention from healthcare providers.

Accompanying the rising life expectancy is the growth in healthcare costs, which are increasing faster than the economic growth in most places [3]. In developed countries, healthcare spending ranges from 11% to 18% of GDP with the aging workforce being a key driver of these increasing costs. In recently developed countries such as Brazil and China, this figure is only 5%—10% but is expected to significantly increase in the next decade [4]. The total global healthcare costs are expected to rise from 8.4 trillion USD in 2015 to 18.3 trillion USD in 2030 [5]. In parallel, measures for lost productivity incurred due to chronic diseases are estimated at 47 trillion USD over the same period. Indeed, such loss of productivity as a result of aging and disease is a major reason for cost pressures of governments and healthcare providers. There is a need for a major

Artificial Intelligence in Healthcare
DOI: https://doi.org/10.1016/B978-0-12-818438-7.00001-0

change in the healthcare system over the coming years to cope with the increasing health-related demands from the aging population.

1.1.1 The rising need for technology

With the increasing healthcare spending, a rise in a health-conscious population has also been observed, which can provide opportunities for better health management and reduced healthcare spending. This includes consumer goods and services to increase health and wellness such as healthy nutrition, fitness, and meditation retreats. Other increasing trends include wearable and mobile health technologies. Technology is indeed changing the way patients interact with healthcare and how the healthcare system understands the patients, making the system more like a consumer market.

With the rise of electronic medical records (EMRs), personalized genomics, lifestyle and health data, and the capacity for better and faster analysis of data, digital trends are profoundly changing the healthcare system. This is also leading to breakthroughs in treatments based on correlations found from the collection and integration of healthcare data. The use of these data in large volumes is relatively new to the healthcare sector compared with other well-established sectors such as financial services. Recently, many digital stakeholders are seeking to disrupt the healthcare system by taking a technological approach to healthcare data. For instance, Google is building systems biology programs and analytical tools and applying these in areas such as digital pathology [6]. Furthermore, companies like Garmin, Fitbit (Google), and Apple are using information including heart rate and sleep data from their smart watches to predict an overall state of health of an individual [7].

Technology development is disrupting healthcare in many ways with one of the recent developments that has made the largest impact being the digitization of medical records, the EMRs.

The EMR provides a systematized, digital collection of patient health information, which can be shared across healthcare settings (Fig. 1.1). This includes notes and information collected by the clinicians in the office, clinic or hospital and contains the patient's medical history, diagnoses, and treatment plans. One can imagine that for each patient a substantial amount of healthcare data is accumulated over the course of their life, which can potentially be used to obtain a better understanding of medical conditions, diagnoses, and treatments. With the increasing adoption of EMRs and health monitoring technology, it is now possible to pull data together for individual patients and consolidate these to identify the needs of a subject to work

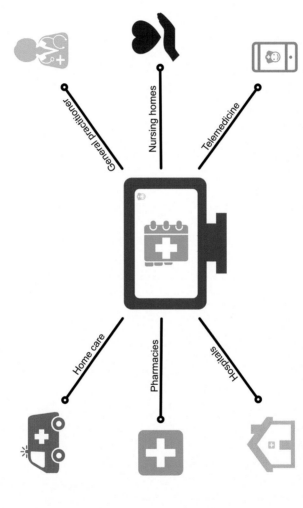

Figure 1.1 Electronic medical records can be acquired from multiple sources. Each of these sources provides a significant amount of data that could immensely improve the healthcare system overall. EMRs can make this process faster, more accurate, and precise.

toward a more unified goal for the entire population [8]. Although there is a complex transition toward such consolidation from the "old"/established system of data logging to the new digitalized world, this is believed to lead to high value for patients along the line [9].

1.1.2 New models in healthcare

With the increasing spending and burden on healthcare services and the recent technological developments, the healthcare system is ready for disruptions, from the basic understanding and differentiation of "health" and "care" to the processes of managing health. It is time to replace the old models with new ideas that can make healthcare more cost-efficient and improve healthcare standards across the entire system. Studies by McKinsey & Company have identified a few important characteristics among some of the new care models that are nowadays presented by various organizations [10].

Increased process management: Standardization of operational and clinical processes, increased use of new technologies and analytics including self-management and remote monitoring of patients, and more focus on performance.

Focus on people: Strong leadership and effective people development and improved workforce models such as optimum use of highly qualified clinicians and displacement of less-skilled tasks to new professions such as health coaches.

Focus on patients: Motivating patients to take an active role with self-management and larger differentiation of healthcare services depending on patient needs and desired outcomes.

Wider scale: Increasing operating scale to support expansion of healthcare services and ensure improved performance management.

The new care models require new business models from the healthcare sector including the healthcare providers and industry. A several number of changes and trends are observed in the healthcare industry over the recent years to comply with the changing environment and the rapid technological developments. Medical device companies are transforming into service entities where lab and remote management constitute important ways to engage with users and drive extra revenue. Pharmaceutical companies are focusing more on service and added value of products and an increasing number of mergers and acquisitions have taken place over the past years. Healthcare providers are also expanding their value chain by going into post discharge monitoring of patients thereby making the distinction less clear between health and care. Moreover, many rather "unusual" partnerships are beginning to form in the healthcare sector; for instance, partnerships between technology and pharma companies (e.g., Google with Novartis, Sanofi and Pfizer; IBM with Merck, GSK and Pfizer; Apple with J&J; Microsoft with AstraZeneca and Novartis), between device and pharma (glucose monitoring, others), payers and providers and also nongovernmental medical organizations with biotech companies (ORUK and Renovos) [11,12]. In some parts of the world, insurers are also becoming active in the practice of prevention and offer loyalty programs and lower premiums as incentives for healthy behavior to reach the goal of reduction in payouts and a more health-conscious population where focus is on prevention instead of treatment or a cure. It is critical that pharma and medical device industry follow suit in making society healthier even if that may take away some of the financial incentives seen from the current market and business model. There are new opportunities to financially benefit from a health-centered system and moving away from the disease-centered picture of healthcare [13].

Indeed, we are currently at a point with excellent opportunities to transition from a reactive to a more proactive healthcare view. Whereas today's healthcare system is mainly reactionary, providing healthcare based on requests and demands from patients, tomorrow's system may be based on health maintenance and health solutions. The value- and outcome-based system changes the way healthcare positions itself with regard to the patient.

1.2 Value-based treatments and healthcare services

The primary mission of healthcare is to create/maintain a healthier/ healthy society and improve the lives of its citizens. Short-term objectives including universal access to healthcare and profit optimization are distractions compared with the overall mission. Healthcare is primarily a service of importance to patients, and healthcare should, thus, be patient-centered rather than being provider-centered. Current input-based approach (patients seen, drugs or devices sold) is being replaced with an output-based approach (health outcomes for patients). With an input-based payment system there is an incentive to do as little as possible and receive maximum compensation, whereas with the output-based payment system there is an incentive to optimize productivity and maximize benefits. This is more advantageous for patients, for the healthcare system and for society, assuming a higher output and reduced future costs.

1.2.1 Value-based healthcare

The value-based healthcare approach serves to increase healthcare out-comes for patients in a more efficient way compared with the current volume-based approach (Fig. 1.2). Currently, the healthcare system rewards volume over outcomes or performance, even though the perfor-mance of healthcare is of major importance. Public and private institutions are trying to change this volume-based focus (also known as the legacy system) toward a more value-based focus. Value can be defined as patient health outcomes generated per unit of currency spent. Delivering high value to patients measured by health outcomes is the main purpose of healthcare and it is believed that value is the only goal that can bring together the interests of all participants in the healthcare environment and prevent reduced healthcare services. Such a value-based system will require a major restructuring in healthcare management and delivery, including organizational structures and payment models [14].

One of the primary considerations with a value-based healthcare approach is defining the value of importance to patients. It is often perceived by healthcare practitioners that increasing services such as the number of office visits is equivalent to good healthcare. However, good healthcare out-comes are not reflected by more visits, tests, or procedures but by better value and health status. For payment-based outputs or patient value and sys-tem value, it is necessary to be able to measure the clinical outcomes. This is

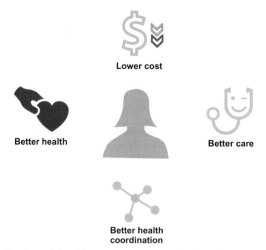

Figure 1.2 A value-based healthcare system can lead to better outcomes for the patient and ultimately a better society.

not the case for the current healthcare system where clinical outcomes are not required to the same degree. With the value-based system, it is also important to have an appropriate method for reporting and evaluating risk-adjusted outcomes for individual health conditions together with costs to assess outcomes. Such outcomes can be multifactorial with parameters such as recovery time, side effects, complications, follow-up treatments, and not only based on mortality and morbidity. They should also be determined for a full cycle of care and should be related and weighed alongside current condition and existing diagnoses of the patients. An outcome setup of three tiers was proposed by Porter and Lee [15].

Tier 1: Health status, including mortality and functional status.
Tier 2: Nature of care and recovery, including duration and readmission.
Tier 3: Sustainability of health.

Here, the aim should be to cover all three tiers. For instance, an extended 5-year survival rate does not provide the full picture of the patient's health status and should include assessment of potential complications, risk of readmission, pain and level of independent functioning [16].

1.2.2 Increasing health outcomes

Today, it is possible to collect data on clinical activities and health status of patients before and after treatment. This occurs in part through the EMR system, where following patients over a period results in better outcomes over time. Further, using these public datasets, health status of a patient population can be calculated and analyzed with health risks assessed for each individual [17]. Many countries in Europe have embraced digitization of medical records. According to a European Commission survey, Denmark, United Kingdom, and the Netherlands have in 2013 reached a patient record digitization of over 80%. This high level of digitization has facilitated national efforts for pilot studies and numerous collaborations, focused on patient outcomes. Denmark as an example included a digital approach consisting of remote monitoring and video consultations allowing GPs to efficiently coordinate care. In the United Kingdom and Spain, incentives to reduce unplanned readmissions have provided a push for improving effectiveness around patient discharge and remote monitoring [18].

Another payment strategy is the creation of bundled reimbursements for medical conditions. This would, for instance, include a single price for a full care cycle of an acute condition and a time-based price for care of a chronic condition or for preventive care.

Integration and collaboration among healthcare providers including clinics and hospitals is an essential aspect of the value-based healthcare approach while still allowing room for competition. Patients should have the option to choose the best or the most fitting healthcare providers based on their reported value and outcomes, and not just be assigned to a physician by the system but should have the option to select a physician or change their physician at a later stage if they choose to do so. This would take us a step closer to the concept of patient-centered care [19].

1.2.3 Patient-centered care (the patient will see you now)

Patient-centered care is a healthcare approach where the individual is placed at the center and their preferences, needs, and values are considered in all clinical decisions and are acknowledged as essential for their well-being. Patient-centered care encourages the healthcare professionals to have a two-way relationship with the patient by emphasizing what is preferred by her and generally provide support and care.

Studies have demonstrated that patient-centered care improves the satisfaction of patients as well as the quality of care and health outcomes which can reduce costs [20].

The patient-centered care approach can increase participation of the patients in the whole care path with shared responsibilities and decision-making in a constructive and collaborative manner. Instead of regarding the patient as a passive component in the healthcare process, a more contractual view is needed where the patient is seen as an active player who is an important part in the decision-making process. Patient-reported outcome measures are already being used by healthcare providers in many countries to determine how effectively the treatment improves patient quality of life [21]. The patients' involvement in the decision-making process has shown to enhance their adherence to treatment plans as well as result in improved patient satisfaction and health outcomes [22].

1.2.4 Personalized medicine

The potential of personalized treatments for transforming clinical practice has been a topic of discussion over the last decade with much of its hype yet to be realized aside from a few success stories. Personalized medicine can be broadly categorized either as drugs with associated diagnostics or advanced diagnostics for screening and risk identification [23].

Personalized medicine is believed to provide a significant value for the healthcare system. For instance, in some conditions such as rheumatoid arthritis, a large number of nonresponders to certain drug therapies will be identified by applying markers that predict treatment response (Fig. 1.3). Advanced personalized diagnostics will also provide value via solutions for data interpretation, decision-making, and analytics.

Advances in gene sequencing have led to a decrease in cost of whole-genome sequencing and it is predicted that sequencing will be common practice for clinical diagnostics in the next few years. Although not every indication will benefit from the added information from gene sequencing, those that are related to genetics are likely to accrue substantial value from being able to pinpoint cause of disease, genotype classification, prevention and treatment options [24].

Until now, oncology has been the focus for personalized medicine due to the obvious variation among the cancer patient populations and, hence, the high potential for personalized therapy. Other disease areas also have potential for personalized medicine including immunology-related conditions (e.g., transplants and autoimmune disorders), CNS disorders, and pediatric diseases, among others [25]. Indeed, this method is now being thought out for conditions that were previously thought incurable. One such condition is amyotrophic lateral sclerosis (ALS). ALS is a devastating motor neuron disease that manifests itself in a progressive manner whereby the death of upper and lower motor neurons leads to an eventual paralysis of the patient. Furthermore, it is estimated that up to 10% of patients with ALS have familial form of the disease [26]. While studies on humans have been rather

Figure 1.3 Personalized medicine can help tailor the treatment path to accommodate for the needs of individual patients.

limited, personalized medicine can allow researchers to categorize patients together based on their characteristics. This categorization can be achieved by utilizing certain biomarkers and perhaps with identification of similar mutations shared between patients. It is hypothesized that gene therapy holds great potential for personalized medicine and that some successful early tests have been conducted on animal models that show immense promise for those suffering from this fatal condition [27].

The adoption and reimbursement of personalized medicine and diagnostics is different from country to country with some adopting a more restricted policy for new options. These treatments are more expensive than conventional broader spectrum approach and will have to demonstrate a clear advantage for healthcare payers to cover these for patients in need of high complexity products for their condition.

We are facing an aging population and a rise in healthcare expenditures, and although the WHO has suggested a definition of health, people across different populations perceive health and healthcare differently. To provide the best care, it is necessary to agree on health and health outcomes. The value-based healthcare approach leads us to consider not only the cost of care but also emphasize more on the outcomes. This approach needs to be brought to life by implementing an appropriate payment system, with the use of proper measures for outcomes and an increased use of information technology to deliver healthcare and encourage patient participation. Both the value-based approach and patient-centered care work toward humanizing the healthcare business compared with the present business commodity view of healthcare services. Personalized medicine adds a further element of individualized care to the system, making healthcare an even more personal experience with precise treatments and the goal of providing an increased quality of life.

1.3 Increasing data volumes in healthcare

During the last decades, we have witnessed a boom in the volume of data produced in the greater healthcare environment. Pharmaceutical companies and academic institutions have been accumulating data from years of research and development into various databases. Payers and healthcare providers have collected large amounts of data from patient records and have digitized and consolidated these in EMRs and other systems. Further, governments and other public entities have begun to open their large library of healthcare knowledge including data from past clinical trials and data from insurance programs. Finally, advances in technology have made it easier to collect and process data from different sources [28]. A combination of these events has resulted in the generation of large volumes of data, and a variety of these can be employed for healthcare outcomes including improved diagnostics, healthcare decisions, treatments, and rehabilitation.

Despite the increased generation of data, most pharmaceutical and medical device companies are slow in operationalizing digital technology compared with companies in other sectors, such as the financial and automotive sector. The hesitation to invest more in digital transformation can be linked to a limited understanding of how implementation of new technologies in their product lines can create substantial business value. Further, there is a lack of digital talent among the workforce who also understand the industry, and there is generally a lack of focus among senior leadership, which is frequently observed in pharma and medical device companies [29]. It is largely acknowledged that such technological changes in the healthcare industry must be initiated and led from the top management of the individual organizations. This ensures that technological transitions are performed in a coordinated manner across the entire organization, providing a well-designed digital infrastructure.

There is a need for new and improved big data tools and technologies that can be employed to manage the growing healthcare data, and tools that can be used to extract these in libraries and find correlations between disease, prognosis, patients, and populations. Further, there is also a need for tools that can be used to link the data with diagnostics and treatments to integrate the new knowledge acquired with the existing healthcare ecosystem. The healthcare sector is highly regulated and large changes are neither fast nor easy to implement but we can expect a steady transition towards an increased collection, integration, and application of data across the whole sector [30].

1.3.1 Big data and data accumulation

Big data is defined as large datasets where the Log (n^*p) exceeds 7 with volumes of complex, variable, and high velocity data [31].

Big data in healthcare refers to complex datasets with unique features beyond large volume and differentiates itself from small data in a few ways. Data as we knew it before, or traditional data, is data which is created in the application of generating data. Such information is created by health systems or purchased as a product from digital services. Here, data is often a by-product of the many individual applications [32]. Big data, on the other hand, is characterized as any collection of data that is so large in terms of both volume and complexity. This is often difficult to process with conventional data-processing tools and therefore requires immense processing capacity running in parallel on numerous servers. The data gathered is often unstructured and full of errors, requiring it to be filtered, organized, and validated to be further analyzed [33]. Big data typically employs a great amount of statistically oriented tools including machine learning and predictive modeling to make sense of the data. Therefore, processing big data into useful healthcare-related information has the potential to transform the current practice of clinical care.

Big data can create value for healthcare stakeholders by providing transparency (data accessibility, processing time), improving performance (analyze variability, quick identification of root causes), segmenting populations to customized actions (creating specific segments in patient population based on data to tailor services and match their needs), and replacing/supporting human decision with automation (leveraging data, predictive and prescriptive algorithms, data correlations) [34].

Global revenues from big data and business analytics are estimated to grow to USD 187 billion by 2020. Many organizations are planning business solutions that utilize such data to give them a competitive edge in the market. Healthcare is one of the sectors that is expected to experience the fastest revenue growth in the next years based on the application of big data and analytics [35]. Currently, big data generally refers to data collected from established sources such as clinical registries, administrative claim records, EMRs, clinical trials as well as medical equipments (e.g., imaging), patient-reported data, internet data from search engines, social media, and geolocation services. Big data is generated and delivered across the whole healthcare ecosystem including medical research (drug discovery and clinical trials), prediction and prevention (patient profile and predictive models), patient experience and care (telemedicine, health apps, and personalized care), and everyday life (wearables, gene sequencing, online diagnostics) [36] (Fig. 1.4). An online diagnostics service such as WebMD gathers 212 million unique visitors per month and is among the top 50 most visited web pages in the United States, while even more people utilize Google's search engine as a platform to search for causes of their symptoms and worries [37]. These internet-based platforms generate enormous amounts of data from their users, which can be of great value

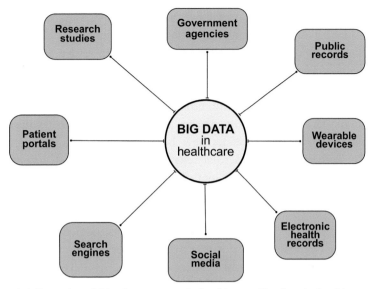

Figure 1.4 Examples of big data sources in healthcare. Big data in healthcare is an accumulation of data collected over a long period of time. The analysis of this data can help leverage the patient experience and well-being. It is often very difficult to approach big data using conventional methods that are rather slow, expensive, and inefficient.

for companies that want to establish correlations based on such data. It is of course also an asset to the companies and organizations that are collecting or in some way managing the data as they can extract information for their own use or sell the information to third parties [38].

1.3.2 Data generation sources

Data can be generated from various sources either intentionally or unintentionally. As mentioned above, EMRs, health insurance records, public health libraries, and clinical trial databases are large sources of data generation which span a large patient population and history. There are, however, many other smaller data generation sources including data collected using various devices and tools by the individual patients or at the clinical setting.

Medical equipments are becoming more digital and connected. This gradual change provides increasing streams of data that can be used in a processed state in EMRs or additional information could be extracted from the raw data collected from integration of different equipment. In particular, imaging-based diagnostics technologies have experienced complete digitization and typically incorporate various image processing tools to render the images for diagnostic purposes. These images showing a patient's anatomy and physiology could be reused for other health-related applications or used for diagnosis of a separate condition if such data were shared between technology platforms [39].

Telemedicine and remote monitoring have also experienced accelerated growth recently based on the greater availability of sensors, wireless, and internet technology. It provides new healthcare opportunities for underserved communities such as rural regions far from medical expertise, for instance, using a virtual care approach, and could also help to alleviate imbalances in demand for access to physicians between partnering locations. There is an increasing amount of consumer electronics and wearables, as well as digitized medical monitoring and diagnostic devices with tests that produce a substantial amount of data over long periods. Such data accumulation has also been ignited from patients beginning to actively use monitoring tools and taking ownership of their condition.

Much of the new data will come from devices that continuously or semi-continuously generate a large amount of information from several devices for individuals. Nearly everyone owns a smart phone and/or other smart devices that record data that can be considered relevant for health purposes including location-based environmental conditions, pedometer and mobility, sleep patterns, etc. and follows people through and between whole life cycles of health conditions (Fig. 1.5). Many people also make use of devices for health

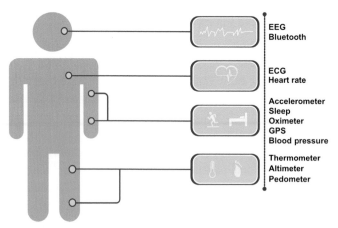

Figure 1.5 Utilizing the latest technology and sensors, it is possible to extract useful health information about the current state of our bodies. This allows individuals to lead healthier, more productive, and active lives.

or fitness, these are classified as wearables. Global sales of wearable devices were estimated to be as high as 274 million units in 2016 [40]. Among consumers, fitness bands are still the most popular type of wearable, and people cite health as a top motivator for using the new wearable technologies. Smart watches are also popular, especially in the US.

Finally, mobile apps are the most significant new source of data for healthcare self-management. There were 318,000 mhealth apps available in the top mobile app stores worldwide in 2018 with more than 200 new mhealth apps being added each day [41]. Currently, the demand for these apps does not meet the excessive supply of apps. Many of the apps are considered irrelevant and most apps have not been through any regulatory review process. With hundreds of apps available for various medical conditions, it is difficult for patients to browse and select the right app for their needs. However, there is a growing demand for health apps and their adoption is driven by advances in software and analytics. Additionally, the average waiting times for access to healthcare and the convenience of on-demand services from smartphones/ tablets play a crucial role in the growth of these devices [42].

Data storage also plays an important role in big data processing. With the increasing volumes of data, there is a need for efficient ways to store data. Cloud servers are suitable for this purpose as they provide high elasticity and powerful computing on an enormous scale.

1.3.3 Big data types

Big data is characterized by its high variety in the type of data and data source utilized as well as its applications. Especially within the healthcare sector, a large variety of data is acquired including new types of what did not exist prior to the emergence of smart devices. Indeed, when we consider data, variety may be as important a feature as the volume. This is due to the integration of many sources of data for better understanding and ultimately making sense of all the input [43]. These sources of data can be collected in a variety of formats, including text, numbers, diagrams/graphs, video, audio, and a mixture of these. Examples of these data formats and sources include Images (medical imaging data), unstructured text (narratives, physician notes, reports, summaries), streaming (monitors, implants, wearables, smartphones), social media (Facebook and Twitter), structured data (NLP annotations, EMR data sources), and dark data (server logs, account information, emails, documents) [44].

Data can generally be acquired in the format that is structured, semistructured, or unstructured and is acquired from either primary (EMRs, clinical decision support systems, etc.) or secondary sources (insurance companies, pharmacies, laboratories, etc.). One of the major complexities of big data is that it is mostly unstructured qualitative or incongruent. This makes it difficult to organize and process the data. Data preprocessing via text mining, image analysis, or other forms of processing is required to give structure to the data and extract relevant information from the raw data. There are numerous techniques for such purposes including statistical parsing, computational linguistics for data preprocessing, entity recognition, and term frequency for extraction with text data such as clinical notes [45]. The types of data most frequently used for big data applications include clinical data (incl. EMR data), research publications, reference materials, clinical trial data/references, genomic data, streamed data from monitoring devices/apps, web and social media data, and organizational and administrational data such as financial transactions [46].

1.4 Analytics of healthcare data (machine learning and deep learning)

Increasingly, large volumes of healthcare data are collected and there is a need for powerful analytical methods to process and interpret these data in order to make sense and create value. Big data is currently considered too large and heterogeneous to be stored and used.

Whereas big data as an entity is useless, the processing of big data to make predictions or decisions using artificial intelligence (AI) has the potential to transform the current practice of clinical care. AI techniques have the potential to extract valuable information from big data and use it for advancing healthcare.

1.4.1 Machine learning

Machine learning (ML) is a subdiscipline of AI and constitutes a group of techniques used to solve data analysis by finding patterns among the data. These patterns provide the opportunity for understanding complex health situations (e.g., identify disease risk factors) or predicting future health outcomes (e.g., predict disease prognosis). ML draws competences from various areas of science including data science, statistics, and optimization. This tool is different from traditional programming in that it uses learning algorithms as opposed to traditional algorithms, and it utilizes mathematical models driven by probability and statistics and invokes predictions using available data to extrapolate unavailable data (Fig. 1.6). Learning algorithms can learn

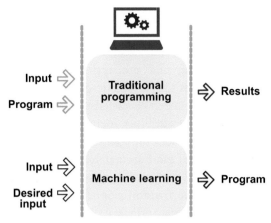

Figure 1.6 The distinction between traditional programming and machine learning.

from data and by feeding this (input) together with the desired results we can obtain the learning algorithm [47,48].

ML can be classified into three learning types: (1) supervised learning, (2) unsupervised learning, and (3) semisupervised learning.

Supervised learning uses datasets with prelabeled outcomes to train algorithms in solving classification and regression problems. It uses input variables to predict a defined outcome and is well-suited for regression problems, but it is demanding in time due to manual data labeling (supervision) and requires a large amount of data [49]. The most common types of supervised learning algorithms include decision tree, Naïve Bayes, and Support Vector Machine. Decision trees group attributes by value-based sorting and is used for data classification purposes. Naïve Bayes is mainly used for classifying text data and employs methods based on probability of events. Support Vector Machine is also used for classification of data and uses principle of margin calculation, a way of assessing the distance between classes. Supervised learning is suitable for predictive modeling by building relationships between the patient traits (as input) and the outcome of interest (as output) [50]. Classifiers are the predominantly used ML algorithms for healthcare applications and are, for instance, used for suggesting possible diagnoses, identifying a certain group of patients, differentiating between classes of documents, and defining abnormal levels for various lab results [44].

Unsupervised learning uses datasets with unlabeled inputs and lets the algorithms extract patterns, features, and structure from the data, autonomously. It uses previously learned features to recognize data it is introduced to. It is used for detecting hidden patterns in the data and the objective is to find this pattern without human feedback [51]. Unsupervised learning is useful for feature extraction and clustering. Common algorithms used include K-Means Clustering and principal component analysis (PCA). K-Means Clustering involves the automatic grouping of data into clusters (k distinct clusters) based on shared characteristics. The algorithm is typically provided with a given set of attributes for each item and the number of clusters to be used. It then divides the items into combinations of attributes that fit the number of groups most accurately. PCA involves the reduction of data from the perspective of data dimension to make data computation faster [52]. It uses orthogonal transformations to represent variables which are potentially correlated using principal components that are linearly uncorrelated. Some of the more common applications of clustering approaches within healthcare include risk management for clinical outcomes and health insurance reimbursements, identification of similar patients with complex or rare diseases and population health management via grouping of patients [53].

Semisupervised learning is a mixture of supervised and unsupervised learning where only a subset of the data output is labeled, as fully labeled data is research intensive to acquire. It can be useful when unlabeled data is already available and labeling the data is tedious. Semisupervised learning methods include Generative Models, Self-Training, and Transductive Support Vector Machine. Generative models use mixed distributions such as Gaussian mixture models as identifiers for prediction. Self-training is a model where classifiers are first trained with labeled data and subsequently fed with unlabeled data and predictions are then added in the training set [54].

There are numerous classifications, learning models, and concepts within the field of ML, too many to cover in this short introduction. Yet, there are a few concepts that are important to mention here including artificial neural networks (ANNs), natural language processing (NLP), and deep learning (DL).

ANNs are models based on the brain and biological neurons and are used to perform computational problems. Similar to biological neurons, artificial neurons receive inputs from other neurons and each input carries a weight, which determines the importance of the input. Collectively, the artificial neurons comprise an ANN with numerous connections between neurons. One can picture a neural network as an extension of linear regression to capture complex nonlinear relationships between input variables and an outcome. ANNs can be trained using both supervised and unsupervised learning in the process of optimizing its prediction accuracy [55]. NLP is a technology that enables computers to understand and process human language. The process of reading and understanding language is complex as most languages do not follow logical and consistent rules and can be used in various ways. NLP leverages a selection of specialized ML tools to characterize textual information by building a pipeline, whereby the problem is broken up into smaller pieces that can be solved separately. In the healthcare space, it is used to extract structured data from unstructured clinical text data that is often hard for clinicians or professionals to process. NLP is regarded as one of the most developed of the healthcare analytics applications. It has demonstrated that it can bring value by extracting data from various text-based sources including medical records, clinical notes, and social media. For instance, a large part of clinical information is in the form of narrative text from physical examinations, lab reports and discharge summaries. It is important to note that these are unstructured and incomprehensible to computer programs. Here, NLP is useful in extracting the relevant information from the narrative [56].

1.4.2 Deep learning

DL is a specific type of ML and can be regarded as a modern version of the ANN technique (Fig. 1.7). It is like a neural network but with many layers of abstraction rather than a direct input to output. The advances in computing power, the availability of more and larger datasets, and the introduction of new data formats have enabled DL [57].

Exploration of complex nonlinear patterns can be optimized using DL. This is often characterized by the numerous hidden layers to enable handing of data with high complexity and different structures. It is driven by a type of unsupervised learning that can be performed locally at each level of abstraction [58]. Algorithms used in DL are special in the way that they can autonomously generate features of interest in input data once they are fed with sufficient training examples (usually several million examples of data points). For instance, this could be feeding the system with millions of x-ray images, each labeled with the desired answer, such as the presence of a nodule/tumor, and once sufficiently trained, the algorithm can easily recognize a potential nodule in an image.

DL is particularly suitable for sequential and unstructured data such as images, audio, and video, where it can discover intricate structure in large datasets. It plays an important role in image recognition (medical

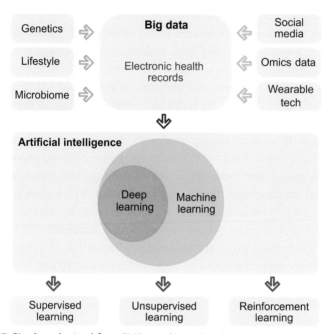

Figure 1.7 Big data obtained from EMRs can be analyzed and studied by AI (ML and DL).

diagnostics, facial recognition), speech recognition (voice assistants), mobile apps, machine vision, and robots [59]. Commonly used DL algorithms for healthcare include convolution neutral networks (CNNs), recurrent neural networks (RNNs), and deep reinforcement learning.

CNNs typically have numbered nodes with a special connectivity pattern between its neurons based on the numbers and are especially good at dealing with unstructured and sequential data. They play an important role in image/audio/video recognition as well as NLP. The term convolution stems from convolution layer, a linear transformation that preserves spatial information in its dimensionality.

RNNs are sequence-based and play an important role in NLP, video processing, and other processes. These networks use the sequential information and perform the same recurrent task for each element of the sequence and capture the output information in their internal memory [60].

Reinforcement neural networks are based on trial and error, much like learning in humans. This method differs from supervised learning in that inputs and outputs are not initially introduced.

At its core, all the above technology are here to serve a single purpose: Increase the quality of life for humans. From conditions that affect multiple organs, infections, and unwanted side effects of medications, the ability of healthcare workers to precisely pinpoint and extrapolate the trajectory of a condition will lead to significantly better outcomes for both the healthcare system and ultimately the patient. It is worth stating that much of healthcare around the globe will have to deal with increasing complexities such as an increase in diversity in the clinical measurements, heterogeneity in clinical phenotypes, or the lack of specific biomarkers for often fatal conditions. We believe that the implementation of the right policies that encompasses the usage of these tools can reduce the burden on the healthcare system and help thousands with novel, high precision solutions.

1.5 Conclusions/summary

Today, our lifestyles, the global demographics, and our needs as individuals are rapidly changing and more people need healthcare than ever before. Healthcare costs are also becoming increasingly expensive and with increased demands and the technological developments, major changes in the healthcare value chain and business models are on their way to disrupt the healthcare system as we know it. It is time to change the way we see health and disease in a static sense and instead view life as a dynamic process where the maintenance of health is followed from far before symptoms of any sort start to appear. Technological developments within data science have begun having an impact on the healthcare ecosystem but are yet to show their full potential. The digitization of the healthcare system in the last decade has led to an ever-increasing source for data production. The amount of data generated is increasing exponentially and collection of healthcare data from various sources is rapidly on the rise. Collection, management, and storage of all this data is very costly and may not be of value unless converted into insights that can be used by the healthcare ecosystem. For maximum optimization and usefulness, the data needs to be analyzed, interpreted, and acted upon. ML and DL allow for the generation of new knowledge from all the collected data and can help reduce the cost/time-based burden on all healthcare systems via learning, better prediction, and diagnosis. These tools serve to improve the overall clinical workflow from the preventive and diagnostic phase to the prescriptive and restorative phase in health management.

References

[1] Available from: <https://www.who.int/ageing/publications/global_health.pdf>.
[2] Available from: <https://apps.who.int/iris/bitstream/handle/10665/186463/9789240694811_eng.pdf?sequence = 1>.
[3] Elton J, O'Riordan A. Healthcare disrupted: next generation business models and strategies. John Wiley & Sons; 2016.
[4] WHO report on healthcare spending. Available from: <https://apps.who.int/iris/rest/bitstreams/1165184/retrieve>.
[5] Available from: <https://www2.deloitte.com/content/dam/Deloitte/global/Documents/Life-Sciences-Health-Care/gx-lshc-hc-outlook-2018.pdf>.
[6] Available from: <https://www.mckinsey.com/industries/pharmaceuticals-and-medical-products/our-insights/using-data-and-technology-to-improve-healthcare-ecosystems>.
[7] Available from: <https://www.cnbc.com/2019/05/24/apple-acquires-asthma-detection-start-up-tueo-health.html> and <https://synapse.koreamed.org/DOIx.php?id = 10.4258/hir.2015.21.4.315>.
[8] Available from: <https://www.accenture.com/_acnmedia/accenture/conversion-assets/dotcom/documents/global/pdf/dualpub_22/accenture-global-report-acn-doctors-study-sg.pdf>.
[9] Available from: <https://www.mckinsey.com/industries/healthcare-systems-and-services/our-insights/digitizing-healthcare-opportunities-for-germany>.
[10] Available from: <https://www.mckinsey.com/industries/healthcare-systems-and-services/our-insights/new-models-of-healthcare>.
[11] Danzon PM. Differential pricing of pharmaceuticals: theory, evidence and emerging issues. PharmacoEconomics. 2018;36(12):1395−405.
[12] Available from: <https://futureworlds.com/renovos-announced-as-first-startup-to-secure-investment-from-uk-orthopaedic-research-charity/>.
[13] McMahon Jr LF, Tipirneni R, Chopra V. Health System Loyalty Programs: an innovation in customer care and service. JAMA 2016;315:863−4. Available from: https://doi.org/10.1001/jama.2015.19463.
[14] Porter ME. Value-based health care delivery. Ann Surg 2008;248(4):503−9.
[15] Porter ME, Lee TH. The strategy that will fix healthcare. Harv Bus Rev 2013;91:24.
[16] Putera I. Redefining health: implication for value-based healthcare reform. Cureus. 2017;9(3).
[17] Available from: <https://www.cdc.gov/nchs/data_access/ftp_data.htm>.
[18] Hamel MB, Blumenthal D, Collins SR. Health care coverage under the Affordable Care Act—a progress report. N Engl J Med 2014;371(3):275−81.
[19] Hillary W, Justin G, Bharat M, Jitendra M. Value based healthcare. Adv Manag 2016;9(1):1.
[20] Manderscheid R, Kathol R. Fostering sustainable, integrated medical and behavioral health services in medical settings. Ann Intern Med 2014;160(1):61−5.
[21] Available from: <https://assets.kpmg/content/dam/kpmg/pdf/2016/01/value-based-organizations.pdf>.
[22] Collins S, Britten N, Ruusuvuori J, Thompson A. Understanding the process of patient participation. Patient participation in health care consultations: a qualitative perspective. McGraw-Hill Education; 2007. p. 3−21.
[23] Available from: <https://www.mckinsey.com/industries/pharmaceuticals-and-medical-products/our-insights/mckinsey-personalized-medicine-compendium-the-path-forward>.
[24] Baudhuin LM, Biesecker LG, Burke W, Green ED, Green RC. Predictive and precision medicine with genomic data. Clin Chem 2020;66(1):33−41.
[25] Selt F, Deiß A, Korshunov A, Capper D, Witt H, van Tilburg CM, et al. Pediatric targeted therapy: clinical feasibility of personalized diagnostics in children with relapsed and progressive tumors. Brain Pathol 2016;26(4):506−16.

[26] Picher-Martel V, Valdmanis PN, Gould PV, Julien JP, Dupré N. From animal models to human disease: a genetic approach for personalized medicine in ALS. Acta Neuropathol Commun 2016;4(1):70.

[27] Thomsen GM, Gowing G, Latter J, Chen M, Vit JP, Staggenborg K, et al. Delayed disease onset and extended survival in the SOD1G93A rat model of amyotrophic lateral sclerosis after suppression of mutant SOD1 in the motor cortex. J Neurosci 2014;34(47):15587–600. Available from: https://doi.org/10.1523/jneurosci.2037-14.2014.

[28] Available from: <https://www.mckinsey.com/industries/healthcare-systems-and-services/our-insights/the-big-data-revolution-in-us-health-care>.

[29] Available from: <https://www.mckinsey.com/business-functions/digital-mckinsey/our-insights/four-keys-to-successful-digital-transformations-in-healthcare>.

[30] Wang Y, Kung L, Wang WY, Cegielski CG. An integrated big data analytics-enabled transformation model: application to health care. Inf Manag 2018;55(1):64–79.

[31] Baro E, et al. Toward a literature-driven definition of big data in healthcare. Biomed Res Int 2015;2015:639021.

[32] Dinov ID. Volume and value of big healthcare data. J Med Stat Inform 2016;4:pii:3.

[33] Raghupathi W, Raghupathi V. Big data analytics in healthcare: promise and potential. Health Inf Sci Syst 2014;2(1):3.

[34] Based on analysis by McKinsey big data: the next frontier for innovation, competition and productivity. Available from: <https://www.mckinsey.com/business-functions/mckinsey-digital/our-insights/big-data-the-next-frontier-for-innovation>; 2011.

[35] Olavsrud T. Big data and analytics spending to hit USD 187 billion. Available from: <https://www.cio.com/article/3074238/big-data-and-analytics-spending-to-hit-187-billion.html>; 2016.

[36] Available from: <https://med.stanford.edu/content/dam/sm/sm-news/documents/StanfordMedicineHealthTrendsWhitePaper2017.pdf>.

[37] WebMD focuses on content, social media as users flock to mobile. MobiHealth News. August 2015. Available from: <https://www.mobihealthnews.com/45771/webmd-focuses-on-content-social-media-as-users-flock-to-mobile>.

[38] IHTT. Transforming healthcare through big data strategies for leveraging big data in the healthcare industry. Available from: <http://ihealthtran.com/wordpress/2013/03/iht%C2%B2-releases-big-data-research-report-download-today/>; 2013.

[39] Li KC, Marcovici P, Phelps A, Potter C, Tillack A, Tomich J, et al. Digitization of medicine: how radiology can take advantage of the digital revolution. Acad Radiol 2013;20(12):1479–94.

[40] Gartner says worldwide wearable devices sales to grow 18.4 percent in 2016. Gartner. February 2016. Available from: <https://www.gartner.com/en/newsroom/press-releases/2016-02-02-gartner-says-worldwide-wearable-devices-sales-grow-18-percent-in-2016>.

[41] Available from: <https://www.iqvia.com/insights/the-iqvia-institute/reports/the-growing-value-of-digital-health>.

[42] Rathbone AL, Prescott J. The use of mobile apps and SMS messaging as physical and mental health interventions: systematic review. J Med Internet Res 2017;19(8):e295.

[43] Bean, R. Variety, not volume, is driving big data initiatives; 2016.

[44] Natarajan P, Frenzel JC, Smaltz DH. Demystifying big data and machine learning for healthcare. CRC Press; 2017.

[45] Dinov ID. Methodological challenges and analytic opportunities for modeling and interpreting big healthcare data. GigaScience. 2016;5(1):12.

[46] Priyanka K, Kulennavar N. A survey on big data analytics in healthcare. Int J Comput Sci Inf Technol 2014;5(4):5865–8.

[47] Goodfellow I, Bengio Y, Courville A. Deep learning. MIT Press; 2016.

[48] Domingos P. The master algorithm: How the quest for the ultimate learning machine will remake our world. Basic Books; 2015.

[49] Buchlak QD, Esmaili N, Leveque JC, Farrokhi F, Bennett C, Piccardi M, et al. Machine learning applications to clinical decision support in neurosurgery: an artificial intelligence augmented systematic review. Neurosurg Rev 2019;1−9.

[50] Wiens J, Shenoy ES. Machine learning for healthcare: on the verge of a major shift in healthcare epidemiology. Clin Infect Dis 2017;66(1):149−53.

[51] Holzinger A. Interactive machine learning for health informatics: when do we need the human-in-the-loop? Brain Inform 2016;3(2):119−31.

[52] Dey A. Machine learning algorithms: a review. Int J Comput Sci Inf Technol 2016;7 (3):1174−9.

[53] Gao B, Zhang L, Wang H. Clustering of major cardiovascular risk factors and the association with unhealthy lifestyles in the Chinese adult population. PLoS One 2013;8(6):e66780.

[54] Jiang F, Jiang Y, Zhi H, Dong Y, Li H, Ma S, et al. Artificial intelligence in healthcare: past, present and future. Stroke Vasc Neurol 2017;2(4):230−43.

[55] Gurney K. An introduction to neural networks. CRC Press; 2014.

[56] Friedman C, Elhadad N. Natural language processing in health care and biomedicine. In Biomedical informatics. Springer, London; 2014. pp. 2552−2584.

[57] LeCun Y, Bengio Y, Hinton G. Deep learning. Nature. 2015;521(7553):436.

[58] Bengio Y. Learning deep architectures for AI. Found Trends Mach Learn 2009;2 (1):1−27.

[59] Krittanawong C, Zhang H, Wang Z, Aydar M, Kitai T. Artificial intelligence in precision cardiovascular medicine. J Am Coll Cardiol 2017;69(21):2657−64.

[60] Mandic DP, Chambers J. Recurrent neural networks for prediction: learning algorithms, architectures and stability. John Wiley & Sons, Inc; 2001.

CHAPTER 2

The rise of artificial intelligence in healthcare applications

Adam Bohr[1] and Kaveh Memarzadeh[2]
[1]Sonohaler, Copenhagen, Denmark
[2]ChemoMetec, Lillerød, Denmark

2.1 The new age of healthcare

Big data and machine learning are having an impact on most aspects of modern life, from entertainment, commerce, and healthcare. Netflix knows which films and series people prefer to watch, Amazon knows which items people like to buy when and where, and Google knows which symptoms and conditions people are searching for. All this data can be used for very detailed personal profiling, which may be of great value for behavioral understanding and targeting but also has potential for predicting healthcare trends. There is great optimism that the application of artificial intelligence (AI) can provide substantial improvements in all areas of healthcare from diagnostics to treatment. There is already a large amount of evidence that AI algorithms are performing on par or better than humans in various tasks, for instance, in analyzing medical images or correlating symptoms and biomarkers from electronic medical records (EMRs) with the characterization and prognosis of the disease [1].

The demand for healthcare services is ever increasing and many countries are experiencing a shortage of healthcare practitioners, especially physicians. Healthcare institutions are also fighting to keep up with all the new technological developments and the high expectations of patients with respect to levels of service and outcomes as they know it from consumer products including those of Amazon and Apple [2]. The advances in wireless technology and smartphones have provided opportunities for on-demand healthcare services using health tracking apps and search platforms and have also enabled a new form of healthcare delivery, via remote interactions, available anywhere and anytime. Such services are relevant for underserved regions and places lacking specialists and help reduce costs and prevent unnecessary exposure to contagious illnesses at the clinic.

Artificial Intelligence in Healthcare
DOI: https://doi.org/10.1016/B978-0-12-818438-7.00002-2

Telehealth technology is also relevant in developing countries where the healthcare system is expanding and where healthcare infrastructure can be designed to meet the current needs [3]. While the concept is clear, these solutions still need substantial independent validation to prove patient safety and efficacy.

The healthcare ecosystem is realizing the importance of AI-powered tools in the next-generation healthcare technology. It is believed that AI can bring improvements to any process within healthcare operation and delivery. For instance, the cost savings that AI can bring to the healthcare system is an important driver for implementation of AI applications. It is estimated that AI applications can cut annual US healthcare costs by USD 150 billion in 2026. A large part of these cost reductions stem from changing the healthcare model from a reactive to a proactive approach, focusing on health management rather than disease treatment. This is expected to result in fewer hospitalizations, less doctor visits, and less treatments. AI-based technology will have an important role in helping people stay healthy via continuous monitoring and coaching and will ensure earlier diagnosis, tailored treatments, and more efficient follow-ups.

The AI-associated healthcare market is expected to grow rapidly and reach USD 6.6 billion by 2021 corresponding to a 40% compound annual growth rate [4].

2.1.1 Technological advancements

There have been a great number of technological advances within the field of AI and data science in the past decade. Although research in AI for various applications has been ongoing for several decades, the current wave of AI hype is different from the previous ones. A perfect combination of increased computer processing speed, larger data collection data libraries, and a large AI talent pool has enabled rapid development of AI tools and technology, also within healthcare [5]. This is set to make a paradigm shift in the level of AI technology and its adoption and impact on society.

In particular, the development of deep learning (DL) has had an impact on the way we look at AI tools today and is the reason for much of the recent excitement surrounding AI applications. DL allows finding correlations that were too complex to render using previous machine learning algorithms. This is largely based on artificial neural networks and compared with earlier neural networks, which only had 3−5 layers of connections, DL networks have more than 10 layers. This corresponds to simulation of artificial neurons in the order of millions.

There are numerous companies that are frontrunners in this area, including IBM Watson and Google's Deep Mind. These companies have shown that their AI can beat humans in selected tasks and activities including chess, Go, and other games. Both IBM Watson and Google's Deep Mind are currently being used for many healthcare-related applications. IBM Watson is being used to investigate for diabetes management, advanced cancer care and modeling, and drug discovery, but has yet to show clinical value to the patients. Deep Mind is also being looked at for applications including mobile medical assistant, diagnostics based on medical imaging, and prediction of patient deterioration [6,7].

Many data and computation-based technologies have followed exponential growth trajectories. The most known example is that of Moore's law, which explains the exponential growth in the performance of computer chips. Many consumer-oriented apps have experienced similar exponential growth by offering affordable services. In healthcare and life science, the mapping of the human genome and the digitization of medical data could result in a similar growth pattern as genetic sequencing and profiling becomes cheaper and electronic health records and the like serve as a platform for data collection. Although these areas may seem small at first, the exponential growth will take control at some point. Humans are generally poor at understanding exponential trends and have a tendency to overestimate the impact of technology in the short-term (e.g. 1 year) while underestimating the long-term (e.g. 10 years) effect.

2.1.2 Artificial intelligence applications in healthcare

It is generally believed that AI tools will facilitate and enhance human work and not replace the work of physicians and other healthcare staff as such. AI is ready to support healthcare personnel with a variety of tasks from administrative workflow to clinical documentation and patient outreach as well as specialized support such as in image analysis, medical device automation, and patient monitoring.

There are different opinions on the most beneficial applications of AI for healthcare purposes. Forbes stated in 2018 that the most important areas would be administrative workflows, image analysis, robotic surgery, virtual assistants, and clinical decision support [8]. A 2018 report by Accenture mentioned the same areas and also included connected machines, dosage error reduction, and cybersecurity [9]. A 2019 report from McKinsey states important areas being connected and cognitive devices, targeted and personalized medicine, robotics-assisted surgery, and electroceuticals [10].

In the next sections, some of the major applications of AI in healthcare will be discussed covering both the applications that are directly associated with healthcare and other applications in the healthcare value chain such as drug development and ambient assisted living (AAL).

2.2 Precision medicine

Precision medicine provides the possibility of tailoring healthcare interventions to individuals or groups of patients based on their disease profile, diagnostic or prognostic information, or their treatment response. The tailor-made treatment opportunity will take into consideration the genomic variations as well as contributing factors of medical treatment such as age, gender, geography, race, family history, immune profile, metabolic profile, microbiome, and environment vulnerability. The objective of precision medicine is to use individual biology rather than population biology at all stages of a patient's medical journey. This means collecting data from individuals such as genetic information, physiological monitoring data, or EMR data and tailoring their treatment based on advanced models. Advantages of precision medicine include reduced healthcare costs, reduction in adverse drug response, and enhancing effectivity of drug action [11]. Innovation in precision medicine is expected to provide great benefits to patients and change the way health services are delivered and evaluated.

There are many types of precision medicine initiatives and overall, they can be divided into three types of clinical areas: complex algorithms, digital health applications, and "omics"-based tests.

Complex algorithms: Machine learning algorithms are used with large datasets such as genetic information, demographic data, or electronic health records to provide prediction of prognosis and optimal treatment strategy.

Digital health applications: Healthcare apps record and process data added by patients such as food intake, emotional state or activity, and health monitoring data from wearables, mobile sensors, and the likes. Some of these apps fall under precision medicine and use machine learning algorithms to find trends in the data and make better predictions and give personalized treatment advice.

Omics-based tests: Genetic information from a population pool is used with machine learning algorithms to find correlations and predict treatment responses for the individual patient. In addition to genetic information, other biomarkers such as protein expression, gut microbiome, and metabolic profile are also employed with machine learning to enable personalized treatments [12].

Here, we explore selected therapeutic applications of AI including genetics-based solutions and drug discovery.

2.2.1 Genetics-based solutions

It is believed that within the next decade a large part of the global population will be offered full genome sequencing either at birth or in adult life. Such genome sequencing is estimated to take up 100–150 GB of data and will allow a great tool for precision medicine. Interfacing the genomic and phenotype information is still ongoing. The current clinical system would need a redesign to be able to use such genomics data and the benefits hereof [13].

Deep Genomics, a Healthtech company, is looking at identifying patterns in the vast genetic dataset as well as EMRs, in order to link the two with regard to disease markers. This company uses these correlations to identify therapeutics targets, either existing therapeutic targets or new therapeutic candidates with the purpose of developing individualized genetic medicines. They use AI in every step of their drug discovery and development process including target discovery, lead optimization, toxicity assessment, and innovative trial design.

Many inherited diseases result in symptoms without a specific diagnosis and while interpreting whole genome data is still challenging due to the many genetic profiles. Precision medicine can allow methods to improve identification of genetic mutations based on full genome sequencing and the use of AI.

2.2.2 Drug discovery and development

Drug discovery and development is an immensely long, costly, and complex process that can often take more than 10 years from identification of molecular targets until a drug product is approved and marketed. Any failure during this process has a large financial impact, and in fact most drug candidates fail sometime during development and never make it onto the market. On top of that are the ever-increasing regulatory obstacles and the difficulties in continuously discovering drug molecules that are substantially better than what is currently marketed. This makes the drug innovation process both challenging and inefficient with a high price tag on any new drug products that make it onto the market [14].

There has been a substantial increase in the amount of data available assessing drug compound activity and biomedical data in the past few years. This is due to the increasing automation and the introduction of new experimental techniques including hidden Markov model based text to speech synthesis and parallel synthesis. However, mining of the large-scale chemistry data is needed to efficiently classify potential drug compounds and machine learning techniques have shown great potential [15]. Methods such as support vector machines, neural networks, and random forest have all been used to develop models to aid drug discovery since the 1990s. More recently, DL has begun to be implemented due to the increased amount of data and the continuous improvements in computing power. There are various tasks in the drug discovery process where machine learning can be used to streamline the tasks. This includes drug compound property and activity prediction, de novo design of drug compounds, drug–receptor interactions, and drug reaction prediction [16].

The drug molecules and the associated features used in the in silico models are transformed into vector format so they can be read by the learning systems. Generally, the data used here include molecular descriptors (e.g., physicochemical properties) and molecular fingerprints (molecular structure) as well as simplified molecular input line entry system (SMILES) strings and grids for convolutional neural networks (CNNs) [17].

2.2.2.1 Drug property and activity prediction

The properties and activity on a drug molecule are important to know in order to assess its behavior in the human body. Machine learning-based techniques have been used to assess the biological activity, absorption, distribution, metabolism, and excretion (ADME) characteristics, and physicochemical properties of drug molecules (Fig. 2.1). In recent years, several libraries of chemical and biological data including ChEMBL and PubChem have become available for storing information on millions of molecules for various disease targets. These libraries are machine-readable and are used to build machine learning models for drug discovery. For instance, CNNs have been used to generate molecular fingerprints from a large set of molecular graphs with information about each atom in the

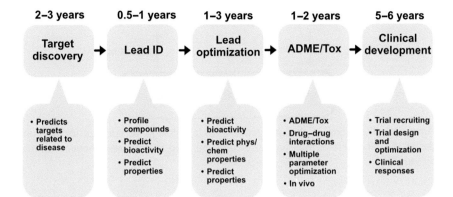

Figure 2.1 Machine learning opportunities within the small molecule drug discovery and development process.

molecule. Neural fingerprints are then used to predict new characteristics based on a given molecule. In this way, molecular properties including octanol, solubility melting point, and biological activity can be evaluated as demonstrated by Coley et al. and others and be used to predict new features of the drug molecules [18]. They can then also be combined with a scoring function of the drug molecules to select for molecules with desirable biological activity and physiochemical properties. Currently, most new drugs discovered have a complex structure and/or undesirable properties including poor solubility, low stability, or poor absorption.

Machine learning has also been implemented to assess the toxicity of molecules, for instance, using DeepTox, a DL-based model for evaluating the toxic effects of compounds based on a dataset containing many drug molecules [19]. Another platform called MoleculeNet is also used to translate two-dimensional molecular structures into novel features/descriptors, which can then be used in predicting toxicity of the given molecule. The MoleculeNet platform is built on data from various public databases and more than 700,000 compounds have already been tested for toxicity or other properties [20].

2.2.2.2 De novo design through deep learning

Another interesting application of DL in drug discovery is the generation of new chemical structures through neural networks (Fig. 2.2). Several DL-based techniques have been proposed for molecular de novo design. This also includes protein engineering involving the molecular design of proteins with specific binding or functions.

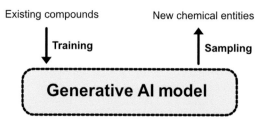

Figure 2.2 Illustration of the generative artificial intelligence concept for de novo design. Training data of molecular structures are used to emit new chemical entities by sampling.

Here, variational autoencoders and adversarial autoencoders are often used to design new molecules in an automated process by fitting the design model to large datasets of drug molecules. Autoencoders are a type of neural network for unsupervised learning and are also the tools used to, for instance, generate images of fictional human faces. The autoencoders are trained on many drug molecule structures and the latent variables are then used as the generative model. As an example, the program druGAN used adversarial autoencoders to generate new molecular fingerprints and drug designs incorporating features such as solubility and absorption based on predefined anticancer drug properties. These results suggest a substantial improvement in the efficiency in generating new drug designs with specific properties [21]. Blaschke et al. also applied adversarial autoencoders and Bayesian optimization to generate ligands specific to the dopamine type 2 receptor [22]. Merk et al. trained a recurrent neural network to capture a large number of bioactive compounds such as SMILES strings. This model was then fine-tuned to recognize retinoid X and peroxisome proliferator-activated receptor agonists. The identified compounds were synthesized and demonstrated potent receptor modulatory activity in in vitro assays [23].

2.2.2.3 Drug–target interactions

The assessment of drug–target interactions is an important part of the drug design process. The binding pose and the binding affinity between the drug molecule and the target have an important impact on the chances of success based on the in silico prediction. Some of the more common approaches involve drug candidate identification via molecular docking, for prediction and preselection of interesting drug–target interactions.

Molecular docking is a molecular modeling approach used to study the binding and complex formation between two molecules. It can be used to find interactions between a drug compound and a target, for example a receptor, and predicts the conformation of the drug compound in the binding site of the target. The docking algorithm then ranks the interactions via scoring functions and estimates binding affinity. Popular commercial molecular docking tools include AutoDock, DOCK, Glide, and FlexX. These are rather simple and many data scientists are working on improving the prediction of drug—target interaction using various learning models [24]. CNNs are found useful as scoring functions for docking applications and have demonstrated efficient pose/affinity prediction for drug—target complexes and assessment of activity/inactivity. For instance, Wallach and Dzamba build AtomNet, a deep CNN to predict the bioactivity of small molecule drugs for drug discovery applications. The authors showed that AtomNet outperforms conventional docking models in relation to accuracy with an AUC (area under the curve) of 0.9 or more for 58% of the targets [25].

Current trends within AI applications for drug discovery and development point toward more and more models using DL approaches. Compared with more conventional machine learning approaches, DL models take a long time to train because of the large datasets and the often large number of parameters needed. This can be a major disadvantage when data is not readily available. There is therefore ongoing work on reducing the amount of data required as training sets for DL so it can learn with only small amounts of available data. This is similar to the learning process that takes place in the human brain and would be beneficial in applications where data collection is resource intensive and large datasets are not readily available, as is often the case with medicinal chemistry and novel drug targets. There are several novel methods being investigated, for instance, using a one-shot learning approach or a long short-term memory approach and also using memory augmented neural networks such as the differentiable neural computer [17].

2.3 Artificial intelligence and medical visualization

Interpretation of data that appears in the form of either an image or a video can be a challenging task. Experts in the field have to train for many years to attain the ability to discern medical phenomena and on top of that have to actively learn new content as more research and information presents itself. However, the demand is ever increasing and there is a significant shortage of experts in the field. There is therefore a need for a fresh approach and AI promises to be the tool to be used to fill this demand gap.

2.3.1 Machine vision for diagnosis and surgery

Computer vision involves the interpretation of images and videos by machines at or above human-level capabilities including object and scene recognition. Areas where computer vision is making an important impact include image-based diagnosis and image-guided surgery.

2.3.1.1 Computer vision for diagnosis and surgery

Computer vision has mainly been based on statistical signal processing but is now shifting more toward application of artificial neural networks as the choice for learning method. Here, DL is used to engineer computer vision algorithms for classifying images of lesions in skin and other tissues. Video data is estimated to contain 25 times the amount of data from high-resolution diagnostic images such as CT and could thus provide a higher data value based on resolution over time. Video analysis is still premature but has great potential for clinical decision support. As an example, a video analysis of a laparoscopic procedure in real time has resulted in 92.8% accuracy in identification of all the steps of the procedure and surprisingly, the detection of missing or unexpected steps [26].

A notable application of AI and computer vision within surgery technology is to augment certain features and skills within surgery such as suturing and knot-tying. The smart tissue autonomous robot (STAR) from the Johns Hopkins University has demonstrated that it can outperform human surgeons in some surgical procedures such as bowel anastomosis in animals. A fully autonomous robotic surgeon remains a concept for the not so near future but augmenting different aspects of surgery using AI is of interest to researchers. An example of this is a group at the Institute of Information Technology at the Alpen-Adria Universität Klagenfurt that uses surgery videos as training material in order to identify a specific intervention made by the surgeon. For example, when an act of dissection or cutting is performed on the patient's tissues or organs, the algorithm recognizes the likelihood of the intervention as well as the specific region in the body [27]. Such algorithms are naturally based on the training on many videos and could be proven very useful for complicated surgical procedures or for situations where an inexperienced surgeon is required to perform an emergency surgery. It is important that surgeons are actively engaged in the development of such tools ensuring clinical relevance and quality and facilitating the translation from the lab to the clinical sector.

2.3.2 Deep learning and medical image recognition

The word "Deep" refers to the multilayered nature of machine learning and among all DL techniques, the most promising in the field of image recognition has been the CNNs. Yann LeCun, a prominent French computer scientist introduced the theoretical background to this system by creating LeNET in the 1980s, an automated handwriting recognition algorithm designed to read cheques for financial systems. Since then, these networks have shown significant promise in the field of pattern recognition.

Similar to radiologists that during the medical training period have to learn by constantly correlating and relating their interpretations of radiological images to the ground truth, CNNs are influenced by the human visual cortex, where image recognition is initiated by the identification of the many features of the image. Furthermore, CNNs require a significant amount of training data that comes in the form of medical images along with labels for what the image is supposed to be. At each hidden layer of training, CNNs can adjust the applied weights and filters (characteristics of regions in an image) to improve the performance on the given training data.

Briefly and very simply (Fig. 2.3), the act of convolving an image with various weights and creating a stack of filtered images is referred to as a convolutional layer, where an image essentially becomes a stack of filtered images. Pooling is then applied to all these filtered images, where the original stack of images becomes a smaller representation of themselves and all negative values are removed by a rectified linear unit (ReLU). All these operations are then stacked on top of one another to create layers, sometimes referred to as Deep stacking. This process can be repeated multiple times and each time the image gets filtered more and relatively smaller. The last layer is referred to as a fully connected layer where every value assigned to all layers will contribute to what the results will be. If the system produces an error in this final answer, the gradient descent can be applied by adjusting the values up and down to see how the error changes relative to the right answer of interest. This can be achieved by

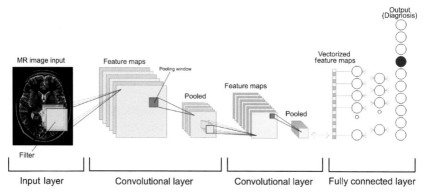

Figure 2.3 The various stages of convolutional neural networks at work. *Adapted from Lundervold AS, Lundervold A. An overview of deep learning in medical imaging focusing on MRI. Z Med Phys. 2019;29:102−27.*

an algorithm called back propagation that signifies "learning from mistakes." After learning a new capability from the existing data, this can be applied to new images and the system can classify the images in the right category (Inference), similar to how a radiologist operates [28].

2.3.3 Augmented reality and virtual reality in the healthcare space

Augmented and virtual reality (AR and VR) can be incorporated at every stage of a healthcare system. These systems can be implemented at the early stages of education for medical students, to those training for a specific specialty and experienced surgeons. On the other hand, these technologies can be beneficial and have some negative consequences for patients.

In this section, we will attempt to cover each stage and finally comment on the usefulness of these technologies.

2.3.3.1 Education and exploration
Humans are visual beings and play is one of the most important aspects of our lives. As children the most important way for us to learn was to play. Interaction with the surroundings allowed us to gain further understanding of the world and provided us with the much-needed experience. The current educational system is limited and for interactive disciplines such as medicine this can be a hindrance. Medicine can be visualized as an art form and future clinicians are the artist. These individuals require certain

skills to fulfill the need for an ever-evolving profession. Early in medical school, various concepts are taught to students without them ever experiencing these concepts in real life. So game-like technologies such as VR and AR could enhance and enrich the learning experience for future medical and health-related disciplines [29]. Medical students could be provided with and taught novel and complicated surgical procedures, or learn about anatomy through AR without ever needing to interact or involve real patients at an early stage or without ever needing to perform an autopsy on a real corpse. These students will of course be interacting with real patients in their future careers, but the goal would be to initiate the training at an earlier stage and lowering the cost of training at a later stage.

For today's training specialists, the same concept can be applied. Of course, human interaction should be encouraged in the medical field but these are not always necessary and available when an individual is undergoing a certain training regimen. The use of other physical and digital cues such as haptic feedback and photorealistic images and videos can provide a real simulation whereby learning can flourish and the consequences and cost of training are not drastic (Fig. 2.4).

In a recent study [30], two groups of surgical trainees were subjected to different methods for Mastoidectomy, where one group ($n = 18$) would go through the standard training path and the other would train on a freeware VR simulator [the visible ear simulator (VES)]. At the end of the training, a significant improvement in surgical dissection was

Figure 2.4 Virtual reality can help current and future surgeons enhance their surgical abilities prior to an actual operation. (Image obtained from a video still, OSSOR VR).

observed for those who trained with VR. For real-life and precise execution, AR would be more advantageous in healthcare settings. By wearing lightweight headsets (e.g., Microsoft HoloLens or Google Glass) that project relevant images or video onto the regions of interest, the user can focus on the task without ever being distracted by moving their visual fields away from the region of interest.

2.3.3.2 Patient experience

Humans interact with their surroundings with audiovisual cues and utilize their limbs to engage and move within this world. This seemingly ordinary ability can be extremely beneficial for those who are experiencing debilitating conditions that limit movement or for individuals who are experiencing pain and discomfort either from a chronic illness or as a side effect of a treatment. A recent study, looking at the effect of immersive VR for patients who had suffered from chronic stroke patients, found this technology to be contributing positively to the state of patients. During the VR experience, the patients are asked to grab a virtual ball and throw it back into the virtual space [31]. For these patients, this immersive experience could act as a personal rehabilitation physiotherapist who engages their upper limb movement multiple times a day, allowing for possible neuroplasticity and a gradual return of normal motor function to these regions.

For others, these immersive technologies could help cope with the pain and the discomfort of their cancer or mental health condition. A study has shown that late-stage adult cancer patients can use this technology with minimum physical discomfort and in return benefit from an enhanced relaxed state, entertainment, and a much-needed distraction [32]. These immersive worlds provide a form of escapism with their artificial characters and environments, allowing the individual to interact and explore the surrounding while receiving audiovisual feedback from the environment, much like all the activities of daily living.

2.4 Intelligent personal health records

Personal health records have historically been physician-oriented and often have lacked patient-related functionalities. However, in order to promote self-management and improve the outcomes for patients, a patient-centric personal health record should be implemented. The goal is to allow ample freedom for patients to manage their conditions, while freeing up time for the clinicians to perform more crucial and urgent tasks.

2.4.1 Health monitoring and wearables

For millennia individuals relied on physicians to inform them about their own bodies and to some extent, this practice is still applied today. However, the relatively new field of wearables is changing this. Wearable health devices (WHDs) are an upcoming technology that allow for constant measurement of certain vital signs under various conditions. The key to their early adoption and success is their application flexibility—the users are now able to track their activity while running, meditating, or when underwater. The goal is to provide individuals with a sense of power over their own health by allowing them to analyze the data and manage their own health. Simply, WHDs create individual empowerment (Fig. 2.5).

At first look, a wearable device might look like an ordinary band or watch; however, these devices bridge the gap between multiple scientific disciplines such as biomedical engineering, materials science, electronics, computer programming, and data science, among many others [33]. It would not be an exaggeration to refer to them as ever-present digital health coaches, as increasingly it is encouraged to wear them at all times in order to get the most out of your data. Garmin wearables are a good example of this, with a focus on being active, they cover a vast variety of sports and provide a substantial amount of data on their Garmin connect application where users can analyze and observe their daily activities. These are increasingly accompanied by implementation of gamification.

Gamification refers to utilization of game design elements for nongame-related applications. These elements are used to motivate and drive users to reach their goals [34]. On wearable platforms, data gathered from daily activities can serve as competition between different users on the platform. Say, that your average weekly steps are around 50,000 steps.

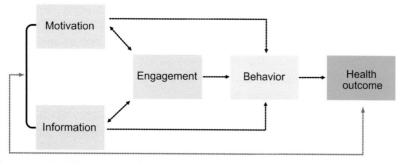

Figure 2.5 Health outcome of a patient depends on a simple yet interconnected set of criteria that are predominantly behavior dependent.

Here, based on specific algorithms, the platform places you on a leader-board against individuals whose average weekly steps are similar to yours or higher, with the highest ranking member exceeding your current average weekly steps. As a result of this gamified scenario, the user can push themselves to increase their daily activities in order to do better on the leaderboard and potentially lead a healthier life. While the gamification aspect of wearables and their application could bring benefits, evidence of efficacy is scarce and varies widely with some claiming that the practice might bring more harm than good.

Remote monitoring and picking up on early signs of disease could be immensely beneficial for those who suffer from chronic conditions and the elderly. Here, by wearing a smart device or manual data entry for a prolonged period, individuals will be able to communicate to their healthcare workers without the need of disrupting their daily lives [35]. This is a great example of algorithms collaborating with healthcare professionals to produce an outcome that is beneficial for patients.

2.4.2 Natural language processing

Natural language processing (NLP) relates to the interaction between computers and humans using natural language and often emphasizes on the computer's ability to understand human language. NLP is crucial for many applications of big data analysis within healthcare, particularly for EMRs and translation of narratives provided by clinicians. It is typically used in operations such as extraction of information, conversion of unstructured data into structured data, and categorization of data and documents.

NLP makes use of various classifications to infer meaning from unstructured textual data and allows clinicians to work more freely using language in a "natural way" as opposed to fitting sequences of text into input options to serve the computer. NLP is being used to analyze data from EMRs and gather large-scale information on the late-stage complications of a certain medical condition [26].

There are many areas in healthcare in which NLP can provide substantial benefits. Some of the more immediate applications include [36]

1. Efficient billing: extracting information from physician notes and assigning medical codes for the billing process.
2. Authorization approval: Using information from physician notes to prevent delays and administrative errors.

3. Clinical decision support: Facilitate decision-making for members of healthcare team upon need (for instance, predicting patient prognosis and outcomes).
4. Medical policy assessment: compiling clinical guidance and formulation appropriate guidelines for care.

One application of NLP is disease classification based on medical notes and standardized codes using International Statistical Classification of Diseases and Related Health Problems (ICD). ICD is managed and published by the WHO and contains codes for diseases and symptoms as well as various findings, circumstances, and causes of disease. Here is an illustrative example of how an NLP algorithm can be used to extract and identify the ICD code from a clinical guidelines description. Unstructured text is organized into structured data by parsing for relevant clauses followed by classification of ICD-10 codes based on frequency of occurrence. The NLP algorithm is run at various thresholds to improve classification accuracy and the data is aggregated for the final output (Fig. 2.6).

2.4.3 Integration of personal records

Since the introduction of EMRs, there have been large databases of information on each patient, which collectively can be used to identify healthcare trends within different disease areas. The EMR databases contain the history of hospital encounters, records of diagnoses and interventions, lab test, medical images, and clinical narratives. All these datasets can be used to build predictive models that can help clinicians with diagnostics and various treatment decision support. As AI tools mature it will be possible to extract all kinds of information such as related disease effects and correlations between historical and future medical events [37]. The only data often missing is data from in between interventions and between hospital visits when the patient is well or may not be showing symptoms. Such data could help to construct an end-to-end model of both "health" and "disease" for studying long-term effects and further disease classifications.

Although the applications of AI for EMRs are still quite limited, the potential for using the large databases to detect new trends and predict health outcomes is enormous. Current applications include data extraction from text narratives, predictive algorithms based on data from medical tests, and clinical decision support based on personal medical history. There is also great potential for AI to enable integration of EMR data with various health applications. Current AI applications within healthcare

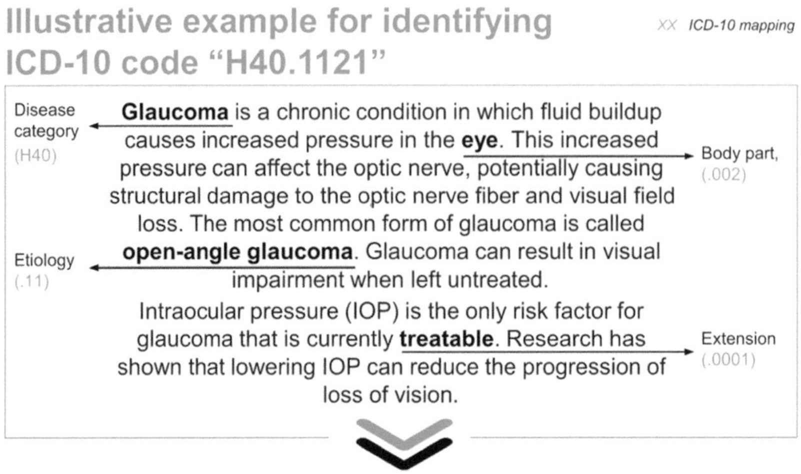

Illustrative example for identifying ICD-10 code "H40.1121"

xx ICD-10 mapping

Disease category (H40)

Glaucoma is a chronic condition in which fluid buildup causes increased pressure in the **eye**. This increased pressure can affect the optic nerve, potentially causing structural damage to the optic nerve fiber and visual field loss. The most common form of glaucoma is called **open-angle glaucoma**. Glaucoma can result in visual impairment when left untreated.

Body part, (.002)

Etiology (.11)

Intraocular pressure (IOP) is the only risk factor for glaucoma that is currently **treatable**. Research has shown that lowering IOP can reduce the progression of loss of vision.

Extension (.0001)

Text from clinical guidance extract	ICD-10 nomenclature	ICD-10
Glaucoma	Disease category	H40
Open angle	Etiology	0.11
Eye	Body part	0.002
Treatable	Extension	0.0001

= H40.1121 (Glaucoma/Primary open-angle/ Left eye/Mild stage)

Figure 2.6 Example of ICD-10 mapping from a clinical guidelines' description [36].

are often standalone applications, these are often used for diagnostics using medical imaging and for disease prediction using remote patient monitoring [38]. However, integrating such standalone applications with EMR data could provide even greater value by adding personal medical data and history as well as a large statistical reference library to make classifications and predictions more accurate and powerful. EMR providers such as Cerner, Epic, and Athena are beginning to add AI functionality such as NLP in their systems making it easier to access and extract data held in their libraries [39]. This could facilitate the integration of, for instance, Telehealth and remote monitoring applications with EMR data and the data integration transfer could even go both ways including the addition of remote monitoring data in the EMR systems.

There are many EMR providers and systems globally. These use various operating systems and approaches with more than a thousand EMR providers operating in the United States alone. Integration of EMR records on their own poses a great challenge and interoperability of these systems is important to obtain the best value from the data. There are various international efforts in gathering EMR data across countries including Observational Health Data Science and Informatics (OHDSI), who have consolidated 1.26 billion patient records from 17 different countries [40]. Various AI methods have been used to extract, classify, and correlate data from EMRs but most generally make use of NLP, DL, and neural networks.

DeepCare is an example of an AI-based platform for end-to-end processing of EMR data. It uses a deep dynamic memory neural network to read and store experiences and in memory cells. The long short-term memory of the system models the illness trajectory and healthcare processes of users via a time-stamped sequence of events and in this way allows capturing long-term dependencies [41]. Using the stored data, the framework of DeepCare can model disease progression, support intervention recommendation, and provide disease prognosis based on EMR databases. Studying data from a cohort of diabetic and mental health patients it was demonstrated that DeepCare could predict the progression of disease, optimal interventions, and assessing the likelihood for readmission [37].

2.5 Robotics and artificial intelligence-powered devices

There are numerous areas in healthcare where robots are being used to replace human workforce, augment human abilities, and assist human healthcare professionals. These include robots used for surgical procedures such as laparoscopic operations, robotic assistants for rehabilitation and patient assistance, robots that are integrated into implants and prosthetic, and robots used to assist physicians and other healthcare staff with their tasks. Some of these devices are being developed by several companies especially for interacting with patients and improving the connection between humans and machines from a care perspective. Most of the robots currently under development have some level of AI technology incorporated for better performance with regard to classifications, language recognition, image processing, and more.

2.5.1 Minimally invasive surgery

Although many advances have been seen in the area surrounding surgery measured by the outcomes of surgical procedures, the main practice of surgery still remains a relatively low-tech procedure for the most part using hand tools and instruments for "cutting and sewing." Conventional surgery relies greatly on sensing by the surgeon, where touching allows them to distinguish between tissues and organs and often requires open surgery. There is an ongoing transformation within surgical technology and focus has especially been placed in reducing the invasiveness of surgical procedure by minimizing incisions, reducing open surgeries, and using flexible tools and cameras to assist the surgery [42]. Such minimally invasive surgery is seen as the way forward, but it is still in an early phase with many improvements to be made to make it "less of a big deal" for patients and reduce time and cost. Minimal invasive surgery requires different motor skills compared with conventional surgery due to the lower tactile feedback when relying more on tools and less on direct touching. Sensors that provide the surgeon with finer tactile stimuli are under development and make use of tactile data processing to translate the sensor input into data or stimuli that can be perceived by the surgeon. Such tactile data processing typically makes use of AI, more specifically artificial neural networks to enhance the function of this signal translation and the interpretation of the tactile information [43]. Artificial tactile sensing offers several advantages compared with physical touching including a larger reference library to compare sensation and standardization among surgeons with respect to quantitative features, continuous improvement, and level of training.

An example where artificial tactile sensing has been used includes screening of breast cancer, as a replacement for clinical breast examination to complement medical imaging techniques such as x-ray mammography and MRI. Here, the artificial tactile sensing system was built on data from reconstruction of mechanical tissue measurements using a pressure sensor as reference data. During training of the neural network, the weight of the input data adjusts according to the desired output [44]. The tactile sensory system can detect mass calcifications inside the breast tissue based on palpation of different points of the tissue and comparing with different reference data, and subsequently determine whether there are any significant abnormalities in the breast tissue. Artificial tactile sensing has also been used for other applications including assessment of liver, brain, and submucosal tumors [45].

2.5.2 Neuroprosthetics

Our species has always longed for an eternal life, in the ancient Vedic tradition there exists a medicinal drink that provides "immortality" for those who drink it. The Rig Veda, which was written some 5000 years ago, comments: "We drank soma, we became immortal, we came to the light, we found gods." This is similar in ancient Persian culture, where a similar legendary drink is called Hoama in the Zoroastrain sacred book, Avesta [46,47]. This longing for "enhancement" and "augmentation" has always been with us, and in the 21st century we are gradually beginning to move towards making some past myths into reality. In this section, we will cover some recent innovations that can utilize AI to assist and allow humans to function better. Most research in this area is to assist individuals with preexisting conditions and have not been implemented in normal functioning humans for the sake of human augmentation; however, this can perhaps change in the coming years.

Neuroprosthetics are defined as devices that help or augment the subject's own nervous system, in both forms of input and output. This augmentation or stimulation often occurs in the form of an electrical stimulation to overcome the neurological deficiencies that patients experience.

These debilitating conditions can impair hearing, vision, cognitive, sensory or motor skills, and can lead to comorbidities. Indeed, movement disorders such as multiple sclerosis or Parkinson's are progressive conditions that can lead to a painful and gradual decline in the above skills while the patient is always conscious of every change. The recent advances in brain machine interfaces (BMIs) have shown that a system can be employed where the subjects' intended and voluntary goal-directed wishes (electroencephalogram, EEG) can be stored and learned when a user "trains" an intelligent controller (an AI). This period of training allows for identification of errors in certain tasks that the user deems incorrect, say that on a computer screen, a square is directed to go left and instead it goes to right and also in a situation where the BMI is connected to a fixed robotic hand, the subject directs the device to go up and the signals are interpreted as a down movement. Correct actions are stored, and the error-related brain signals are registered by the AI to correct for future actions. Because of this "reinforcement learning," the system can potentially store single to several control "policies," which allow for patient personalization [48]. This is rather similar to the goals of the company Neuralink which aims to bring the fields of material science, robotics, electronics, and neuroscience together to try and solve multifaceted health problems [49].

While in its infancy and very exploratory, this field will be immensely helpful for patients with neurodegenerative diseases who will increasingly rely on neuroprostheses throughout their lives.

2.6 Ambient assisted living

With the aging society, more and more people live through old age with chronic disorders and mostly manage to live independently up to an old age. Data indicates that half of people above the age of 65 years have a disability of some sort, which constitutes over 35 million people in the United States alone. Most people want to preserve their autonomy, even at an old age, and maintain control over their lives and decisions [50]. Assistive technologies increase the self-dependencies of patients, encouraging user participation in Information and Communication Technology (ICT) tools to provide remote care services type assistance and provide information to the healthcare professionals. Assistive technologies are experiencing rapid growth, especially among people aged 65−74 years [51]. Governments, industries, and various organizations are promoting the concept of AAL, which enables people to live independently in their home environment. AAL has multiple objectives including promoting a healthy lifestyle for individuals at risk, increasing the autonomy and mobility of elderly individuals, and enhancing security, support, and productivity so people can live in their preferred environment and ultimately improve their quality of life. AAL applications typically collect data through sensors and cameras and apply various artificially intelligent tools for developing an intelligent system [52]. One way of implementing AAL is using smart homes or assistive robots.

2.6.1 Smart home

A smart home is a normal residential home, which has been augmented using different sensors and monitoring tools to make it "smart" and facilitate the lives of the residents in their living space. Other popular applications of AAL that can be a part of a smart home or used as an individual application include remote monitoring, reminders, alarm generation, behavior analysis, and robotic assistance.

Smart homes can be useful for people with dementia and several studies have investigated smart home applications to facilitate the lives of dementia patients. Low-cost sensors in an Internet of Things (IoT) architecture can be a useful way of detecting abnormal behavior in the home. For instance, sensors are placed in different areas of the house including the bedroom, kitchen, and bathroom to ensure safety. A sensor can be placed on the oven and detect the use of the cooker, so the patient is reminded if it was not switched off after use. A rain sensor can be placed by the window to alert the patient if the window was left open during rain. A bath sensor and a lamp sensor can be used in the bathroom to ensure that they are not left on [53].

The sensors can transmit information to a nearby computing device that can process the data or upload them to the cloud for further processing using various machine learning algorithms, and if necessary, alert relatives or healthcare professionals (Fig. 2.7). By daily collection of patient data, activities of daily living are defined over time and abnormalities can be detected as a deviation from the routine. Machine learning algorithms used in smart home applications include probabilistic and discriminative methods such as Naive Bayes classifier and Hidden Markov Model, support vector machine, and artificial neural networks [54].

In one example, Markov Logic Network was used for activity recognition design to model both simple and composite activities and decide on

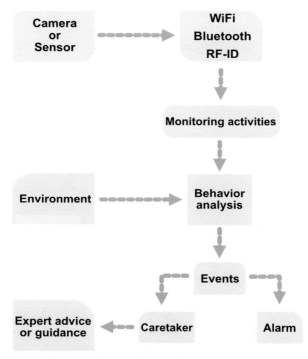

Figure 2.7 Process diagram of a typical smart home or smart assistant setup.

appropriate alerts to process patient abnormality. The Markov Logic Network used handles both uncertainty modeling and domain knowledge modeling within a single framework, thus modeling the factors that influence patient abnormality [55]. Uncertainty modeling is important for monitoring patients with dementia as activities conducted by the patient are typically incomplete in nature. Domain knowledge related to the patient's lifestyle is also important and combined with their medical history it can enhance the probability of activity recognition and facilitate decision-making. This machine learning-based activity recognition framework detected abnormality together with contextual factors such as object, space, time, and duration for decision support on suitable action to keep the patient safe in the given environment. Alerts of different importance are typically used for such decision support and can, for instance, include a low-level alarm when the patient has forgotten to complete a routine activity such as switching off the lights or closing the window and a high-level alarm if the patient has fallen and requires intervention by a caretaker. One of the main aims of such activity monitoring approaches, as well as other monitoring tools, is to support healthcare practitioners in identifying symptoms of cognitive functioning or providing diagnosis and prognosis in a quantitative and objective manner using a smart home system [56]. There are various other assistive technology devices for people with dementia including motion detectors, electronic medication dispensers, and robotic devices for tracking.

2.6.2 Assistive robots

Assistive robots are used to support the physical limitations of the elderly and dysfunctional people and help them by assisting in daily activities and acting as an extra pair of hands or eyes. Such assistive robots can help in various activities such as mobility, housekeeping, medication management, eating, grooming, bathing, and various social communications. An assistive robot named RIBA with human-type arms was designed to help patients with lifting and moving heavy things. It has been demonstrated that the robot is able to carry the patient from the bed to a wheelchair and vice versa. Instructions can be provided to RIBA either by using tactile sensors using a method known as tactile guidance to teach by showing [57].

The MARIO project (Managing active and healthy Aging with use of caring Service robots) is another assistive robot which has attracted a lot of attention. The project aims to address the problems of loneliness, isolation, and dementia, which are commonly observed with elderly people.

This is done by performing multifaceted interventions delivered by service robots. The MARIO Kompaï companion robot was developed with the objective to provide real feelings and emotions to improve acceptance by dementia patients, to support physicians and caretakers in performing dementia assessment tests, and promote interactions with the end users. The Kompaï robot used for the MARIO project was developed by Robosoft and is a robot containing a camera, a Kinect motion sensor, and two LiDAR remote sensing systems for navigation and object identification [58]. It further includes a speech recognition system or other controller and interface technologies, with the intention to support and manage a wide range of robotic applications in a single robotic platform similar to apps for smartphones. The robotic apps include those focused on cognitive stimulation, social interaction, as well as general health assessment. Many of these apps use AI-powered tools to process the data collected from the robots in order to perform tasks such as facial recognition, object identification, language processing, and various diagnostic support [59].

2.6.3 Cognitive assistants

Many elderly people experience a decline in their cognitive abilities and have difficulties in problem-solving tasks as well as maintaining attention and accessing their memory. Cognitive stimulation is a common rehabilitation approach after brain injuries from stroke, multiple sclerosis or trauma, and various mild cognitive impairments. Cognitive stimulation has been demonstrated to decrease cognitive impairment and can be trained using assistive robots.

Virtrael is one of such cognitive stimulation platforms and serves to assess, stimulate, and train various cognitive skills that experience a decline in the patient. The Virtrael program is based on visual memory training and the project is carried out by three different key functionalities: configuration, communication, and games. The configuration mode allows an administrator to match the patient with a therapist and the therapist to configure the program for the patient. The communication tool allows communication between the patient and the therapist and between patients., The games are intended to train cognitive skills of the patient including memory, attention, and planning (Fig. 2.8) [60].

2.6.4 Social and emotional stimulation

One of the first applications of assistive robots and a commonly investigated technology is companion robots for social and emotional

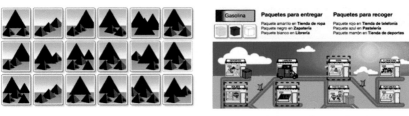

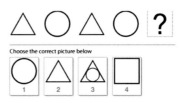

(A) Choose the pictures with two pyramids facing the sun

(B) Plan the delivery of a set of packages

(C) Search for the missing objects and move them to their correct place

(D) Choose the next picture in the series

Figure 2.8 Example of games used for training cognitive skills of patients [60].

stimulation. Such robots assist elderly patients with their stress or depression by connecting emotionally with the patient with enhanced social interaction and assistance with various daily tasks. The robots vary from being pet-like robots to more peer-like and they are all interactive and provide psychological and social effects. The robotic pet PARO, a baby seal robot, is the most widely used robotic pet and carries various sensors to sense touch, sounds, and visual objects [61]. Another robot is the Mario Kampäi mentioned earlier, which focuses on assisting elderly patients with dementia, loneliness, and isolation. Yet, another companion robot Buddy, by Blue Frog Robotics, assists elderly patients by helping with daily activities such as reminders about medication and appointments, as well as using motion sensors to detect falls and physical inactivity. Altogether, studies investigating cognitive stimulation seem to demonstrate a decrease in the rate of cognitive decline and progression of dementia.

2.7 The artificial intelligence can see you now

AI is increasingly becoming an integral part of all our lives. From smartphones to cars and more importantly our healthcare. This technology will continue to push boundaries and certain norms that have been dormant and accepted as the status quo for hundreds of years, will now be directly challenged and significantly augmented.

2.7.1 Artificial intelligence in the near and the remote

We believe that AI has an important role to play in the healthcare offerings of the future. In the form of machine learning, it is the primary capability behind the development of precision medicine, widely agreed to be a sorely needed advance in care. Although early efforts at providing diagnosis and treatment recommendations have proven challenging, we expect that AI will ultimately master that domain as well. Given the rapid advances in AI for imaging analysis, it seems likely that most radiology and pathology images will be examined at some point by a machine. Speech and text recognition are already employed for tasks like patient communication and capture of clinical notes, and their usage will increase.

The greatest challenge to AI in these healthcare domains is not whether the technologies will be capable enough to be useful, but rather ensuring their adoption in daily clinical practice. For widespread adoption to take place, AI systems must be approved by regulators, integrated with EHR systems, standardized to a sufficient degree that similar products work in a similar fashion, taught to clinicians, paid for by public or private payer organizations, and updated over time in the field. These challenges will ultimately be overcome, but they will take much longer to do so than it will take for the technologies themselves to mature. As a result, we expect to see limited use of AI in clinical practice within 5 years and more extensive use within 10 years.

It also seems increasingly clear that AI systems will not replace human clinicians on a large scale, but rather will augment their efforts to care for patients. Over time, human clinicians may move toward tasks and job designs that draw on uniquely human skills like empathy, persuasion, and big-picture integration. Perhaps the only healthcare providers who will risk their careers over time may be those who refuse to work alongside AI.

2.7.2 Success factors for artificial intelligence in healthcare

A review by Becker [62] suggests that AI used in healthcare can serve clinicians, patients, and other healthcare workers in four different ways. Here, we will use these suggestions as inspirations and will expand on

their contribution toward a successful implementation of AI in healthcare: (Fig. 2.9)
1. Assessment of disease onset and treatment success.
2. Management or alleviation of complications.
3. Patient-care assistance during a treatment or procedure.
4. Research aimed at discovery or treatment of disease.

2.7.2.1 Assessment of condition

Prediction and assessment of a condition is something that individuals will demand to have more control over in the coming years. This increase in demand is partly due to a technology reliable population that has grown to learn that technological innovation will be able to assist them in leading healthy lives. Of course, while not all answers lie in this arena, it is an extremely promising field.

Mood and mental health-related conditions are immensely important topic in today's world and for good reason. According to the WHO, one in four people around the world experiences such conditions and as a result can accelerate their path toward ill-health and comorbidities. Recently, machine learning algorithms have been developed to detect words and intonations of an individual's speech that may indicate a mood disorder. Using neural networks, an MIT-based lab has conducted research onto the detection of early signs of depression using speech. According to the researchers, the "model sees sequences of words/speaking style" and decides whether these emerging patterns are likely to be seen in individuals with and without depression [63]. The technique employed by the researchers is often referred to as a sequence modeling, where model sequences of audio and text from patients with and without depression are fed to the system and as these accumulate, various text

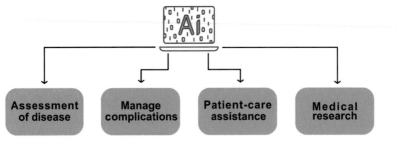

Figure 2.9 The likely success factors depend largely on the satisfaction of the end users and the results that the AI-based systems produce.

patterns could be paired with audio signals. For example, words such as "low", "blue," and "sad" can be paired with more monotone and flat audio signals. Additionally, the speed and the length of pauses can play a major role in detection of individuals experiencing depression. An example of this can be seen in Fig. 2.10 where within a period of 60 seconds and based on the tone and words used, it is possible to measure an estimated emotion.

2.7.2.2 Managing complications

The general feeling of being unwell and its various complications that accompany mild illnesses are usually well tolerated by patients. However, for certain conditions, it is categorically important to manage these symptoms as to prevent further development and ultimately alleviate more complex symptoms. A good example for this can be seen in the field of infectious diseases. In a study published in the journal of trauma and acute care surgery, researchers think that by understanding the microbiological niches (biomarkers) of trauma patients, we could hold the key to future wound infections and therefore can allow healthcare workers to take the necessary arrangements to prevent the worst outcome [64]. Machine learning techniques can

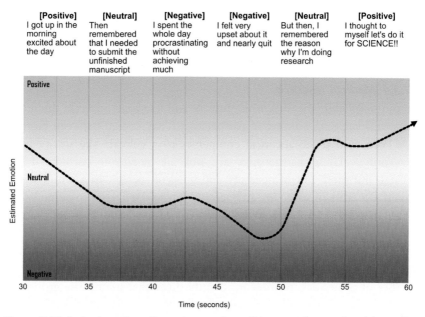

Figure 2.10 Early detection of certain mood conditions can be predicted by analyzing the trend, tone of voice, and speaking style of individuals.

also contribute toward the prediction of serious complications such as neuropathy that could arise for those suffering from type 2 diabetes or early cardiovascular irregularities. Furthermore, the development of models that can help clinicians detect postoperative complications such as infections will contribute toward a more efficient system [65].

2.7.2.3 Patient-care assistance

Patient-care assistance technologies can improve the workflow for clinicians and contribute toward patient's autonomy and well-being. If each patient is treated as an independent system, then based on the variety of designated data available, a bespoke approach can be implemented. This is of utmost importance for the elderly and the vulnerable in our societies. An example of this could be that of virtual health assistants that remind individuals to take their required medications at a certain time or recommend various exercise habits for an optimal outcome. The field of Affective Computing can contribute significantly in this arena. Affective computing refers to a discipline that allows the machine to process, interpret, simulate, and analyze human behavior and emotions. Here, patients will be able to interact with the device in a remote manner and access their biometric data, all the while feeling that they are interacting with a caring and empathetic system that truly wants the best outcome for them. This setting can be applied both at home and in a hospital setting to relieve work pressure from healthcare workers and improve service.

2.7.2.4 Medical research

AI can accelerate the diagnosis process and medical research. In recent years, an increasing number of partnerships have formed between biotech, MedTech, and pharmaceutical companies to accelerate the discovery of new drugs. These partnerships are not all based on curiosity-driven research but often out of necessity and need of society. In a world where certain expertise is rare, research costs high and effective treatments for certain conditions are yet to be devised, collaboration between various disciplines is key. A good example of this collaboration is seen in a recent breakthrough for antibiotic discovery, where the researchers devised/trained a neural network that actively "learned" the properties of a vast number of molecules in order to identify those that inhibit the growth of *E. coli*, a Gram negative bacterial species that is notoriously hard to kill [66]. Another example is the recent research carried out regarding the pandemic of COVID-19 all around the world. Predictive Oncology, a precision medicine company has announced that they are launching an AI platform to accelerate the production of new diagnostics and vaccines, by using more than 12,000 computer simulations per machine. This is combined with other efforts to employ DL to find molecules that can interact with the main proteases (M^{pro} or $3CL^{pro}$) of the virus, resulting in the disruption of the replication machinery of the virus inside the host [67,68].

2.7.3 The digital primary physician

As you walk into the primary care physician's room, you are greeted by the doctor. There is an initial eye to eye contact, then an exchange of pleasantries follows. She further asks you about your health and how she can be of help. You, the patient, have multiple medical problems: previous presence of sciatica, snapping hip syndrome, high cholesterol, an above-average blood pressure, and chronic sinusitis. However, because of the limited time that you have with the doctor, priorities matter [69]. You categorize your own conditions and tend to focus on the most important to you, the chronic sinusitis. The doctor asks you multiple questions about the condition and as you are explaining your symptoms, she types it all in your online record, does a quick examination, writes a prescription, and says to come back in 6 weeks for further examination. For your other conditions, you probably need to book a separate appointment unless you live in a country that designates more than 20 minutes per patient.

The above scenario is the normal routine in most countries. However, despite the helpfulness of the physician, it is not an ideal system and it is likely that if you were in the position of the above patient, you will walk away dissatisfied with the care received. The frustration with such systems has led to an immense pressure on the health workers and needs to be addressed. Today, there are numerous health-related applications that utilize and combine the power of AI with that of a remote physician to answer some of the simple questions that might not warrant a physical visit to the doctors.

2.7.3.1 Artificial intelligence prequalification (triage)

Prior to having access to an actual doctor, trained AI bots can qualify whether certain symptoms warrant an actual conversation with a physician. Many questions are asked of the patient and based on each response; the software encourages the user to take specific actions. These questions and answers are often vigorously reviewed by medical professionals at each stage to account for accuracy. In important cases, a general response of "You should see a doctor" is given and the patient is directed to book an appointment with a primary care physician.

2.7.3.2 Remote digital visits

The unique selling point for these recent innovations is that they allow remote video conversations between the patient and the physician. Normally, the patient books an appointment for a specific time, often during the same day. This provides them with ample time to provide as much information as possible for the physician responsible to review and carefully analyze the evidence before talking to the patient. The information can be in the form of images, text, video, and audio. This is extremely encouraging and creative as many people around the world lack the time and resources to visit a physician and allows remote work for the physician.

2.7.3.3 The future of primary care

In a recent study, when asked about the future of AI on primary care, while acknowledging its potential benefits, most practitioners were extremely skeptical regarding it playing a significant role in the future of the profession. One main pain point refers to the lack of empathy and the ethical dilemma that can occur between AI and patients [70]. While this might be true for today, it is naive to assume that this form of technology will remain dormant and will not progress any further. Humanity prefers streamlining and creative solutions that are effective and take less out of our daily lives. Combine this with the ever-increasing breakthroughs in the field of smart healthcare materials [71] and AI, one could envisage patients managing most of their own conditions at home and when necessary get in touch with a relevant healthcare worker who will refer them to more specialized physicians who could tend to their needs. It is also very important to note that at the time of an epidemic, an outbreak, natural or manmade disaster, or simply when the patient is away from their usual dwelling, a technology that allows humans to remotely interact and solve problems will have to become a necessity. At the time of writing (Early 2020), the threat of a SARS-COV-2 epidemic looms over many countries and is expanding at an unprecedented rate. World experts speculate that the infection rate is high and has the potential to remain within a population and cause many fatalities in many months to come. It is therefore essential to promote remote healthcare facilities/technologies and to have permanent solutions in place to save lives in order to reduce any unnecessary burden or risk on both healthcare workers and patients alike.

References

[1] Miller DD, Brown EW. Artificial intelligence in medical practice: the question to the answer? Am J Med 2018;131(2):129−33.
[2] Kirch DG, Petelle K. Addressing the physician shortage: the peril of ignoring demography. JAMA 2017;317(19):1947−8.
[3] Combi C, Pozzani G, Pozzi G. Telemedicine for developing countries. Appl Clin Inform 2016;07(04):1025−50.
[4] Bresnick J. Artificial intelligence in healthcare market to see 40% CAGR surge; 2017.
[5] Lee K-F. AI superpowers: China, Silicon Valley, and the new world order. 1st ed. Houghton Mifflin Harcourt; 2019.
[6] King D, DeepMind's health team joins Google Health.
[7] Hoyt RE, Snider D, Thompson C, Mantravadi S. IBM Watson Analytics: automating visualization, descriptive, and predictive statistics. JMIR Public Health Surveill 2016;2(2):e157.
[8] Marr B. How is AI used in healthcare—5 powerful real-world examples that show the latest advances. Forbes; 2018.
[9] Kalis B, Collier M, Fu R. 10 promising AI applications in health care. Harvard Business Review; 2018.
[10] Singhal S, Carlton S. The era of exponential improvement in healthcare? McKinsey Co Rev.; 2019.
[11] Konieczny L, Roterman I. Personalized precision medicine. Bio-Algorithms Med-Syst 2019; 15.
[12] Love-Koh J, et al. The future of precision medicine: potential impacts for health technology assessment. Pharmacoeconomics 2018;36(12):1439−51.
[13] Kulski JK. Next-generation sequencing—an overview of the history, tools, and 'omic' applications; 2020.
[14] Hughes JP, Rees S, Kalindjian SB, Philpott KL. Principles of early drug discovery. Br J Pharmacol 2011;162(6):1239−49.
[15] Ekins S, et al. Exploiting machine learning for end-to-end drug discovery and development. Nat Mater 2019;18(5):435−41.
[16] Zhang L, Tan J, Han D, Zhu H. From machine learning to deep learning: progress in machine intelligence for rational drug discovery. Drug Discov Today 2017;22 (11):1680−5.
[17] Lavecchia A. Deep learning in drug discovery: opportunities, challenges and future prospects. Drug Discov Today 2019;24(10):2017−32.
[18] Coley CW, Barzilay R, Green WH, Jaakkola TS, Jensen KF. Convolutional embedding of attributed molecular graphs for physical property prediction. J Chem Inf Model 2017;57(8):1757−72.
[19] Mayr A, Klambauer G, Unterthiner T, Hochreiter S. DeepTox: toxicity prediction using deep learning. Front Environ Sci 2016;3:80.
[20] Wu Z, et al. MoleculeNet: a benchmark for molecular machine learning. Chem Sci 2018;9.
[21] Kadurin A, Nikolenko S, Khrabrov K, Aliper A, Zhavoronkov A. druGAN: an advanced generative adversarial autoencoder model for de novo generation of new

molecules with desired molecular properties in silico. Mol Pharm 2017;14 (9):3098−104.

[22] Blaschke T, Olivecrona M, Engkvist O, Bajorath J, Chen H. Application of generative autoencoder in de novo molecular design. Mol Inform 2018;37 (1−2):1700123.

[23] Merk D, Friedrich L, Grisoni F, Schneider G. De novo design of bioactive small molecules by artificial intelligence. Mol Inform 2018;37.

[24] Shi T, et al. Molecular image-based convolutional neural network for the prediction of ADMET properties. Chemom Intell Lab Syst 2019;194:103853.

[25] Wallach HA, Dzamba MI. AtomNet: a deep convolutional neural network for bio-activity prediction in structure-based drug discovery. arXiv 2015;.

[26] Hashimoto DA, Rosman G, Rus D, Meireles OR. Artificial intelligence in surgery: promises and perils. Ann Surg 2018;268:70−6.

[27] Petscharnig S, Schöffmann K. Learning laparoscopic video shot classification for gynecological surgery. Multimed Tools Appl 2018;77:8061−79.

[28] Lundervold AS, Lundervold A. An overview of deep learning in medical imaging focusing on MRI. Z Med Phys 2019;29:102−27.

[29] Chien CH, Chen CH, Jeng TS. An interactive augmented reality system for learning anatomy structure. In: Proceedings of the International MultiConference of Engineers and Computer Scientists 2010, IMECS 2010; 2010. http://www.iaeng.org/publication/IMECS2018/

[30] Frendø M, Konge L, Cayé-Thomasen P, Sørensen MS, Andersen SAW. Decentralized virtual reality training of mastoidectomy improves cadaver dissection performance: a prospective, controlled cohort study. Otol Neurotol 2020;41(4).

[31] Lee SH, Jung HY, Yun SJ, Oh BM, Seo HG. Upper extremity rehabilitation using fully immersive virtual reality games with a head mount display: a feasibility study. PM R 2020;12:257−62.

[32] Baños RM, et al. A positive psychological intervention using virtual reality for patients with advanced cancer in a hospital setting: a pilot study to assess feasibility. Support Care Cancer 2013;21:263−70.

[33] Dias D, Cunha JPS. Wearable health devices—vital sign monitoring, systems and technologies. Sensors (Basel) 2018;18(8):2414.

[34] Johnson D, Deterding S, Kuhn K-A, Staneva A, Stoyanov S, Hides L. Gamification for health and wellbeing: a systematic review of the literature. Internet Interv 2016;6:89−106.

[35] Athilingam P, Labrador MA, Remo EFJ, Mack L, San Juan AB, Elliott AF. Features and usability assessment of a patient-centered mobile application (HeartMapp) for self-management of heart failure. Appl Nurs Res 2016;32:156−63.

[36] Rangasamy ASS, Nadenichek R, Rayasam M. Natural language processing in healthcare; 2018.

[37] Pham T, Tran T, Phung D, Venkatesh S. Predicting healthcare trajectories from medical records: a deep learning approach. J Biomed Inform 2017;69:218−29.

[38] Rojahn K, et al. Remote monitoring of chronic diseases: a landscape assessment of policies in four European countries. PLoS One 2016;11:e0155738.

[39] Davenport TH, Hongsermeier TM, Mc Cord KA. Using AI to improve electronic health records. Harvard Business Review; 2018.

[40] Wang F, Casalino LP, Khullar D. Deep learning in medicine—promise, progress, and challenges. JAMA Intern Med 2019;179:293−4.

[41] Pham T, Tran T, Phung D, Venkatesh S. DeepCare: a deep dynamic memory model for predictive medicine. In: Lecture Notes in Computer Science (including subseries Lecture Notes in Artificial Intelligence and Lecture Notes in Bioinformatics). Springer; 2016.

[42] Konstantinova J, Jiang A, Althoefer K, Dasgupta P, Nanayakkara T. Implementation of tactile sensing for palpation in robot-assisted minimally invasive surgery: a review. IEEE Sens J 2014;14.

[43] Naeini FB, et al. A novel dynamic-vision-based approach for tactile sensing applications. IEEE Trans Instrum Meas 2019;1.

[44] Naidu AS, Naish MD, Patel RV. A breakthrough in tumor localization: combining tactile sensing and ultrasound to improve tumor localization in robotics-assisted minimally invasive surgery. IEEE Robot Autom Mag 2017;24.

[45] Madani N, Mojra A. Quantitative diagnosis of breast tumors by characterization of viscoelastic behavior of healthy breast tissue. J Mech Behav Biomed Mater 2017;68:180—7.

[46] Simha RK. How Russian scientists cracked the secret of a Vedic ritual drink; 2017.

[47] David O. Scientific verification of vedic knowledge: archaeology online.

[48] Iturrate I, Chavarriaga R, Montesano L, Minguez J, del Millán JR. Teaching brain-machine interfaces as an alternative paradigm to neuroprosthetics control. Sci Rep 2015;5(1):13893.

[49] Musk E. An integrated brain-machine interface platform with thousands of channels. J Med Internet Res 2019;21(10):e16194.

[50] Roberts AW, Ogunwole SU, Blakeslee L, Rabe MA, The population 65 years and older in the United States: 2016; 2018.

[51] Anderson WL, Wiener JM. The impact of assistive technologies on formal and informal home care. Gerontologist 2015;55:422—33.

[52] Barnay T, Juin S. Does home care for dependent elderly people improve their mental health? J Health Econ 2016;45:149—60.

[53] Demir E, Köseoğlu E, Sokullu R, Şeker B. Smart home assistant for ambient assisted living of elderly people with dementia. Procedia Comp Sci 2017;113:609—14.

[54] Fahad LG, Ali A, Rajarajan M. Learning models for activity recognition in smart homes. In: Information science and applications. Berlin: Springer; 2015. p. 819—26.

[55] Gayathri KS, Easwarakumar KS. Intelligent decision support system for dementia care through smart home. Procedia Comp Sci 2016;93:947—55.

[56] Nef T, et al. Evaluation of three state-of-the-art classifiers for recognition of activities of daily living from smart home ambient data. Sensors (Basel), 15. 2015. p. 11725—40.

[57] Joseph A, Christian B, Abiodun AA, Oyawale F. A review on humanoid robotics in healthcare. In: MATEC Web of Conferences; 2018. https://www.matec-conferences.org/

[58] D'Onofrio G, et al. MARIO Project: validation and evidence of service robots for older people with dementia. J Alzheimers Dis 2019;68:1587—601.

[59] Koumakis L, Chatzaki C, Kazantzaki E, Maniadi E, Tsiknakis M. Dementia care frameworks and assistive technologies for their implementation: a review. IEEE Rev Biomed Eng 2019;12:4—18.

[60] Garcia-Alonso J, Fonseca C, editors. Gerontechnology: First International Workshop. In: First international workshop on gerotechnology. Springer; 2018.

[61] Vitanza A, D'Onofrio G, Ricciardi F, Sancarlo D, Greco A, Giuliani F. Assistive robots for the elderly: innovative tools to gather health relevant data. In: Data science for healthcare: methodologies and applications. Springer; 2019.

[62] Becker A. Artificial intelligence in medicine: what is it doing for us today? Health Policy Technol 2019;9:198—205.

[63] Matheson R. Model can more naturally detect depression in conversations. MIT News; 2018.

[64] Dente CJ, et al. Towards precision medicine: accurate predictive modeling of infectious complications in combat casualties. J Trauma Acute Care Surg 2017;83(4).

[65] Hu Z, et al. Accelerating chart review using automated methods on electronic health record data for postoperative complications. AMIA Annu Symp Proc 2016;2016:1822—31.

[66] Stokes JM, et al. A deep learning approach to antibiotic discovery. Cell 2020;180:688—702.

[67] Zhang H, Saravanan KM, Yang Y, Hossain MT, Li J, Ren X, et al. Deep learning based drug screening for novel coronavirus 2019-nCov. Preprints 2020;.

[68] Zhang L, et al. Crystal structure of SARS-CoV-2 main protease provides a basis for design of improved α-ketoamide inhibitors. Science 2020;eabb3405.

[69] Irving G, et al. International variations in primary care physician consultation time: a systematic review of 67 countries. BMJ Open 2017;7:e017902.

[70] Blease C, Kaptchuk TJ, Bernstein MH, Mandl KD, Halamka JD, Desroches CM. Artificial intelligence and the future of primary care: exploratory qualitative study of UK general practitioners' views. J Med Internet Res 2019;21:e12802.

[71] Zang Y, Zhang F, Di CA, Zhu D. Advances of flexible pressure sensors toward artificial intelligence and health care applications. Mater Horiz 2015;2:140—56.

CHAPTER 3

Drug discovery and molecular modeling using artificial intelligence

Henrik Bohr
Chemical Engineering, Technical University of Denmark, Lyngby, Denmark

3.1 Introduction. The scope of artificial intelligence in drug discovery

In the present chapter, we discuss the use of machine learning (ML) as a tool of artificial intelligence, AI, for development of pharmaceutical drugs [1]. This entails the issue of drug discovery and drug design of new drug molecules. Pharmaceutical drug discovery requires handling and analyzing large databases of chemical compounds and therefore means to scan over a huge amount of chemical information. This can be achieved by using machine learning techniques (MLTs) that can operate and reach their goal in a relatively short time. In drug discovery, what matters is not only processing a large amount of data in a short time but also an ability to correlate or associate different data to molecular structure and/or properties. These tasks can be done by MLTs, for example, artificial neural networks (ANNs), that greatly help to develop design of drugs. In the next section, we shall give a list of various MLTs [2] that are of great use in biotechnology for drug development.

As mentioned above, ML is a necessary tool for handling a large amount of chemical data in drug development. However, the type of data related to molecular structures and function that is needed for drug development is often highly nonlinear in nature [3] when it comes to assessment and prediction of drug effects. This means that a little change in the molecular structure of the drug can change the effect of the drug dramatically. This nonlinearity is also typical for assessment of what mushrooms that are edible or lethal although having almost identical structure. This is opposed to linear incrementation/extrapolation or linear regression methods [4], where a little change in the structure is followed by a small effect

Artificial Intelligence in Healthcare
DOI: https://doi.org/10.1016/B978-0-12-818438-7.00003-4

unlike the effect of being alive or dead. The nonlinearity is an ability that can exist in some MLT, for example, in the synapse function of neural networks.

Still the most obvious reason for making use of ML or any other automated computational method in biotechnology or drug development is the ability to scan large datasets in a short amount of time and pick out a particular number or pattern that is optimal with respect to a certain performance.

3.1.1 Areas in which machine learning techniques are applied in biotechnology

Drug design and discovery has usually, before the massive use of AI and MLT, been utilizing either trial and error methods or it has used hand-fitting by starting with a known compound, which is then changed chemically and tested experimentally in rodents for toxicity and effect. Today, the drug industry is instead making strong use of AI and computational methods where the drug process and drug fitting are being simulated in computer systems with the help of software packages.

The areas of application of MLT in Biotech [2,3] are typically (1) screening of large libraries of ligands, (2) virtual screening or high-throughput screening, (3) structure of potential ligands and their pharmacophoric values as a drug, (4) quantitative structure–activity relationship (QSAR) studies, (5) Biological functionality of drug receptors and docking of drugs to them aided by molecular dynamics (MD) simulations, and (6) expected toxicity properties of given drugs. In all these areas, especially ANN is a valuable tool often together with MD programs (see Fig. 3.1). In general, applications of MLT in medicine have been famous for the diagnosis of diseases in patients and also in the analysis of tissue samples in pathology. Below is given an example of point 5 about docking of a drug molecule.

The main aim of drug docking to a receptor molecule is to be able to see how well a given drug molecule fits into a specific receptor. In terms of a quantitative measure, the aim is to be able to predict the binding affinity of that drug to a given receptor. In such a search, it is implicit that the most effective drug for a given receptor is the one that binds most strongly to the receptor. However, there are cases where a drug that binds poorly to a receptor can have a strong effect as a medication and vice versa. Of course, the effect of a drug has to be qualified. The effect can be the potency to inhibit another biological function. There are, for instance,

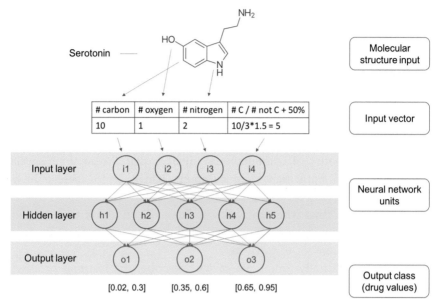

Figure 3.1 Graphical description of the training and operation of an ANN for chemical drug design. The basic figure for the principles of drug design using AI with ANN. On the top is shown the chemical structure of a potential drug molecule, Serotonin. The data for this are in the first row distributed into categories and assembled as an input vector. This is then coded in the input layer of a neural network shown with neurons in three layers where the hidden layer is next and with the output layer in the bottom. The architecture of the network is a multilayer perceptron described in Section 3.2.2. The design procedure is depicted to the right as a flow diagram and described in detail in Section 3.7 as the main story of drug design by AI.

cases where a drug is binding to another than the usual binding site and still has a strong effect. There is also the case of the drug dermorphine, which binds 10,000 times stronger than ordinary morphine but with no beneficial effect for the consumer. It is, however, a highly potent opioid agonist for the mu–opioid receptor [5].

There are other aims in the drug docking effort being, for instance, finding the binding site on the receptor molecule for a given class of drugs. However, having a given specific drug the main aim in the drug docking effort is to find or predict the binding affinity for that drug on a particular receptor molecule.

In the usual programs for general molecular docking of a ligand to a protein, MD programs are used, as explained in a later chapter, to find the optimal binding of the ligand drug to a protein being either a protein receptor or an oligopeptide binding protein as described in [4].

Besides methods in drug docking, there are ML applications for other drug discovery areas such as in High-Throughput Screening, Virtual Screening, QSAR studies, and Docked Complexes analysis that all are connected to drug docking.

3.2 Various types of machine learning in artificial intelligence

ML is here a matter of computational methods that also in some special cases have been implemented in computer hardware to achieve high-speed data processing for that particular purpose.

The type of tasks where ML plays a large role is a task that contains complex problems such as biosystem involving a large set of data and described by a large set of variables.

There are basically four types of ML described by learning processes [4−6]:

1. Supervised learning. Supervision often done by a person handling labeled data arranged in training set of inputs (descriptors) to be withhold with correct output values.
2. Unsupervised learning. Here, the data are unlabeled and there is no supervisor or teacher with a known prescription but the machine has to guess a prescription from the data patterns.
3. Semisupervised learning. A smaller set of labeled data are within a larger set of unlabeled data, where a pattern has to be guessed from the few data and being less costly.
4. Reinforced learning. This is a system with a continuous learning and aimed at using the interaction with the environment and to a given set of data.

The common models of algorithms for the supervised learning are Bayes systems [7], linear regression [8], and neural networks [9], among few others. They are the most popular for handling biochemistry data.

For the unsupervised learning, the most popular algorithm has been the associated memory model of Kohonen, the so-called Kohonen feature map [10]. In the case of semisupervised learning, the algorithms [11] similar to the ones for supervised learning have been used. For the reinforced learning [12], the algorithms can be Q-learning and deep learning (DL) networks [13].

ML methods applied to drug discovery will often involve large datasets and ML is necessary for handling and surveying a large amount of data and these data are a necessity for drug development. On the other hand, MLTs can also be useful in the opposite situation of sparse datasets where machine network methods [13] are employed. Here, auxiliary data with perhaps even a few data points were proposed as a variant for one-shot learning. In this case, the human brain has been an inspiration for overcoming the problem with very little data by, for example, similarity measures with multiple neural networks to infer more data. As for most cases, MLTs are strong and needed when large data are to be surveyed and processed. Later, more details will be given about data mining (DM) with ML in a separate chapter. In a later section, various types of ANN models will be discussed as useful tools for drug design and development.

3.2.1 Artificial neural networks as tools in drug discovery

ANNs are one of the most used methods of ML. They are modeled after the biological brain and were developed as a technique in the 1940s almost 80 years ago [14]. In the following, we shall be reviewing the most used architectures of these neural networks that are basically based on and classified as Associated Memory models. These models have been successful in simulating brain processes and have shown great application in science and engineering [15].

The Associated Memory neural networks are analogous, and actually isomorphic, to spin systems in condensed matter physics. The network systems contain procedures for doing cognitive tasks such as learning.

The basic use of ANN in the AI technology is that of classification and generalization.

An ANN system is defined by the following [11]:

1. State variables n_i, $n_i = 0,1$ are associated with each neuron element i and evolved in time as $n_i(t + 1) = f(\sum_j W_{ij} n_j(t) + \theta_i)$.
2. The neuron elements are connected to each other and to each connection between two elements i, j is associated with a real-valued weight, W_{ij}.
3. A real-valued bias θ_i (threshold fct.) is associated with each neuron element.
4. A transfer function $f_i(n_i, W_{ij})$, with $(i \neq j)$ is defined for each neuron element i and which determines the state of the neuron element as a function of its bias, the weight of the incoming connections and the states of each element joined to it by these links. The transfer function is usually chosen to be a sigmoidal function.

The neuron elements are usually just called neurons, or nodes, the connections are called synapses and the bias known as the activation threshold. The time development or evolution of the neuron states in the network is described by the same equation as in the spin systems of solids.

The training of the neural network is often made from memory of various applications. This is by the same interaction (learning) using the Hebbs rule [3] where the interaction terms or the weights, W_{ij}, are formed by stored patterns, n_i^k, that is, $W_{ij} = 1/N \sum_{k=1}^{p} n_i^k n_j^k$.

3.2.2 Architecture of artificial neural network for drug discovery applications

The various types of ANN listed below, which are also the most used ones in drug discovery applications, are classified by their architecture or by the way the neuron elements are connected, and they are all governed by the same evolution equation.

1. The first ANN is the fully connected associated memory network, or sometimes called the Random neural network, where all neurons are connected to each other with often no specific input neurons but where the neuron states are started with random values. This network is useful for modeling various features of the biological brain, as demonstrated in [16].

2. The ANN that was important in the new development of the neural network revolution was the Hopfield network, which is an on-layer neural network with as many neurons as input signals and connected to active neurons giving output. This network is capable of making autoassociations for forming or regenerating pictures from corrupted data. It is similar to 2-D Ising spin models.

3. The most well-known architecture is that of the perceptron network, where there is a distinctive layer of neurons with input values and connected to neurons processed with data from the input and connected to output neurons (see Fig. 3.2). Usually the perceptron networks are used for only two layers of neurons, the input and the output layers with weighted connections going from input to output neurons and not in between neurons in the same layer.

4. Multilayer perceptron networks are perhaps the most popular ANN with hidden layers of neurons that are connected only to neurons in upper layers or to neurons in layers like in Fig. 3.2. These ANNs are capable of performing recall and extrapolation of any type of logical problems.

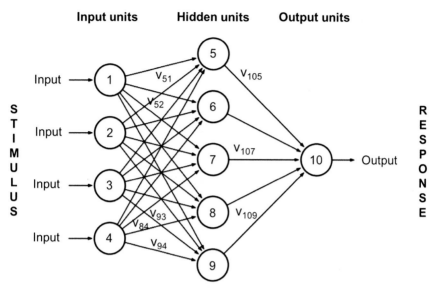

Figure 3.2 Architecture of three-layer feedforward network called the multilayer perceptron network. Each layer is depictured vertically as a set of neurons drawn as circular units with connection lines from the input units (left) to the units in the next layer, with hidden units to, finally, the output units at the right side. Thus the information flow is unidirectional depictured by arrows flowing from left to right and with weight factors V_{ij} attach to each connection line. The neuron units are numbered and so their synaptic connections by numbers describing what are connected.

5. Recurrent neural networks are ANN with feedback loop so the information that in ordinary perceptron networks go forward to the output neuron now also can flow backwards. These networks are optimized with fixed points which are similar to random networks. Thus it is harder to train.

6. Jury networks. Here, one uses several independent ANNs where the majority results are chosen as the result for the output values for the entire network systems. This is not a particular architecture but rather a procedure for improving the reliability of the output.

7. The Kohonen feature map network with no unique information stream like in the perceptron and where the network is unsupervised as opposed to supervised perceptron. It is similar (isomorphic) to Hopfield networks and thus to Ising spin systems. The way the network is laid out makes it useful for classifying molecular reactions in chemistry.

The first of these networks is usually trained with backpropagation, error-correcting networks where the difference between the actual output neuron values and the correct target values is propagated backwards and used for adjusting the weight parameters to obtain the optimal performance by minimizing the error function. The most employed ANN for drug discovery is networks under class 4, 5, 6, and 7. In the training of the ANN, an important concept is that of Hebbian learning, also discussed later, which is a type of reinforced learning.

ANN has first of all been used in drug discovery as a tool for gene search in the huge gene databases such as the GenBank is the NIH genetic sequence database, an annotated collection of all publicly available DNA sequences.

3.2.3 Artificial neural network methods for structure prediction in proteins from their sequence

One of the earliest and most popular applications in biotechnology was training ANN on protein structures and their sequences to predict structures that are new to the training set from their corresponding sequences [15]. Hence, the input to the network is a sequence and the output is some 2-D or 3-D structural information about the protein molecule. Therefore the input could consist of a 20-letter amino acid code while the output could be, for example, a 2-letter code of a particular secondary structure, for example, helix, being present or not present.

In binary code:

The input of amino acid could, for example, for amino acid alanine, be 10000000000000000000 as the middle position, i, read into a window of seven amino acids accounting for the interaction and moved along the sequence. The output of secondary structure of Helix/noHelix could be 1 or 0.

The network is trained on known structures from their sequence to predict novel structures on known sequences. In counting the correctness score, a random one will be 50% with no systematic correlation between input and output, while a score of, say, 60%, indicates a detected correlation. The study in [15] reported up to 70% where of course the training and test set contain low homology in sequence space.

As mentioned in the previous section about ANN architecture, the most popular one here has been the multilayer Perceptron network in which the basic element of the ANN, the neurons, is processing units that produce output from a characteristic, nonlinear function (often a sigmoidal function) of a weighted sum of input data (see Fig. 3.2).

3.2.4 Artificial neural network methods for spectroscopy in biomedicine

In this section, we mention again another application of ANN and applied to spectroscopy.

The task of the ANN for this study has been to classify and identify tissue classes through infrared images from infrared spectra. These ANNs were part of a larger software package, CytoSpec NeuroDeveloper 2.5. The special ability of the ANN methods is to correlate and classify input data with output data through a carefully, arranged training that can be either supervised or unsupervised [14]. The ANNs used for the present type of problem are multilayered feedforward perceptron neural networks [15] (Fig. 3.3). In the following, the most common architecture of these networks is reviewed.

The training process as described above is presenting a set of selected, nonhomologous input data and then adjusting the variable interconnecting synaptic weights such that the output neurons are producing the output values close to the ones given in the training set. This is achieved by the minimization procedure of the backpropagation, error-correcting algorithm and should make the network able to predict new graphical output on the basis of new infrared spectra. Hence, the network can generate new infrared images from combinations of infrared spectra as input. To make more sharp distinction in the classification of spectra

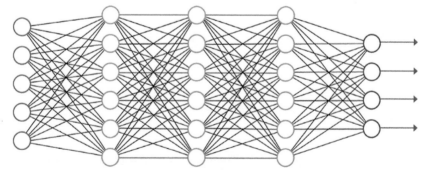

Figure 3.3 Schematic picture of a massive feedforward network for deep learning. In this picture, a very complexed feedforward neural network is drawn schematically with a large number of hidden layer neurons to be trained and advantageously reduced to avoid overfitting using algorithms such as DropConnections [4]. The training of such a complex ANN is possible because of new powerful computer hardware and software.

corresponding to specific types of tissue, one has to make a Hierarchical Cluster Analysis in several spectral ranges for the input data.

Regarding the statistics: The input data of absorption spectra have a precision of classification around 10%, which corresponds to 1 over the number of peak assignments. For the output data of tissue images, the accuracy is 1 over the number of classes of distinct images being around 10, so again the precision is around 10% and of course a different error measure than that of the previous section.

3.3 Molecular modeling and databases in artificial intelligence for drug molecules

Molecular modeling of drug molecules is usually done using MD calculations based on classical mechanics [17] and where the objective is to find the structures of the conformational states of a drug that have optimal effects. Often there is a drug that has to fit to a given drug receptor of known structure and in that such modeling or computations are appropriately based on classical MD [18]. Recently, in the last decade there has been a desire in the field of computational molecular biology to get electronic structures of molecules for understanding their biochemical function. For such tasks, quantum mechanical (QM) techniques are required to calculate and construct the electronic structures and effects of the molecular orbitals.

However, such electronic calculations are not yet possible to carry out for larger molecular structures such as drug receptor molecules, due to their electronic complexity but works reasonably for peptides and smaller oligonucleotides [19] and provides valuable information and methods for drug discovery and development. These computational techniques, classical as well as quantal, can certainly be considered as ML tools and involves huge computer processing like the ANNs and often use programs of system packages [18,19]. In Section 3.4, application of MD and QM is discussed further.

3.3.1 Databases for the training sets in drug discovery

The use of neural network for biomolecular technology is, as in general, very dependent on the extraction of data from databases. In each field of application, one will have to find good descriptors of the target values one will have to extract from the databases. In the case of design of drug molecules binding to receptors [20], the molecular binding data for the

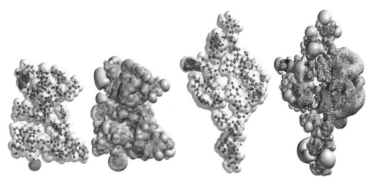

Figure 3.4 Graphical plot of the electronic structures of two oligonucleotides (13-mers) with different chirality. The molecules are shown as potential surfaces derived from quantum mechanics calculations of molecular wave functions. The two molecules contain several hundred atoms and have the same sequence of nucleotides but with a slight change in arrangement of two of the atoms. This minor change of symmetry realized in quantum mechanics alters the molecular structure and thus the chemistry. Each of the two molecules is shown by their electronic density (transparent) to the left shown also with their atoms and their electrostatic potential surface (solid is negative potential) to the right. The highest occupied electronic orbitals are blue/red and the lowest unoccupied are orange. These quantum electronic details are important information when one has to assess a drug molecule for its properties.

training of ML systems can be obtained from the online network of the ligand–receptor binding database, bindingdb.org [21]. The chemical descriptor for the ligand molecule in the database is often a picture of the molecule to be used as input data in the ML system and being correlated to a binding strength for a given receptor molecule. Thus the input to the ML system is structural data of the ligand molecule and output data are binding strength data as shown schematically in Fig. 3.4 [20].

3.3.2 Database mining

DM is an important ingredient in the AI technology, especially drug development. This is also due to the available software tools such as Waikato Environment for Knowledge Analysis (WEKA) for analyzing large databases. A book has been written about that [22]. In that book, basic ML and DM methods are reviewed. That basically includes (1) decision trees as in C4.5, (2) instance-based learning, (3) rules for classification and association, (4) vector machine techniques, (5) Bayes classifiers, (6) clustering algorithm as in K-means, (7) prediction from linear regression trees and model trees and issues of (8) data-cleansing and boosting or stacking. We shall here discuss especially three of the methods: C4.5, K-means, and Support Vector machines.

C4.5 uses a decision tree on data that have already been classified. New data are allocated to a class on the basis of the attributes of the object. In the case of drug molecules, it could be bonding structures.

K-means creates groups from the set of objects so the members of a group are similar. It is a kind of cluster analysis where the center of each cluster is found by the most typical member. It is important with a metric measuring the difference between the members of the space of clusters like.

Support Vector Machines. This method is similar to C4.5 but without using decision trees. It can learn to classify into, for example, two classes by a hyperplane separating the objects in the two classes. It could be balls of two different colors. A crucial point or problem here is to work in higher dimensions.

For a field like drug development, DM contains crucial tools and is essential for finding important leads for a drug to a given receptor molecule. The methods of DM would not have had the impact in the past due to limited drug databases, that is, few and small, but today drug search and drug libraries are huge and contain modern computer techniques and resources [23].

3.4 Computational mechanics ML methods in molecular modeling

ML can also involve and actually constitute computer simulation of molecular systems and in that sense molecular modeling can be considered as an ML program package. Below in the following subchapters, the use of classical and quantum mechanics is also described.

3.4.1 Molecular dynamics simulation for drug development

The MLTs for employing MD have been developed during the last part of the last century and are involving large computational systems that are of great use in drug development. Essentially the MD methods are about solving the classical equations of motion for all atoms in the molecular system including solvents. The computational systems of AI consist of large computer program packages such as NAMD, CHARMM, and What-if [17,18]. To make comparisons to ANN, the training of the MD system is a matter of adjusting the force field in accordance with the data of structural chemistry. The input consists of atomic composition of the molecules and output is then the computed molecular structures. The MD simulation techniques basically consist of solving Newtonian equations of motion and integration of observables in time developments.

3.4.2 Computations with quantum mechanical techniques for drug development

In the previous section, the ML algorithms make use of classical mechanics equations. This section describes the ML algorithms that include QM equations. The ML for this application of drug development on the molecular level consists first of a chemistry construction program package, Spartan [21] where the molecule is constructed from its sequence data. Next, the optimization is carried out with the quantum mechanics equations in the Hartree-Fock (HF) scheme. These procedures are necessary for obtaining and describing the electronic structures and properties of the atoms and the molecule of the drug.

Electronic properties of biomolecules such as proteins, DNA, and drug molecules are important for studying the effects of drugs on a receptor and for understanding their biological function. These electronic properties of proteins and of the oligonucleotides with respect to the molecular orbitals and the electrostatic potential surface can be obtained by the employment of QM calculations made within the DFT or HF optimization systems. In this optimization, an energy function is optimized with respect to structural, geometric parameters of electrostatic energy configurations. Here, electron dynamics is separated from nuclear motions under the Born-Oppenheimer approximation. Finding the minimized energy configuration means that nuclear coordinates and electron orbitals are obtained as a wave function in the ground state of the molecule. If the structures seem realistic (correct bond lengths, etc.) and the optimization steps have converged providing a unique energy function this and the wave function are describing the target molecule at the atomic level. The wave function in this approach is constructed from a basis set of Gaussian functions.

In a particular case of a molecule of oligonucleotides of small DNA/RNA segments, the quantum calculations were possible [22,24] meaning that convergence could be reached for systems of many hundreds of atoms and in a matter of a few hundred of CPU hours.

The electronic properties of the oligonucleotides, both with respect to their molecular orbitals and their electrostatic surfaces, are advantageously obtained by using QM HF optimization employing the basis set of Gaussian wave functions. This is because the force field of the oligonucleotides is more constrained than in the case of peptides. In the present approach, a large energy functional can be optimized with respect to the electrostatic energy configuration solely on the basis of the electronic density by separating electron and nuclear motions. Once the energy-minimized electronic structure is obtained, that is, the nuclear coordinates and electron orbitals in the ground state, the electrostatic surfaces and the HOMO (highest occupied molecular orbital) and LUMO (lowest unoccupied molecular orbital) bonding system can be calculated providing, for example, information about electron donor/acceptor sites.

Fig. 3.4 shows the case study of a set of two oligonucleotides as phosphorothioates with specific chiral structures depending on whether the sulfur atom is left or right of the oxygen atom being either R or S, respectively, and depending on what nucleotide has what chiral symmetry. The oligonucleotide can be described as RSRRRRSR and with sequence of nucleotides CACACTCC. In Fig. 3.4, the oligonucleotide 8-mer is shown in the full R state configuration to the left and to the right as the full S state configuration both shown in atomic resolution and shown with an electrostatic surface structure where the negative surface is shown in a solid representation and the positive potential as a mesh representation. The HOMO state is blue/red surface and the LUMO is transparent. The all-R state is globular in structure to the left and the all-S state to the right is stretched out helical. The important point here is that the two oligomers (all-R or all-S configurations) are very different in appearance although having the same sequence but with different chiral symmetry. Such structures can only be calculated by the use of quantum mechanics in an ML program system with chemistry molecular setup and with quantal optimization. The helical structure turned out to be the basis for a better drug than the globular structure [22].

3.5 Drug characterization using isopotential surfaces

In the previous section, the construction of isopotential surfaces was mentioned as a useful technique for analyzing the molecular structure and function. Here, we shall discuss how these isopotential surfaces are actually constructed and the meaning of them. To be specific they are equipotential surfaces whose points have the same electrostatic potential relative to the molecular charges and numerically according to an isovalue of a positive test charge outside the molecule. The value of the electrostatic potential is mapped to an electron density surface. The choice for a good isovalue has to do with what kind of bonding is involved in the docking, such as H-bonding with the value of around a few electronvolts (eV). To construct potential surfaces, electric charges and electron densities for the relevant electrons are required. This is a matter of large quantum computations for getting the wave function for the molecule. The docking of a ligand to another protein molecule is very dependent on the electrostatic surface potential of the ligand and the protein both functioning as complementary surfaces.

In another case study, isopotential surfaces are used for drug design in the case of locked nucleic acid modification of oligonucleotide molecules where the chiral properties are studied. The docking of a ligand to another protein molecule is very dependent on the electrostatic surface potential of the various forms of the modified oligonucleotides, even in just a few nucleotides, and plays a role in therapeutic studies. Also here, QM techniques have been important in construction of the isopotential surfaces and in the construction of electronic structures described in [24] and in [25,26], where physical properties have been derived of duplex DNA molecule.

In a similar study, an isopotential surface is constructed for a small peptide, an opioid molecule, (see Fig. 3.5), where the yellow side represents the positive potential and the red side represents the negative potential. The surface is constructed from the isopotential value as 0.2 eV meaning that a positive unit charge is influenced by the potential depicted in the figure. This is a useful information when studying how particles are attracted by the molecule under consideration.

3.6 Drug design for neuroreceptors using artificial neural network techniques

Neuroreceptors are very relevant and obvious areas of application to ANN and other ML methods operating on the molecular level with respect to the ligand, which often is the matter of small molecules of less than 100 atoms. The main obstacle is here to get good chemical descriptors for the relevant ligand molecule and which should be clear enough for training an ANN. Neuroreceptors are popular and important targets for the pharmaceutical industry. In the selection for neuroreceptors, the

Figure 3.5 A picture of the isopotential of a small peptide molecule (Dermorphin). The positive potential surface is shown in yellow and the negative in red. The isopotential surface is an equipotential surface, that is, with the same potential value all over the surface. This isovalue, which defines the position of the surface relative to the molecule, is here set to 0.2 eV, and which is the energy a unit charge (positive) experiences when approaching the surface. The information derived from such electronic surface calculation is important when assessing the binding property of a drug molecule to a receptor or to other molecules.

most important criterion is to have a database of ligand molecules for that particular receptor.

In this review, we shall mention the studies of the following five receptors for Dopamine, Glutamate, Serotonin, Opioid, and Gaba. In the case of dopamine D3 receptor, the crystal structure is used as a template for guiding drug design. This is because it is a target of pharmacotherapeutic interest in a large number of neurological disorders such as schizophrenia, drug addiction, and epileptical disorder [21]. In Table 3.1 is shown the results of applying Deep Belief networks on the four drug molecules glutamate, mu-opioid, serotonin, and serotonin Kd to the corresponding neuroreceptors where serotonin Kd was getting the highest score of up to 92% correctness in a network with two output categories.

We have reported on the results obtained with DL on the serotonin, opioid, and glutamate receptors reported in the following article [24] and discussed in the next section.

Another successful application of neural networks is obtained with Kohonen Networks where the most likely chemical activity sites are

Table 3.1 Performance results of the various datasets and parameters using Soft-Max regression. The first column is giving the dataset of the particular ligands. The second column is giving the output categories involved (often number of output neurons). The third column contains the input vectors (i.e., input parameters) and the last columns give the performance in terms of ratios and scores measured in percent correctness [27].

Dataset	Categorization schemes	Input vectors	Ratio	Scores/[%]
Glutamate	2, 5, 8	2, 12, 16, 4, 14	0.80	80.95, 60, 44
μ-opioid	2, 5, 8	6, 14, 9, 11, 15	0.80	79.31, 44.83, 31.03
Serotonin	2, 5, 8	2, 16, 13, 4, 14	0.80	80.00, 56.00, 40.00
Serotonin k_d	2, 5, 8	2, 8, 1, 11, 7	0.80	92.3, 73.1, 50

predicted on a receptor molecule [1]. The major goal in this application is to optimize the geometry of the chemical molecular substructures in the design molecule. To aid the drug design, one has to get an understanding of the space of chemical interaction of the receptor molecules involved and extract the features about a particular drug's likely place, that is, the drug likeliness. This can be achieved with the help of a Kohonen feature map as shown in [1]. This feature map is constructed by a large set of chemical data that are being classified according to the learning process of the self-organized Kohonen network.

The self-organized map is here used as a statistical data analysis method that with unsupervised learning analyze the input data by using the principles of Hebbian learning. In the textbook by Zupan and Gasteiger [1], Kohonen maps were used to extract features in molecular electrostatic potentials for binding of active compounds to a receptor such as the muscarine receptor where several similar ligands can bind. It has been possible from a lead molecular structure to arrive at the possible site of an active drug by using the Kohonen map. This is done by presenting points of the Van der Waal surfaces of drug molecules and then from these examples to reproduce interaction maps from a Kohonen network and by comparison see which interaction map (of potential surface) best corresponds to the receptor surface.

A feature map of chemical drug likeliness is commonly performed and can be found in numerous literature [1]. This is a good example of how one can use ML for finding and understanding drug—receptor relation and developing drug design with respect to a specific receptor. The Kohonen feature map gives an interesting and informative picture of the chemical interaction space but is not accurate enough for modeling of the molecular structure compared to the quantum structure prediction in drug design. This can be seen in one of the previous chapters about quantum calculations.

3.7 Specific use of deep learning in drug design

DL is a technique consisting of many ML algorithms that operate and make use of ANN with many fully connected hidden layers of processing units. The large number of processing units, or neurons, is carefully trained in DL algorithms to avoid overtraining [27].

Lately, DL has turned out to be a successful tool in training and operating neural networks and other ML methods. The improved algorithms of DL have been developed to avoid overtraining and overfitting when employing neural network systems with a massive number of processing units in many hidden layers. Examples of such algorithms include dropout [4,28], dropconnect [29], and pruning algorithms, and these algorithms can greatly improve the performance of ANN. The pruning technique is a necessary tool for managing large neural networks like those in the living brain. Especially, the DL networks act and rely on the many layers of units that are added during training. Other important improvements in the operation of DL come about by using special ANN other than the usual feedforward ANN, for example, convolutional NN and recurrent NN [4]. One can study the performance of DT and compare it with a simple Soft-Max regression system [27]. Such a system is created and used to benchmark the various ANN in many fields where there are classification problems. This is typical in the field of pharmaceutical databases for drug design where patterns are to be retrieved about chemical structures of molecules on the basis of incomplete information since the real structure of the new drug is unknown and the molecular variability is large.

In Fig. 3.6 is shown the comparison of different neuroreceptors from various databases using Soft-Max regression where performance is measured as a function of categories of the output data. In Fig. 3.1, the full ML, or the ANN network system, is the one that is used for drug design. Here the drug candidates are represented as an input vector that describes

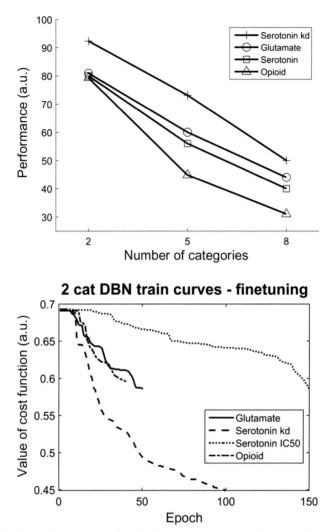

Figure 3.6 The training curves for the deep-learning networks used for drug fitting. The training (finetuning) curves for a two-category network are shown for various neuropeptides where the value of the cost function is shown as a function of the number of training cycles. The serotonin score (effectiveness measured by IC50) is clearly more stable along the training [27].

the chemistry of the ligand molecule. The output values are then the effects of the drug related to the specific ligand molecules in the input. These numbers are taken from the drug database on which the ANN is trained. The ANN is then supposed to predict output values, Kd, IC50, or binding constants, corresponding to new molecules. The predictions can then be judged or tested by database numbers that are novel to the network, that is, not included in the training set. The most successful ANN system for predicting output values for input ligand/molecules was the DL network that used a large number of hidden neurons that were carefully pruned in the training optimization and where Soft-Max regression was used. The results are shown in Table 3.1 for various drug molecules for the neuroreceptors mentioned above [27]. The table contains the three receptor molecules, glutamate, opioid, and serotonine, where the latter is represented by two different outputs, namely, the binding equilibrium constant, Kd, and the bioeffect strength, IC50. The data clearly show that Kd in the case of serotonin is easier to predict being just an equilibrium constant than the IC50 value, which is dependent on many factors. The table also shows, of course, that the two-category network achieves the highest score in percentage of correctness for the ligands compared to the other types of receptors.

3.7.1 Other applications of machine learning in drug development

Other applications of DL in drug discovery have been to handle large libraries of chemical data to achieve properties of particular drug candidates by operating with virtual screening to obtain knowledge of chemical activity for a given molecular structure [30,31] where DL is used for predicting specificity of a given compound.

Particularly in medicinal chemistry molecular properties can be extracted for a given molecular compound and related to the physicochemical bioactivity level so as to target certain properties of a molecule such as its pharmacokinetics properties, for example, absorption, excretion, and toxicity [32,33].

3.8 Possible future artificial intelligence development in drug design and development

Due to the fast development in computer hardware and software, it will be possible in the near future to employ detailed QM calculations for drug molecules. Such development is necessary for obtaining detailed chemical structure information on the atomic level, to be analyzed by computational processes.

An important area of future AI development in medicine is to use real-time spectroscopy and analysis of molecules in the tissue taken during surgery or biopsy of a patient. Several studies have already been carried out in oncosurgery with real-time spectroscopy [34], where AI techniques are necessary for forming a real-time picture of the molecules in the area of operation. This is often just obtained on the basis of sparse data from Infrared Spectroscopy, as in [34], and here AI techniques are important for processing and transforming the data into pictures. This is done with the utilization of ANN, and such application of infrared spectra is described in Section 3.2.4.

In modeling drug receptors, the biggest problem had previously been to obtain atomic and bonding structure information in detail. By special crystallographic techniques for nematic crystals, X-ray diffraction methods could obtain these data for the important membrane-bound drug receptors. However, these data of drug receptors are static. The big thing is here to get a dynamical picture of the drug molecule interacting with the receptor. Here, the quantum computations are relevant since they can aid the analysis and interpretation of the drug interaction process. Today, it is still difficult to simultaneously simulate more than a few hundred atoms, this is less than a full drug-receptor protein molecule embedded in the environment solvent or lipid molecules. However, new development in AI and computer software makes it likely to realize simulation of the drug docking process in several milliseconds and for many thousands of atoms. Such simulations will also be important in biochemical sciences and biotechnology.

References

[1] Zupan J, Gasteiger J. Neural networks in chemistry and drug design. 2nd ed Wiley-VCH; 1999.
[2] Mandlik V, Bejugam PR, Singh S. Application of neural networks in modern drug discovery. Chp. 6 in Artificial neural networks for drug design, delivery and disposition. Elsevier Inc; 2016.
[3] Davidian M, Giltinan DM. Nonlinear models for repeated measurement data: an overview and update. J Agric Biol Environ Stat 2003;8:387–419.
[4] Chen H, Engkvist O, Wang Y, Olivecrona M, Blaschke T. The rise of deep learning in drug discovery. Drug Discov 2018;23:1241–50.
[5] Fu Y, Zhao J, Chen Z. Insight into the molecular mechanics of protein-ligand interaction by molecular dynamics simulation: a case of oligopeptide binding protein. Comput Math Meth Med 2018;2018:3502514.
[6] Melchiorri P, Negri L. The dermorphin peptide family. Gen Pharmacol 1996;27:1099.
[7] Hertz JA, Krogh A, Palmer RG. Introduction to the theory of neural computation. Addison Wesley; 1991.

[8] Kotsiantis SB. Supervised machine learning: a review of classification techniques. Informatica 2007;31:249—68.

[9] Freedman DA. Statistical models: theory and practice. Cambridge University Press; 2009. p. 26.

[10] Kohonen T. Self-organized formation of topologically correct feature maps. Biol Cybern 1982;43(1):59—69.

[11] Haykin S. Neural networks. Prentice-Hall; 1994.

[12] van Otterlo M, Wiering M. Reinforcement learning and markov decision processes. Reinforcement learning. Adaptation, learning, and optimization 2012;vol. 12:3—42.
Bertsekas DP. Dynamic programming and optimal control: approximate dynamic programming, vol. II. Athena Scientific; 2012.

[13] Vinyals O, et al. Matching networks for one shut learning. arXiv 2016;16706.

[14] Mueller B, Reinhardt J. Neural networks; an introduction. Springer-Verlag; 1900.

[15] Bohr H, Bohr J, Brunak S, Cotterill RMJ, Lautrup B, Nørskov L, et al. Protein secondary structure and homology by neural networks. The α-helices in rhodopsin. FEBS Lett 1988;241(1—2):223—8.
Bohr H. Neural networks predicting protein structures [doctoral thesis]. Lyngby: Polytechnical Press; 1997.

[16] McGuire P, et al. Random neural networks show volatility. Neural Syst 2001;9:1—20.

[17] Karplus M, Porter RN. Molecular dynamics. In: Atoms and molecules. Wiley; 1975.

[18] Frisch MJ, et al. Gaussian 09, Revision D.01. Wallingford, CT: Gaussian, Inc; 2013.

[19] Hobza P, Sponer J. Struckture, energetics, and dynamics of the nucleic acid base pairs: noneimpirical an initio calculations. Chem Rev 1999;99:3247—76.

[20] Roothaan CCJ. New development in molecular orbital theory. Rev Mod Phys 1951;23:69—89.

[21] Spartan'14. Irvine, CA: Wavefunction, Inc.

[22] Bohr HG, Irene Shim I, Stein C, Ørum, Hansen HF, Koch T. Eletronic structures of LNA phosphorothioate oligonucleotides. Mol Ther Nucleic Acids 2017;8:428—41.

[23] Geller J. SIGMOD Record, vol. 31, 74—76 (2002) about this subject under the title "Data Mining: Practical Machine Learning Tools and Techniques with Java Implementation" by Witten IH, Frank E. Morgan Kaufmann Publishers; 2000.

[24] Koch T, Shim I, Lindow M, Ørum H, Bohr HG. Quantum mechanical studies of DNA and LNA. Nucleic Acid Ther 2014;24:139—48.

[25] Keck TM, et al. Beyond small molecule SAR — using the dopamine D3 receptor crystal structure to guide drug design. Adv Pharmacol 2014;69:267—300.

[26] Rezac J, et al. Stretched DNA investigated using molecular dynamics and QM calculations. Biophys J 2010;98:101—10.

[27] Agerskov C, et al. Curr Comp Aided Drug Des 2015;11..

[28] Srivastava N, et al. Dropout, a simple way to prevent networks from overfitting. J Mach Learn Res 2014;15:1929—58.

[29] Wan L., et al. Regularization of neural networks using DropConnect. In: Sanjoy J David M. Proceedings of the 30th International Conference on Machine Learning, vol. 28. 2013. p. 1058—1066.

[30] Pereira JC, Caffarena ER, dos Santos CN. Boosting docking-based virtual screening with deep learning. J Chem Inf Model 2016;56:2495—506. Available from: https://doi.org/10.1021/acs.jcim.6b00355.

[31] Zhang R, Li J, Lu J, Hu R, Yuan Y, Zhao Z. Using deep learning for compound selectivity prediction. Curr Comput Aided Drug Des 2016;12:5. Available from: https://doi.org/10.2174/1573409912666160219113250.

[32] Duch W, Swaminathan K, Meller J. Artificial intelligence approaches for rational drug design and discovery. Curr Pharm Des 2007;13:1497—508. Available from: https://doi.org/10.2174/138161207780765954.

[33] Maltarollo VG, Gertrudes JC, Oliveira PR, Honorio KM. Applying machine learning techniques for ADME-Tox prediction: a review. Exp Opin Drug Metab Toxicol 2015;11:259−71. Available from: https://doi.org/10.1517/17425255.2015.980814.

[34] Piva JAAC, et al. Overview of the use of theory to understand IR and Raman spectra and images in biomolecules. Theo Chem Acc 2011;130:1261−73.

CHAPTER 4

Applications of artificial intelligence in drug delivery and pharmaceutical development

Stefano Colombo
Independent Scientist, Leon, Spain

4.1 The evolving pharmaceutical field

The pharmaceutical industry has traditionally been a conservative sector, researching and developing preferentially small-molecule drugs intrinsically featuring (1) stability, (2) adequate potency for therapeutic purposes, and (3) acceptable toxicity for the vast majority of the consumers. Among the most powerful approaches in pharmaceutical R&D is the systematic chemical screening of molecular variants in combinatorial libraries, with the aim to provide novel molecules with positive features to be exploited in healthcare. However, this approach alone is becoming insufficient in providing novel pharmaceutics to the increasingly demanding healthcare industry.

A number of causes can be identified for the limited success of this approach in the R&D of new pharmaceuticals. First, small-molecule drugs, for example, simple molecules easily obtained by chemical synthesis such as penicillin-derived beta-lactam antibiotics, have been largely investigated already for synthetic alternatives, leaving only a few and often problematic new candidates for future development. Further, very stable and potent molecules are already on the market for many therapeutic areas, producing a hard-to-surpass benchmark for new molecules. Next, the ever-harder path through clinical studies and aggressive competition from generic companies make the development of novel molecules less profitable in the absence of defensible product protection strategies, with the exception of a few specific applications or rare-disease niches [1−3].

Nevertheless, monitoring the state of the pharmaceutical industry on the current productivity crisis of novel molecule R&D is misleading. In the last three decades, the pharmaceutical sector has been reinvigorated by

Artificial Intelligence in Healthcare
DOI: https://doi.org/10.1016/B978-0-12-818438-7.00004-6
85

an unprecedented diversity of approaches to the therapeutic development of pharmaceutics, transforming each pitfall encountered by a scientific and technical challenge. The biomolecular drug industry is currently growing very fast, compensating for the crisis of small-molecule drugs and filling the gap left by poorly delivering R&D endeavors. Small-molecule drugs are composed of a relatively low number of atoms forming a functional molecule for which their action, conformation, reactivity, and stability rely on the atomic composition and bonds. Biomolecules are, on the other hand, large entities formed by several molecular units, for example, amino acids for proteins and nucleotides or ribonucleotides for nucleic acids. Their function and stability are determined not only by the atomic features of each unit but also by the supramolecular sequence and spatial conformation.

The use of complex biomolecules as pharmaceutics has already been successfully exploited, leading to blockbuster products. Among these, some immediate examples include insulin and adalimumab (marketed as Humira), which on their own constitute a sizeable share of the current biopharmaceutical market [4,5]. Nevertheless, biomolecules are especially labile and their administration often requires infusion into the bloodstream. As a consequence of the restricted choice of administration routes available, modulating their pharmacokinetics is also hard to achieve [6] (Fig. 4.1). Looking further ahead, nucleic acid–based drugs are just around the corner in the R&D pipeline of many companies, and these will be even more demanding for molecular stabilization and pharmacokinetics modulation [7,8]. In parallel, the enhancement of difficult small-molecule pharmacokinetics by advanced formulations is still a highly relevant goal for many players in the pharmaceutical industry [9−11].

The discipline of modulating molecular features to obtain augmented physicochemical profiles for pharmacological application is generally referred to as drug delivery (Fig. 4.1). In the current R&D industrial landscape, drug delivery represents a multidisciplinary development of the traditional drug formulation sector. Since the establishment of the Nanotechnology National Institute (NNI) in 2000, nanotechnology has been a hot topic in drug delivery, as engineering of nanosized matter was found to be particularly well-suited for interacting with the human body and controlling drug distribution [12−14]. However, by focusing on nanometer engineering, the complexity of the drug delivery systems is increased. Hereafter, Quality-by-Trial approaches based on high-throughput systematic series of attempts became incapable of exploring

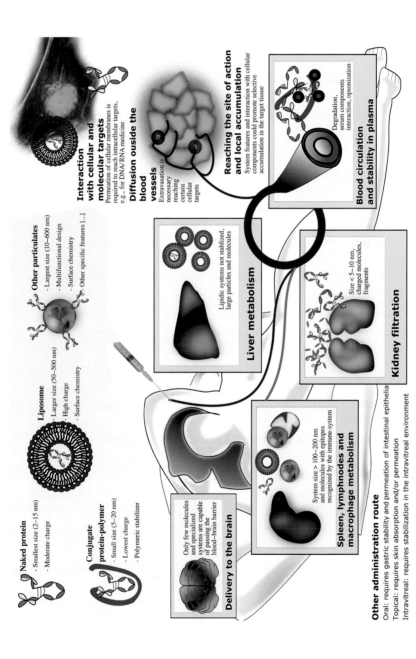

Figure 4.1 Introduction to drug delivery and biological barriers. Active pharmaceutical ingredients need to reach the site of action for being effective. The drug delivery system is engineered to aid active pharmaceutical compound to overcome the biological barrier and improve the absorption, distribution, metabolism, excretion, and toxicity profile of the drug.

the wide design space developing these complex products. Therefore the development and manufacturing of novel products is moving toward quality-focused approaches resolving the escalating complexity of the product quality profile [15]. The passage toward quality-based rational engineering on scientific principles [16,17] is now guiding pharmaceutical R&D and manufacturing innovation, representing one of the most suitable areas for artificial intelligence (AI) implementation.

As boundaries between pharmaceutical sciences, engineering, biotechnology, and applied physics get smaller, fields of computational simulations and bioinformatics have become more influential in pharmaceutical R&D. Adaptive algorithms, such as neural networks, have become increasingly relevant in proteomics and bioinformatics, galvanized by the vast genomics and proteomics projects. More recently, advanced computing methods started to emerge as mature opportunities for pharmaceutical R&D and industrialization units. Today, examples of this technology can be found supporting pharmaceutical discovery, production, and diagnostics [18,19]. Neural networks and other advanced approaches in computing owe their popularity mostly to their early application in scientific niches, such as systems biology and atomic simulations. This is because of the necessity of parsimonious approaches to complex computation, that is, fuzzy logic and heuristic approaches. Stratified adaptive algorithms, such as neural networks, are still currently more widely used than deductive or sensing AI in pharmaceutical R&D, as they readily provide computational power to already existing applications, for example, in silico research and modeling [20]. Thus advanced computational tools such as artificial neural networks have existed in pharmaceutical R&D for a while and display a certain grade of maturity. However, these are most often implemented in highly specialized scientific tasks, for example, molecular design and screenings or computation for process control [18,21−23]. AI intended as a technique capable of inductive or deductive processes or some levels of sensing is still applied to a lesser extent in the pharmaceutical field. One of the reasons for this is that AI is effective when supported by rich, trained, and ordered databases of information to solve problems with extensive data or information-rich environments. Currently, the lack of large standardized datasets is limiting the AI implementation in the pharmaceutical field, resulting in few transferrable AI cases to be imitated and developed [24−26]. Conversely, AI applications in the cosmetic and consumer industry are the leading cases for quality-by-design (QbD)−based product development [27]. Here, large databases of consumers' preference and related product quality profiles are joined to bring together marketing and product research [27].

This chapter aims to explore the involvement and perspective of AI in drug development, delivery, and manufacturing, hence encompassing drug delivery and pharmaceutical product development from inception to quality-by-design production. For this, we will focus preferentially on existing technologies as leading examples. Yet, a certain level of conceptual abstraction is required to fill in the gap between the inception and application of AI.

For this simple purpose, AI in this chapter is defined as a technique with inductive or deductive capability [3] and more specifically categorized as

Passive AI: a method capable of deep learning and adaptation to changing information. The system should be able to summarize the information through a simple output, for example, a suggestion or informed indication. These are very likely commonly used systems in consumer markets, working with a substantially basic cycle of retrieving information, using it to provide the indication, and to some extent revising it to refine the next suggestion.

Predictive AI: represents a subclass of passive AI for structuring and/or modeling data to provide a forecast or a prediction. This type is largely applied in healthcare, finance, and consumer markets. A predictive AI capable of self-creating data (deduction) upon revision and retrieval could show capability for identifying and overcoming common data biases for the evaluation of models. In general, this perspective might enable the progression of AI in fields where the knowledge is potentially lacking and the practical expertise pool is restricted.

Active AI: an AI system capable of sensing, self-probing, and independently exploring the real space (or a simulated version of it), for example, by operating an either simple or complex robotic system. In comparison with the passive approach, the interaction with the physical world is supplying a source of direct verification, self-obtaining the information and revising it. This method could actively fill knowledge gaps. Therefore this designation also extends to factual machine experimental learning activity for those problems with no existing expertise [28–30].

4.2 Drug delivery and nanotechnology

Biotech and pharmaceutical industries have been primarily prompted by the fulfillment of technical requirements, such as mankind mastering genetic modification of cells to produce high-quality biomolecules on an industrial scale.

However, it is hard to imagine obtaining wide therapeutic profit from protein and nucleic acid without the confidence that we can overcome the molecular fragility undermining their biological drug stability and efficacy. That said, most currently humanized monoclonal antibody therapies (the majority of the biopharmaceutical market products) are provided as a liquid suspension or lyophilized powder for injection/infusion of naked proteins. In these products, stabilization by formulation is primarily aimed at ensuring quality retention during manufacturing/marketing/logistics and use [31]. Yet planning the extension of their therapeutic application, the use of naked antibody in intravenous injection, might no longer be sufficient. In fact, it substantially reduces the opportunity of pharmacokinetic control and limits the product life cycle management opportunities, that is, after patent expiration [31]. An example is the development of oral or subcutaneous systems for insulin administration that are aiming to overcome the discomfort and limitation of injections in chronic conditions. Here, engineering is applied to circumvent the natural degradation of the protein drug when it is exposed to the gastric acid environment (for oral delivery) or protease-rich derma (for dermal delivery). It is also used to augment a naturally poor penetration of physical barriers (such as the intestinal endothelium or the dermal layers) that separate the drug from the site of action, for example, the bloodstream [32].

Another goal commonly associated with drug delivery is drug accumulation in the site of action. The controlled local accumulation of drugs by delivering them to the site of action can function as an enhancer of the therapeutic effect with a reduction in toxicity. For example, driving a cytotoxic molecule to preferably accumulate at the tumor mass can increase its biological activity on the tumoral cells, while sparing other healthy tissues from cytotoxic effects. Therefore an equally important aim of contemporary drug delivery research is the improvement of the drug's safety and potency [33,34] (Fig. 4.1).

Drug delivery systems embrace simple molecular conjugates, atom clusters, membranes, particles, microdevices, up to a wide range of macroscale absorption enhancers; with the common aim to confer a different pharmacokinetic profile to the drug payload compared with the naked drug substance [35,36].

Regardless of the specific delivery system's chemistry and mechanism, all delivery systems have a common feature, they have to interact with the biological environment of action, for example, the human body. Therefore having a compatible size has become a component of utmost importance defining the systems' properties. Regardless of the administration route, the actual size for a system to circulate in the circulation system is restricted to the micrometer to nanometer scale.

Consequently, nanotechnology is especially well-featured for delivery systems engineering, because it allows the system to (1) interact with cellular/subcellular structures, (2) molecularly interact with proteins and nucleic acids, (3) escape the immune system, the activation of which could potentially lead to severe side effects. Nevertheless, also the human physiology has adapted itself to manage matters at such scales: blood filtration in the liver and the immune system can quickly clear micrometer particles, and kidney filtration efficiently expels charged molecules of a few nanometers [37−41]. This means that the development of the delivery system has to carefully account for the "absorption, distribution, metabolism, and excretion" (ADME) profile.

Nanomedicine theoretically encompasses any submicron-featured system, including any colloidally stable molecular solution and any submicron synthetic or biological structure (Fig. 4.2). Similarly, drug delivery can be considered to be a stabilization strategy with a positive effect on the molecular pharmacokinetics, thus a vast amount of old-fashioned galenic formulation could be retrofitted to this appellation.

However, a nanomedicine is a drug or therapeutic molecule which in its final formulation actively benefits from nanofeatures improving its pharmacokinetics, as compared with its native form. A monoclonal antibody suspension is a nanosized colloidal dispersion, as its colloidal state only preserves its expected activity rather than improve it. Conversely, the active modulation of an active pharmaceutical compound (API) pharmacokinetics by acting on the colloidal or dimensional feature of API, for example, by conjugating an inactive excipient, represents a drug delivery system. For example, abraxane uses a biomolecule, albumin, to actively accumulate an antitumoral agent, paclitaxel, preferably at the cancer site. More complex systems, such as liposomes, use definite structures in the nano- to micron range to deliver an API payload at the site of action or confer special mechanisms of action to the active molecule (Fig. 4.2).

Ultimately, a nanomedicine describes a system in which nanosized features are actively part of the designed qualities and mechanism of action of the system, and a drug delivery enhanced medicine is a system in which quality attributes that modulate product efficacy are engineered [15,42−44].

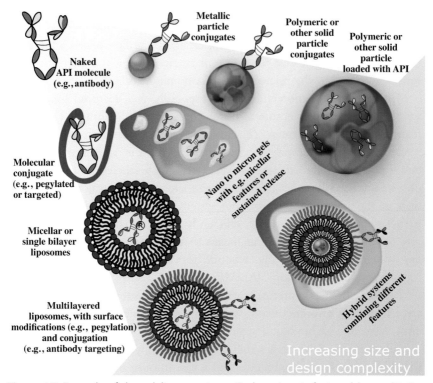

Figure 4.2 Example of drug delivery systems. Each system is featured by qualitative profiles that can be exploited to potentiate the API efficacy or improve its pharmacokinetics.

Therefore to understand modern pharmaceutical technology and drug delivery, a quality-based approach must be introduced, referred to as QbD.

4.3 Quality-by-design R&D

The transition from a systematic series of attempts (a "Trial and Error" approach) to a quality-based rational engineering on scientific principles (referred to as QbD [17,45]) has been crucial for the current pharmaceutical evolution toward advanced drug delivery.

The core of the QbD approach is the definition of quality target product profile (QTPP), which is broadly identified by the essential properties required by the final product to fulfill the intended scope. In a drug delivery system, the QTPP binds the product quality features to its mechanism of action. In this vision, the product quality profile is designed to obtain a certain effect. For example, a specific pharmacokinetics profile with accumulation in a specific tissue. The quality attributes necessary to reach this goal are identified on the basis of acquired knowledge (Fig. 4.3). Next, the manufacturing process is engineered to infuse the product with such qualities. To achieve the QTPP, critical quality attributes (CQAs) are defined. Process parameters and material attributes that have a direct or indirect impact on the CQAs are defined as critical process parameters and critical material attributes (CMAs), respectively. These parameter and attributes are controlled in manufacturing control plans [16,46,47].

As an example, potentiating the effect of an anticancer drug such as doxorubicin for the treatment of solid tumors aims to promote its local accumulation on the highly vascularized cancer matrix. QTPP of the delivery system associated with this development is the local drug accumulation at the tumor site, as opposed to long-term circulation in the bloodstream or liver/kidney excretion. A defined range of material and physical parameters, referred to as attributes, which are potentially critical for obtaining the desired tumor accumulation effect can be readily defined using the scientific knowledge of exploitable mechanisms to reach solid tumor accumulation.

As an example, attributes generally regarded as relevant in mediating a permeability and retention effect are
• Particle size capable of circulating and accumulating selectively in the tumor vasculature.
• Particle charge sufficiently neutral to evade the physiological clearance.
• Sufficient colloidal stability to avoid precipitation and fast clearance.

The optimal ranges defined for size, charge, and other identified parameters constitute the CQA set and typically lay the foundation defining the product design space (PDS).

Material qualities and process parameters are next defined to infuse the final product with the designed CQAs. This is done to ensure reliable production within an optimal subspace of the PDS that is adapted to guarantee robustness during manufacturing.

Next, the manufacturing process is evaluated at each step by attempting to evaluate the impact of each process parameter on the final CQAs, either on the basis of documented knowledge or experimental demonstration. The outcome of this work is the definition of the manufacturing process parameter ranges and the risk mitigation strategies, which are implemented to ensure robust production quality [15,16,48,49].

QTPP

**The product
(a doxorubicin formulation)
induces apoptosis in cancer
cells. As compared to
doxorubicin it is more
effective and less toxic,
as effect of liver accumulation**

The product is intended to
improve the treatment e.g., of liver
cancer. Considering the
patients' needs and **products
on the market** the acceptable
**efficacy, toxicity, and
pharmacokinetics** are set.
The target profile is completed
by a scientifically sound
mechanism of action description

AI Combine medical, marketing,
and cultural sources to
apprise the patients' needs

AI Benchmark the products in
the market to define the
efficacy, toxicity, and kinetics
to be reached during clinical
trial/market

AI Scan scientific and technical
sources for patent opportunities
and mechanism description

CQAs

**Each feature of relevance
is defined in a range,
e.g., minimum-maximum
size, chemical composition
physicochemical state.
The product is designed
accordingly**

Which features of the product
enables the reach of the QTPP,
on the basis of the current
scientific knowledge.
These are to be defined in
specific ranges, constituting
the product design space

AI Scan scientific and technical
sources providing indication
for CQAs definition

AI Aid the definition for range of
CQAs by Design of Experiments
(DoE) and in silico simulations

AI Based on the CQAs defined scout
for alternative delivery systems
or optimized system designs

Manufacturing science & technology
CPPs/ CMAs

**Each material attribute
and process parameters
relevant for infusing the CQAs
in the product is defined
in a range and controlled**

The manufacturing process
is designed for infusing the
CQAs in the product.
Material attributes (CMA)
and process parameters (CPPs)
with direct impact on the
products' CQAs are defined and
controlled in specific ranges,
constituting the process design
space

AI Scan scientific and technical
sources providing indication
of CMA/CPPs for the process
engineering. Provide insight in
material and process attributes

AI Aid the definition for range of
CMAs/CPPs by Design of
Experiments (DoE) and
in silico simulations

AI Aid process control and acquire data
from process analytical technology

AI Use the acquired process data to
perform reporting, support risk
analysis, and suggest optimization
by trend analysis and
in silico process predictions

Figure 4.3 Example schematic of quality-by-design product description and possible
AI implementation.

AI can manage the scanning of big data, making available knowledge for providing informed guesses, for example, helping to identify the CQAs connected with the desired QTPP. Importantly, the full QTPP profile of the product is not only limited to pharmacological or drug delivery features, but any aspect potentially of primary importance for patient safety and use. Therefore heterogeneous sources including toxicity profiles and consumer profile have to be collected and evaluated. For this reason, it is of relevance to note the application of AI to collect information from different databases with the scope of predicting more accurate ADME and toxicity profiles [50].

From this, the product prototype can be designed and characterized. Next, a process capable of manufacturing a product with the defined characteristic has to be implemented. At this stage, AI could also provide support by aiding the process engineering and development toward the development of robust manufacturing. For example, exploring material quality and production parameter ranges can evaluate their impact on the CQAs in an independent manner. This may even reach the potential of supporting the crucial steps of scaling the manufacture process, characterization, and validation [51–53].

4.4 Artificial intelligence in drug delivery modeling

AI can enhance nanosystem design by enabling advanced understanding of the biological environment and thus use this knowledge to create drug products by nanoengineering.

The human body is a complex system, which for drug delivery purposes is often simplified into a compartmental system divided by biological membranes. The concept of biological membranes is related to the physicochemical properties of the barriers separating different biological compartments, hence grouping a very heterogeneous array of epithelia and environments. For example, an orally administered drug is exposed to gastric passage and must achieve the permeation of the intestinal or gastric epithelium to distribute into the bloodstream. Next, it has to reach its target, which can be represented by a tissue, a component of a cellular membrane, or even an intracellular molecule. The permeation of biological barriers can be categorized as passive and active. Passive diffusion through membranes is mediated by physicochemical gradients and generally relies on the molecular feature of the drug. The drug's exposure to the biological environment actively interferes with the molecular features of the drug, complicating the predictive computation in silico of the drug distribution. Furthermore, passive permeation is often inefficient for biologicals and several classes of small molecules, unless mediated by specific drug delivery systems. Active diffusion is mediated by energy-activated cellular systems, such as membrane transport, thus it typically relies on complex biological interactions and much greater energy expenditure. This potentially expands the modeling of the systems to a wider number of specific parameters to compute [54].

The QbD design of a delivery system is tailored to take advantage of a single or a few transport mechanisms which are identified as bottlenecks impacting drug pharmacokinetics, hence simplifying the system studied. For example, the development of the oral formulation of insulin is highly focused on the drug's passage through the intestinal barrier to the bloodstream, as the large majority of the API delivered orally is lost in this passage. One of the limits of this stage of R&D is the predictability of preclinical models. For computationally greedy and complex in silico models, the prediction correctness is typically rooted in correct parametrization [55,56]. Parameters and models are typically obtained from theoretical knowledge (e.g., calculated from physical laws) and experimentally measured.

The quest to identify, develop, and characterize mechanisms, for example, for membrane interactions, in the modeled human environment could greatly profit from the implementation of AI that is able to research and analyze multilayered data. The intelligent literature search for parameters and comparison of models is currently one of the most crucial methods, which is generally left to the specific knowledge and understanding of the research units. An automated system capable of searching, simulating, scoring, and refining the search might greatly contribute to a more systematic model and parameter evaluation. Thus the ability to formulate a supported guess and systemically refine it using the results obtained in simulations in silico or experimental trials would be a consistent improvement, enabling the systematic selection and scoring of simulated models [54,57–59]. System biology databases are not far from possibly being ready to support AI application, providing solid information for the AI training. Currently, a number of examples of neural network applications and software providers, for example, within pharmacokinetics evaluation, already provide a solid background for the implementation of this technology [3,20]. Further examples of advanced AI systems are emerging as tools to incorporate into the model's complex inputs, that is, from drug interactions, phenotypic, chemical, and genomic databases [50,60], leading the way toward personalized medicine perspectives [61].

The next step is using the model to evaluate the possible impact of the drug delivery system on the drug pharmacokinetics, that is, deposition/toxicity. A fundamental decision-making stage is which factors and model should be evaluated and simulated, either to design the quality attributes of the system before experimental trial or to identify the critical attributes inducing the observed experimental results [62].

For example, a company owning a drug delivery technology platform (a liposome) is interested in simulating the impact of their technology providing the long circulation of a number of potential APIs on the market. For each of these APIs, the model should represent the most relevant ADME feature of the APIs compared to the APIs in the liposome, predicting the permeation of the barriers for each drug compared with the others in the liposome (e.g., extravasation in the tumor tissue). This potentially requires studying a complicated system by both experimental and in silico system simulation. An AI tool could help aggregate information from multiple sources and produce indications for these drugs which would greatly benefit the drug delivery system. As an example, the most relevant challenges in the development of a specific API/system could be provided by evaluating molecular, pharmacokinetics, and patient information.

A passive AI could help include new molecular entity features with similar known molecules and actively propose delivery solutions which are predicted to be effective for developing a treatment. Considering that even a minimal increase in bioavailability or local concentration of a potent molecule could determine a great therapeutic improvement, the outcome of accurate mechanistic modeling can have a huge impact on the upcoming development of new products.

Drug repurposing, which is the application of existing therapeutics to new disease, can directly benefit from AI implementation during the early phase of the drug discovery process [3]. However, in many cases, drug development upon formulation or advanced drug delivery might be required for adapting the drug pharmacokinetics or stability to the need of the specific patients.

A hurdle to AI implementation is the availability of databases with consistent information. Considering the eventuality that a large part of today's results might not be sufficiently comparable to enable unbiased evaluation of models and parameters, an active AI capable of sensing and acting to create or verify the current knowledge might help to consolidate the required information for a future AI application. This would in turn enable the development of a passive AI application to digest information and instruct investigators on how to overcome the lack of specific knowledge for the rational design of a product.

The contribution of active AI, recording parameters, and self-supervising experimental results could permit a more rigorous result codification in a knowledge database. Clarifying with an example, an AI controlling a liquid handler, a UV probe, and a pump operate a set of micronized tests, that is, microfluidic lab-on-chip assays for membrane permeation, cellular or enzymatic interactions, and maybe even an organ-on-chip study [63−67]. The AI skims through literature acquiring information on API and delivery systems it is fed with. It can retest the permeability parameters through a membrane or simulated biomembrane, for example, a standardized PAMPA system [68,69]. In this way, it can consolidate the previously reported results in a coherent and documented database; another time the AI will perform a comparative study with a new drug. In the next laboratory, another AI is performing a simulation of glomerular filtration or controlled dissolution, while the enzymatic affinity is tested on a chip in a third lab. The sum of the results is organized in a database. A developer company studying drug delivery systems can interrogate a search for an API that could be of benefit on the basis of their ADME and simulate the impact of drug delivery systems using more precise parameters. Meanwhile, an AI bot could be using the database to recommend a delivery system to an investigator seeking to enhance the intestinal absorption of insulin (Fig. 4.4).

Once the drug's mechanism of action, metabolism, barriers, and delivery strategy is identified, a background for product research and development is laid. From here, the prototyping and development stage can occur as discussed in the next section.

4.5 Artificial intelligence application in pharmaceutical product R&D

Simplifying the research and development process, two main phases are identified: the early phase and the late phase of R&D. Early phase R&D is constituted by the translation of the idea in the design and the realization of a prototype, which is tested bearing in mind the hypothetical mechanism of action as defined in the early research stage [70]. This is infused in the prototype and iteratively refined upon prototype testing.

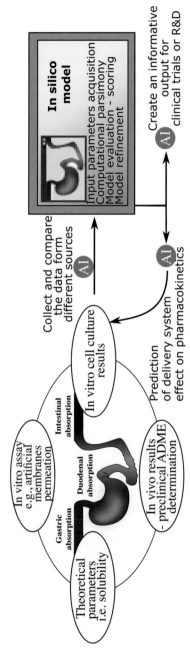

Figure 4.4 Example of AI application in drug delivery in silico simulations.

Simple lab-scale production processes are used to produce volumes to be used in the early product evaluation [16,46,47,71,72]. Besides the product-specific characteristics, the detailed design of the product and correct realization of the prototype with a scalable process are deemed to be the main early phase challenges. Next, early proof of efficacy and safety is necessary, especially in new pharmacological entities and consumer products. The late-stage development is more compelling to the robustness of the actual manufacturing process and scalability development, as the ability to produce increasing volumes of the intended pharmaceutical products at a defined quality can unlock further testing and development opportunities [73]. This is not the only focus of late development, which actually spans until the product life cycle management, however, is considered one of the most crucial aspects (Fig. 4.5).

4.5.1 Artificial intelligence application in prototyping and early development: an example scenario

Regardless of the molecular entity entering the early development stage, the number of administration routes will likely be limited by the molecule's chemical features and application. Building on the initial conditions, a plethora of alternative excipient, devices, and materials produce the initial formulation and drug product prototype [52,74]. Even if the choice of selecting the compounds for screening could be led by a number of relevant objective criteria, for example, predesigned quality target in QbD, typically the investigator's experience and the company's R&D history play a major role. In the same way that a physician who is experienced in successfully treating a disease would be prone to repeatedly prescribing the same therapy to patients with the same diagnosis, scientists tend to favor familiar materials with the most common quality level and from the most common distributor for their work environment. The lack of comparable data makes it challenging, even when there is a clear target quality as a goal [75]. In a practical example: in screening the excipient to formulate an IV antibody product, the investigator is obliged to select a buffering agent and a stabilizer/osmolyte (commonly sugars) and also consider if the addition of an amino acid (or other stabilizers) would improve protein stabilization. Buffering agents are controlled via their pH, osmolarity, and expected manufacturing conditions (e.g., temperature shifts and concentration-dilution steps). Next, the investigator selects a handful of sugars commonly used in formulations, perhaps just focusing on avoiding sugars known to hold potentially severe drawbacks. Several amino acids

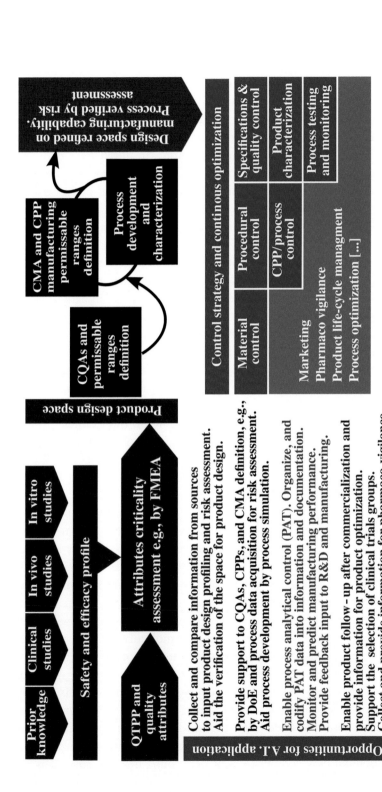

Prior knowledge | **Clinical studies** | **In vivo studies** | **In vitro studies**

Safety and efficacy profile

QTPP and quality attributes

Attributes criticality assessment e.g., by FMEA

Product design space

CQAs and permissable ranges definition

CMA and CPP manufacturing permissable ranges definition

Process development and characterization

Design space refined on manufacturing capability. Process verified by risk assessment

Control strategy and continous optimization

Material control	Procedural control	Specifications & quality control
	CPP/process control	Product characterization
Marketing		Process testing and monitoring

Pharmaco vigilance
Product life-cycle managment
Process optimization [...]

Opportunities for A.I. application

Collect and compare information from sources to input product design profiling and risk assessment.
Aid the verification of the space for product design.

Provide support to CQAs, CPPs, and CMA definition, e.g., by DoE and process data acquisition for risk assessment.
Aid process development by process simulation.

Enable process analytical control (PAT). Organize, and codify PAT data into information and documentation.
Monitor and predict manufacturing performance.
Provide feedback input to R&D and manufacturing.

Enable product follow–up after commercialization and provide information for product optimization.
Support the selection of clinical trials groups.
Collect and provide information for pharmaco-vigilance.

Figure 4.5 Schematic of product development roadmap using quality-by-design approach.

are currently included in formulations of biologics for various purposes. Among these purposes are molecular conformational stabilization and buffering, viscosity reduction, and protection from stress (e.g. oxidative stress). Here, the investigator uses his knowledge to select those that could ensure a useful outcome. Even for formulations featured by a relatively simple composition, varying the amount and quality of each component might result in a costly outcome. So even with a generous budget and an above-the-average laboratory, the decision to only test the commonly used excipients from the habitual provider is not uncommon.

In the above example, most of the choices are driven by past experience or the current opportunity within the research environment. For the investigator, it is often not possible to compare the quality and features of the provider's material as compared to other providers. Similarly, reasons for choosing specific sugars or amino acids for screening are based on the knowledge accessible to the investigator, who might not have access to the raw data or metadata. Further, at this point the investigator only marginally considers consumer-related preferences, except concerns for the safety of the patient. Also, the specific investigator's education, might swing the balance toward certain choices, based on their understanding and acknowledgment of specific risks. Expanding on this example, topical formulations or cosmetics often contain dozens of ingredients in hundreds of alternative variants and quality level, mixed in phases by complex processes, aimed to meet the consumer's subjective pleasure. Here the early phase decisions for material testing in the prototype become extremely difficult [27].

The above scenario is similar to the efforts made by the consumer industry to evaluate data by mixing multiple existing inputs in the early development phase. Passive AI in the pharmaceutical field can aid scientists in selecting the best excipient products and their combinations, in order to create and compare panels of screenings and refine these by active learning based on the outcome results. If learning algorithms can be fed with databases for excipient interaction screenings and molecular stability results, AI could provide systematic suggestions, resulting in stronger evidence-based screenings. Moreover, a deductive AI could identify knowledge gaps in the company/consortium's knowledge by deploying unusual combinations for testing and verification, hence performing calculated deviations. These deviations can turn into useful information to feed the innovation sector of the R&D pipeline with the possibility of discovering patentable compositions. Attempting new combinations or performing negative controls is not unusual among scientists, but they are often operated with a very subjective approach, which makes even worthy observations hard to consolidate in the know-how of many companies.

Going forward, an active AI, such as even a simple robotic arm capable of mixing and measuring a single or few parameters, can actively perform and record experimental arrays. This type of system could even routinely repeat trials with materials from different suppliers to experimentally evaluate the impact of qualitative differences. The innovative effect of such technology would not only be represented by a higher automation in the lab environment, but also the opportunity to produce highly consistent working routines and comparable data across projects, departments, and centers.

All the above examples are potentially valid for advanced formulations, such as nanoformulations, however even for a well-established technology such as liposomes, very little comparative understanding is available, leaving the developmental choices to the investigator's knowledge and consideration, and otherwise inclined toward the most common types of delivery systems. In overcoming such biases, the application of an active AI system might help to speed up the unbiased technology testing and help fill in the available knowledge gaps.

Ultimately, for the early-stage development, it is crucial that the formulated molecule displays stability in relevant conditions to ensure sustainable manufacturing, distribution, and use. While safety and bioactivity of the product are to be fully evaluated in clinical and preclinical stages, stability studies are commonly requested early in pharmaceutical development [76]. At this early stage, the available time and amount of product are limited, and comprehensive long-term stability studies are rarely possible. In this scenario, deep learning and neural networks are commonly applied to predict stability in experimental, accelerated, or exaggerated conditions [20,77].

4.5.2 Artificial intelligence in late-stage development: an example scenario

Late-stage pharmaceutical development in QbD, among other goals, aims to assure the product quality profile from clinical testing to continuous optimization, and toward the product's life cycle management stage. This part of the development pipeline could be different depending on the specific product. Here, the integration of information from different sources is required at all times, for example, the outcomes from R&D,

manufacturing technology, process development, qualification, and quality control. A primary goal of late-stage development is to provide the background for clinical experimentation, aiming to contribute to the safety measures taken for the volunteers and further support for regulatory submission.

Clinical trials typically feature systematic data collection on sizeable groups of volunteers, from relatively homogeneous groups of healthy men to a heterogeneous plethora of patients in the later phases and pharmacovigilance. Basic statistical comparisons by equivalence tests with placebos and benchmark products are deemed to sufficiently depict the product efficacy, at least in the early phases. In the later phases and with pharmacovigilance, the interactions and side effects are actively monitored. The collected data are analyzed and reported to sustain product applications to regulatory agencies and sometimes routed to marketing and product management departments to actively sustain the product life cycle management, by proposing product improvements or device integration. Manufacturers place a lot of effort in producing relevant documentation and investing in continuous optimization of their processes. Today, regulatory communication and revision is a highly resource- and time-demanding task for any company, but also for regulatory agencies and governments. A great deal of effort is taken on the manufacturer's side in the production of supporting documentation that is easily understandable during the agencies' evaluation. On the other side, the clear understanding and meticulous assessment of the provided documentation is a burden on the regulatory agencies. For agencies, this task is becoming harder and harder as new products tend to be very heterogeneous, and thus defy the commonly accepted standards for traditional small molecules. Despite the continuous effort of ICH, EMA, FDA, WHO, and other regulatory agencies to update guidelines for both small molecules and biologics, keeping up with the ever-increasing number of biotech products is a huge challenge. Hence, it is demanded for most companies that a clear, rich, and highly standardized documentation production is in place to support the product's applications. Laboratory information management systems to some extent provide an aid to methodically document the experimental activity for submission. A possible development would be the implementation of AI to gather this information and structure it in reports that are compatible with the submission requirements, that is, as for AI-produced journalism where news reports are elaborated into rich texts [78]. More importantly, the adoption of an associated AI codifying the uptake information on the agency side, for example, by providing knowledge searching and text scanning assistance, could shorten filing and trial times for the agencies. Noticeably, the dialogue between regulatory agencies and submitters is an extremely time-consuming and delicate process, which engages both sides and would be more affordable with AI mediation, by facilitating information sorting and translation. While this might seem to be a farfetched idea, translators and format handlers are a common technology, decreasing the barriers in international regulatory activity and making it easier for information transition across borders. Perhaps the AI-aided interpretation of scientific and industrial documentation prior to regulatory panel interviews could be just a step ahead.

4.5.2.1 Artificial intelligence in manufacturing development and control

Traditional manufacturing processes are typically composed of relatively standardized units and materials. Batches of materials are produced by repeating well-defined and controlled operations. These are designed and developed to be robust and highly repeatable and represent a vast but limited array of elements to choose from.

As an example, common filtration units which are ubiquitous in any sterile production, generally consist of a few alternative commercially available filters defined by the combination of limited materials, pore sizes, and geometries.

Similarly, solvents and plasticizers used in granulation or extrusion units are mostly operating on most used combinations of excipients and reagents. Similarly, most common cryoprotectants and lyoprotectants for dried formulations are extensively used in a majority of these formulations at standardized purity grades.

For all of these materials, databases of chemical and physical attributes are already available, along with the extensive knowledge of materials' interactions. The growth of larger and more information-rich product databases, for example, incorporating manufacturing data, is here to sustain the implementation of passive AI, providing tailored solutions to technological support units or commercial providers to match the client's needs.

Importantly, the AI-assisted process–unit selection would not only represent economic advantages for the providers, as seen in commercial bots, but also can provide advantages in order to design more robust processes for the company. Currently, the primary aim in manufacturing engineering is to create a robust manufacturing process, which could be eventually developed to achieve better productivity. A robust process is expected to consistently deliver products of a defined quality regardless of deviations from controllable conditions. In this context, AI-aided synthetic processes can study materials and parameters and provide a background for the development of tools for process development and characterization [79,80]. In pharmaceutical manufacturing, quality deviations in the product can have severe effects and are therefore tightly controlled and regulated. Regulatory agencies are therefore prone to verifying the process robustness and control when evaluating the acceptance of a product. This assures process robustness in a knowledge-based approach involving extensive data collection. The collected data can highlight the potential risk for any component of the process (e.g., by failure mode and effects analysis [45,81]). Furthermore, the risk assessment produced is analyzed to provide a holistic understanding of the process robustness. This operation might be required for advanced analytical approaches such as multivariate data analysis.

This is particularly crucial, because a failure could jeopardize years of work and investment, especially considering that regulatory agencies might not accept changes in the process without compelling documentation as justification. This is also serious for the manufacturer as their product could be removed from the market. In this scenario, the application of AI in product definition and the collection of relevant material data in support of the process development/characterization might be highly significant. For example, AI is essential for efficiently collecting and organizing a large amount of data during risk-based analysis, and further aiding the continuous data collection and analysis for process validation. While these considerations are equally relevant for any type of pharmaceutical product, the complexity in advanced formulation quality profiles, put further pressure on the material and process assessment, especially when advanced and specialized units are required for controlling the product quality during manufacturing (Fig. 4.6) [45,81].

Further, it is of relevance to discuss the application of AI in process analytical technology (PAT) and continuous manufacturing. In recent years, the development of continuous manufacturing has been supported by the FDA in order to stimulate the current manufacturing processes to evolve toward a more robust and efficient approach. This was also captured in the recent Industry 4.0 action, and continuous manufacturing lines are currently implemented by Janssen, Vertex, and being studied by Novartis [82–86] (Fig. 4.6).

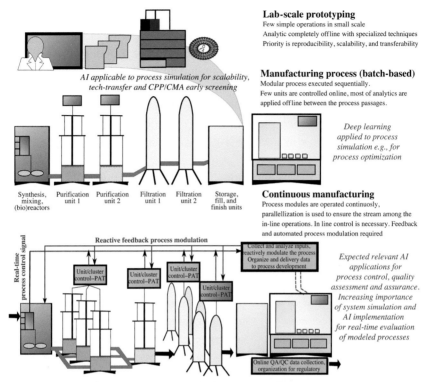

Figure 4.6 Manufacturing stages and batch manufacturing versus continuous manufacturing.

Traditional batch manufacturing processes are denoted by a sequence of finite steps operated in sequence, for each step the reaction/processing volumes are the limiting factors. For example, a defined amount of two reagents are mixed in limited amounts depending on the unit scalability, then if filtration is processed it introduces a limiting factor for the batch volume produced until the process ends with the production of a batch. Therefore the process capability defines the manufacturable volume range for each batch, and for each batch volumes in this defined range are worked stepwise through the discrete procedural stages.

In contrast, continuous manufacturing is defined as a time-resolved process where the processing units are connected and operated simultaneously by modulating the input and output of each unit. As for the above example, two reagents are pumped continuously from a feed to a mixing unit at a defined flow rate and, given the filtration unit capabilities, the input flow is divided among a number of filtration units, these produce a constant volume of intermediate per unit of time. Thus the manufacturing flow continues from the start to the end of the process. In this case, the process is not defined by the volumes processed at each step, but the volumes per unit of time passing in each unit. Examples of AI application for in silico simulation to aid system engineering can be observed in the literature [22], displaying the computational capability of an AI system for in silico predictions of continuous systems. There is very little doubt that AI will become a necessary tool to model, manage, and control continuous processes.

A requirement for a continuous process is the opportunity to control online process efficiency and product quality. Therefore the development of process analytical control technologies is essential to proactively control the process in real time. While in batch manufacturing samples can be taken between the discrete passages and analyzed offline prior to passing to the next step (offline analysis), in continuous manufacturing the process efficiency has to be assessed in real time (online analysis). In the previous example, online control of the pressure across the filter unit of the continuous process is necessary to ensure filter integrity and product quality, and to actively respond if a failure occurs (e.g., a filter loses integrity). In this scenario a fast and precise detection and reaction response is necessary to preserve the production quality. Conversely, in traditional step-by-step manufacturing, a severe and detectable failure like a rupture of a filter typically results in stopping the process and reprocessing the material through the unit, if it is found to be safe. For high-quality complex products, a more precise and reactive multistep control is often necessary, where the failure detection and response might result in complications. This especially applies to nanomedicines which have physical features that impose a high level of control on the processing of the product.

Currently, PAT systems based on UV, NIR, and FTIR are in place to acquire and transmit data continuously; however, actively processing these data is still an open challenge [82,87,88]. AI is potentially capable of crossing the line from simple process tracking and failure reactive response, to proactive process modulation. Shortening the reaction time to detect and react to parameter variability is a primary objective, enabling continuous manufacturing. Reactive changing of parameters is a task that an AI system could manage efficiently; however, its application might not be possible in the near future. This is because a validated process is expected to be well-defined, thus a dynamic system capable of self-optimization is unlikely to be well accepted for process regulation. Eventually, AI systems trained during process development and validation could be able to control the parameters within predefined ranges. This could provide a flexible control system, capable of adapting to various processes and after initial learning being able to carry out the process with high quality. The initial AI learning activity could be implemented by connecting the AI to the multiple PAT outputs and registering the process parameters in real time, until the parameter ranges obtained are fixed and applied to the process control strategy. The training data is then used as support for risk analysis and process characterization. Ultimately, the time-resolved input of multiple PAT associated with several units has to be modeled using a parsimonious computing method capable of deducing correlations between various input parameters. Further, an active AI capable of interacting with the process, for example, changing the parameters systematically, would be able to aid development and optimization.

4.6 Landscape of AI implementation in the drug delivery industry

The lagging acquisition of AI technology in drug delivery makes the future landscape quite uncertain, but room for improvements is clearly visible. The market for drug delivery systems could appear fairly mature in the numbers at first sight, even considering the more restricted area of nanofeatured drug delivery systems where over 300 products or close-to-commercial technologies can be indexed [42−44]. However, the perception of an exhaustive applicative achievement is misleading. Relatively few drug delivery strategies have been implemented in industry, and among these most are represented by relatively simple molecular conjugates with a stabilizing effect with relatively contained impact on the drug's passive distribution [89]. In general, very few innovative drug delivery concepts have been propagated in several applications, and eventually imitated in range and often as part of product life-management plans [15,90]. Technological diffusion by imitation and propagation is a good means by which new technologies are to be widely adopted by companies, but if the rise in new inventions slows as has been observed for the small-molecule R&D crisis, it could be a symptom of issues in the industrialization path.

A sizable portion of innovators in drug delivery are small−medium enterprises. These groups are particularly vulnerable to the escalating risks in the development of new technologies, especially approaching the latest steps in technological readiness development. Here, technological readiness levels (TRLs) can be broadly defined as TRL 1−3 for the preclinical stage of prototyping only requiring small-scale manufacturing, TRL 4−6 for the early to advanced clinical stages requiring some level of GMP production at least at medium scale, and TRL 7−9 for commercial scale and quality manufacturing [73,91].

Considering the possible applications of AI discussed in the previous paragraphs, it is potentially a major game-changer for innovative companies for (1) the early system design and development to moderate the risks upon passage to clinical phases (generally identified in TRL 3−6) and (2) AI implementation in drug delivery smart manufacturing to increase the process robustness, scalability, and transfer (TRL 4 and onward) [73,91,92].

The specialized enterprise network, which is the core of drug delivery R&D, poses some major challenges for AI developers. Highly specialized AI systems have to be customized and developed for specific technological niches. This demands a fast popularization of platforms usable by personnel with average IT/data expertise to be fulfilled by IT experts with an understanding of the specific technological niches. Because of the customization needed, the AI system and toolboxes cannot be black boxes, so it should be possible to control built-in biases and operate a transparent customization. This puts great pressure on developers for simplified command solutions, eventually requiring cutting-edge user interfaces. Alternatively, the development of a generic platform should be flexible enough to create a shared basis for customized development of AI, which would shift the pressure to open platforms.

For high-tech companies, knowledge security is a core asset, thus data safety and AI control ought to be primary concerns for developers. Another question concerns the modality of "forgetting" sensible industrial data once they are acquired by an AI system; it is not clear what the long-term impact of proprietary industrial data survival in AI systems would be.

4.7 Conclusion: the way forward

In conclusion, while the applicability of AI in pharmaceutical technology and drug delivery appears to be rich in promise and opportunity, delays are understandable in adopting this technology within the pharmaceutical industry [93,94].

Curiously, the technological aspects of AI do not appear to be the main restraining factors for applying AI in the pharmaceutical field. With little doubt there are two main nontechnological issues affecting the implementation of AI in the pharmaceutical field: the lack of standardized databases and the conservative regulatory approach adopted in traditional pharmaceutical manufacturing.

For the former, data accumulation and standardization are vital for training AI efficiently, however, pharmaceutical sciences can introduce delays in data codification and standardization compared to other scientific fields. Nevertheless, increasing commitment for initiatives such as the Pistoia Alliance and the action of private pharmaceutical groups are promising [25].

For the latter, an AI system involved in regulating activities such as manufacturing or clinical trials is typically required to be locked to a specific code after training, hence removing the capability of long-term learning. However, since the relatively recent reception of QbD approaches, the pharmaceutical industry has improved, and the recent industry 4.0 action appears to depict a field in rapid transformation. Therefore if an early application of AI would appear, there is a good chance that it might be implemented.

Looking ahead, an important question to ask is which typology of AI would provide a greater benefit to the pharmaceutical field. It could lead to reducing and narrowing the capability of AI to increase productivity (e.g., as a simple tool to enhance robotic operations), or as a passive tool for knowledge/data mining and organization. These scopes are, in fact, already partially fulfilled by deep learning technology, which is not too widely implemented because of the inconsistency of data produced in the pharmaceutical field.

Perhaps the most fruitful experience could be achieved by allowing an interactive AI to explore the space and actively integrate the required knowledge for further application. This can be achieved by a simple robot capable of mixing and measuring to integrate knowledge gaps in incomplete databases and consolidating the current results with verifications.

References

[1] Scannell JW, Blanckley A, Boldon H, Warrington B. Diagnosing the decline in pharmaceutical R&D efficiency. Nat Rev Drug Discov 2012;11:191–200.
[2] Munos B. Lessons from 60 years of pharmaceutical innovation. Nat Rev Drug Discov 2009;8:959–68.
[3] Mak KK, Pichika MR. Artificial intelligence in drug development: present status and future prospects. Drug Discov Today 2019;24:773–80.

[4] Grilo AL, Mantalaris A. The increasingly human and profitable monoclonal antibody market. Trends Biotechnol 2019;37:9−16.

[5] Sarpatwari A, Barenie R, Curfman G, Darrow JJ, Kesselheim AS. The US biosimilar market: stunted growth and possible reforms. Clin Pharmacol Ther 2019;105:92−100.

[6] Wang W. Advanced protein formulations. Protein Sci 2015;24:1031−9.

[7] Daka A, Peer D. RNAi-based nanomedicines for targeted personalized therapy. Adv Drug Deliv Rev 2012;64:1508−21.

[8] Colombo S, Zeng X, Ragelle H, Foged C. Complexity in the therapeutic delivery of RNAi medicines: an analytical challenge. Expert Opin Drug Deliv 2014;11:1481−95. Available from: https://doi.org/10.1517/17425247.2014.927439.

[9] Junghanns JU, Müller RH. Nanocrystal technology, drug delivery and clinical applications. Int J Nanomed 2008;3:295−309.

[10] Jain KK. Nanopharmaceuticals. The handbook of nanomedicine. New York: Springer; 2017. p. 201−71. Available from: <https://doi.org/10.1007/978-1-4939-6966-1_5>.

[11] Kalepu S, Nekkanti V. Insoluble drug delivery strategies: review of recent advances and business prospects. Acta Pharm Sin B 2015;5:442−53.

[12] Dickherber A, Morris SA, Grodzinski P. NCI investment in nanotechnology: achievements and challenges for the future. Wiley Interdiscip Rev Nanomed Nanobiotechnol 2015;7:251−65.

[13] National Research Council of the National Academies. The National Nanotechnology Initiative. Available from: <http://www.nano.gov/node/1071>.

[14] European Commission. Nanotechnology (Phase I report). Available from: <http://ec.europa.eu/DocsRoom/documents/11283/>; 2010.

[15] Colombo S, et al. Transforming nanomedicine manufacturing toward Quality by Design and microfluidics. Adv Drug Deliv Rev 2018;128:115−31. Available from: https://doi.org/10.1016/j.addr.2018.04.004.

[16] Troiano G, Nolan J, Parsons D, Van Geen Hoven C, Zale S. A Quality by Design approach to developing and manufacturing polymeric nanoparticle drug products. AAPS J 2016;18:1354−65.

[17] Trivedi B. Quality by design (QbD) in pharmaceuticals. Int J Pharm Pharm Sci 2012;4:17−29.

[18] Puri M, Pathak Y, Sutariya VK, Tipparaju S, Moreno W. Artificial neural network for drug design, delivery and disposition. Academic Press; 2015. Available from: https://doi.org/10.1016/C2014-0-00253-5.

[19] Vyas M, et al. Artificial intelligence: the beginning of a new era in pharmacy profession. Asian J Pharm 2018;12:72−6.

[20] Hassanzadeh P, Atyabi F, Dinarvand R. The significance of artificial intelligence in drug delivery system design. Adv Drug Deliv Rev 2019;151−152:169−90. Available from: https://doi.org/10.1016/j.addr.2019.05.001.

[21] Sacha GM, Varona P. Artificial intelligence in nanotechnology. Nanotechnology 2013;24.

[22] Wong W, Chee E, Li J, Wang X. Recurrent neural network-based model predictive control for continuous pharmaceutical manufacturing. Mathematics 2018;6:242.

[23] Sable P, Khanvilkar VV. Pharmaceutical applications of artificial intelligence. Int J Pharma Res Health Sci 2018;6:2342−5.

[24] Wise J, et al. The positive impacts of real-world data on the challenges facing the evolution of biopharma. Drug Discov Today 2018;23:788−801.

[25] Taylor D, et al. The Pistoia Alliance Controlled Substance Compliance Service Project: from start to finish. Drug Discov Today 2015;20:175−80.

[26] Roe R. Designing the lab of the future: Robert Roe discusses the role of Pistoia Alliance in creating the lab of the future with Pistoia's Nick Lynch. Sci Comput World 2018;20−2.

[27] Lee CKH, Choy KL, Chan YN. A knowledge-based ingredient formulation system for chemical product development in the personal care industry. Comput Chem Eng 2014;65:40−53.

[28] Schmidhuber J. Deep learning in neural networks: an overview. Neural Netw 2015;61:85−117.

[29] Ha D, Schmidhuber J. World model 2018;. Available from: https://doi.org/10.5281/zenodo.1207631.

[30] Faggella D, Schmidhuber J How robots learn—an interview with Jürgen Schmidhuber, <https://emerj.com/ai-podcast-interviews/how-robots-learn-jurgen-schmidhuber/>; 2019 [accessed 22.04.19].

[31] Cui Y, Cui P, Chen B, Li S, Guan H. Monoclonal antibodies: formulations of marketed products and recent advances in novel delivery system. Drug Dev Ind Pharm 2017;43:519−30.

[32] Wong CY, Martinez J, Dass CR. Oral delivery of insulin for treatment of diabetes: status quo, challenges and opportunities. J Pharm Pharmacol 2016;68:1093−108.

[33] Lammers T, et al. Cancer nanomedicine: is targeting our target? Nat Rev Mater 2016;1:16069.

[34] Hare JI, et al. Challenges and strategies in anti-cancer nanomedicine development: an industry perspective. Adv Drug Deliv Rev 2017;108:25−38.

[35] Tuncer Degim I, Celebi N. Controlled delivery of peptides and proteins. Curr Pharm Des 2007;13:99−117.

[36] Rosen H, Abribat T. The rise and rise of drug delivery. Nat Rev Drug Discov 2005;4:381−5.

[37] Lakkireddy HR, Bazile DV. Nano-carriers for drug routeing—towards a new era. J Drug Target 2019;27:525−41.

[38] National Accademies of Sciences, Engineering, and Medicine. A matter of size: triennial review of the National Nanotechnology Initiative; 2006.

[39] Dowling A, et al. Nanoscience and nanotechnologies: opportunities and uncertainties. Lond R Soc R Acad Eng Rep 2004;46 618−618.

[40] Chang EH, et al. Nanomedicine: past, present and future—a global perspective. Biochem Biophys Res Commun 2015;468:511−17.

[41] Wagner V, Dullaart A, Bock A-K, Zweck A. The emerging nanomedicine landscape. Nat Biotechnol 2006;24:1211−17.

[42] Weissig V, Guzman-Villanueva D. Nanopharmaceuticals (part 2): products in the pipeline. Int J Nanomed 2015;10:1245.

[43] Weissig V, Pettinger T, Murdock N. Nanopharmaceuticals (part 1): products on the market. Int J Nanomed 2014;9:4357.

[44] Etheridge ML, et al. The big picture on nanomedicine: the state of investigational and approved nanomedicine products. Nanomed Nanotechnol Biol Med 2013;9:1−14.

[45] International Conference of Harmonization (ICH). Pharmaceutical quality system Q8-Q12, <http://www.ich.org/products/guidelines/quality/article/quality-guidelines.html> [accessed January 2018].

[46] Rathore AS, Winkle H. Quality by design for biopharmaceuticals. Nat Biotechnol 2009;27:26−34.

[47] Yu LX. Pharmaceutical quality by design: product and process development, understanding, and control. Pharm Res 2008;25:781−91. Available from: https://doi.org/10.1007/s11095-007-9511-1.

[48] Hrkach J, et al. Preclinical development and clinical translation of a PSMA-targeted docetaxel nanoparticle with a differentiated pharmacological profile. Sci Transl Med 2012;4 128ra39-128ra39.

[49] Huang J, et al. Quality by design case study: an integrated multivariate approach to drug product and process development. Int J Pharm 2009;382:23—32.

[50] Cheng F, Zhao Z. Machine learning-based prediction of drug—drug interactions by integrating drug phenotypic, therapeutic, chemical, and genomic properties. J Am Med Inform Assoc 2014;21:e278—86.

[51] Vijayaraghavan V, Garg A, Wong CH, Tai K. Estimation of mechanical properties of nanomaterials using artificial intelligence methods. Appl Phys A Mater Sci Process 2014;116:1099—107.

[52] Gams M, Horvat M, Ožek M, Luštrek M, Gradišek A. Integrating artificial and human intelligence into tablet production process. AAPS PharmSciTech 2014;15:1447—53.

[53] Das PJ, Preuss C, Mazumder B. Artificial neural network as helping tool for drug formulation and drug administration strategies. Artificial neural network for drug design, delivery and disposition. Elsevier Inc; 2015. Available from: <https://doi. org/10.1016/B978-0-12-801559-9.00013-2>.

[54] Bhhatarai B, Walters WP, Hop CECA, Lanza G, Ekins S. Opportunities and challenges using artificial intelligence in ADME/Tox. Nat Mater 2019;18:418—22.

[55] Ehrhardt C, Kim K. Drug absorption studies: in situ, in vitro and in silico models. Springer; 2008.

[56] Siepmann J, Siepmann F. Modeling of diffusion controlled drug delivery. J Control Release 2012;161:351—62.

[57] Vanneschi L. Improving genetic programming for the prediction of pharmacokinetic parameters. Memetic Comput 2014;6:255—62.

[58] Yang S-Y, et al. An integrated scheme for feature selection and parameter setting in the support vector machine modeling and its application to the prediction of pharmacokinetic properties of drugs. Artif Intell Med 2009;46:155—63.

[59] Yu LX, Ellison CD, Hussain AS. Predicting human oral bioavailability using in silico models. Applications of pharmacokinetic principles in drug development. Springer; 2004. p. 53—74. Available from: <https://doi.org/10.1007/978-1-4419-9216-1_3>.

[60] Ryu JY, Kim HU, Lee SY. Deep learning improves prediction of drug—drug and drug—food interactions. Proc Natl Acad Sci USA 2018;115:E4304—11.

[61] Menden MP, et al. Machine learning prediction of cancer cell sensitivity to drugs based on genomic and chemical properties. PLoS One 2013;8:e61318.

[62] Rafienia M, Amiri M, Janmaleki M, Sadeghian A. Application of artificial neural networks in controlled drug delivery systems. Appl Artif Intell 2010;24:807—20.

[63] Cui P, Wang S. Application of microfluidic chip technology in pharmaceutical analysis: a review. J Pharm Anal 2018;9. Available from: https://doi.org/10.1016/J. JPHA.2018.12.001.

[64] Esch EW, Bahinski A, Huh D. Organs-on-chips at the frontiers of drug discovery. Nat Rev Drug Discov 2015;14:248—60.

[65] Kim S, et al. Vasculature-on-a-chip for in vitro disease models. Bioengineering 2017;4:8.

[66] Shuler ML. Organ-, body- and disease-on-a-chip systems. Lab Chip 2017;17:2345—6.

[67] Sims CE, Allbritton NL. Analysis of single mammalian cells on-chip. Lab Chip 2007;7:423—40.

[68] Balogh GT, Müller J, Könczöl A. pH-gradient PAMPA-based in vitro model assay for drug-induced phospholipidosis in early stage of drug discovery. Eur J Pharm Sci 2013;49:81—9.

[69] Avdeef A, et al. Caco-2 permeability of weakly basic drugs predicted with the double-sink PAMPA pKa(flux) method. Eur J Pharm Sci 2005;24:333−49.

[70] Aksu B, et al. A quality by design approach using artificial intelligence techniques to control the critical quality attributes of ramipril tablets manufactured by wet granulation. Pharm Dev Technol 2013;18:236−45.

[71] Rathore AS. Roadmap for implementation of quality by design (QbD) for biotechnology products. Trends Biotechnol 2009;27:546−53.

[72] Xu X, Khan MA, Burgess DJ. A quality by design (QbD) case study on liposomes containing hydrophilic API: I. Formulation, processing design and risk assessment. Int J Pharm 2011;419:52−9.

[73] Eaton MAW, Levy L, Fontaine OMA. Delivering nanomedicines to patients: a practical guide. Nanomed Nanotechnol Biol Med 2015;11:983−92.

[74] Rowe RC, Roberts RJ. Artificial intelligence in pharmaceutical product formulation: knowledge-based and expert systems. Pharm Sci Technol Today 1998;1:153−9.

[75] Khalid MH, Tuszynski PK, Szlek J, Jachowicz R, Mendyk A. From black-box to transparent computational intelligence models: a pharmaceutical case study. Proceedings—2015 13th International Conference on Frontiers of Information Technology (FIT 2015) 2016;114−18. Available from: https://doi.org/10.1109/FIT.2015.30.

[76] Bhirde AA, Chiang MJ, Venna R, Beaucage S, Brorson K. High-throughput in-use and stress size stability screening of protein therapeutics using algorithm-driven dynamic light scattering. J Pharm Sci 2018;107:2055−62.

[77] Ibrić S, Jovanović M, Djurić Z, Parojcić J, Solomun L. The application of generalized regression neural network in the modeling and optimization of aspirin extended release tablets with Eudragit RS PO as matrix substance. J Control Release 2002;82:213−22.

[78] Carlson M. The Robotic Reporter. Digit J 2015;3:416−31.

[79] Segler MHS, Preuss M, Waller MP. Planning chemical syntheses with deep neural networks and symbolic AI. Nature 2018;555:604−10.

[80] Molga EJ, Van Woezik BAA, Westerterp KR. Neural networks for modelling of chemical reaction systems with complex kinetics: oxidation of 2-octanol with nitric acid. Chem Eng Process Process Intensif 2000;39:323−34.

[81] Food and Drug Adminstration. Guidance for industry: Q9 quality risk management, <https://www.fda.gov/downloads/Drugs/Guidances/ucm073511.pdf>; 2016.

[82] Kleinebudde P, Khinast J, Rantanen J. Continuous manufacturing of pharmaceuticals. Wiley; 2017.

[83] Lee SL, et al. Modernizing pharmaceutical manufacturing: from batch to continuous production. J Pharm Innov 2015;10:191−9.

[84] Rantanen J, Khinast J. The future of pharmaceutical manufacturing sciences. J Pharm Sci 2015;104:3612−38.

[85] Dopico M, Gomez A, De la Fuente D, García N, Rosillo R, Puche J. A vision of industry 4.0 from an artificial intelligence point of view. Intl Conf Artif Intell 2016;350:739−40.

[86] Bohr A, Colombo S, Jensen H. Future of microfluidics in research and in the market. Microfluidics for pharmaceutical applications. Elsevier; 2019. p. 425−65. Available from: <https://doi.org/10.1016/B978-0-12-812659-2.00016-8>.

[87] Read EK, et al. Process analytical technology (PAT) for biopharmaceutical products: Part I. concepts and applications. Biotechnol Bioeng 2010;105:276−84.

[88] Aksu B, et al. Strategic funding priorities in the pharmaceutical sciences allied to Quality by Design (QbD) and Process Analytical Technology (PAT). Eur J Pharm Sci 2012;47:402−5.

[89] Park K. Facing the truth about nanotechnology in drug delivery. ACS Nano 2013;7:7442−7.
[90] Venditto VJ, Szoka FC. Cancer nanomedicines: so many papers and so few drugs!. Adv Drug Deliv Rev 2013;65:80−8.
[91] Department of Defense. Technology Readiness Assessment Deskbook; 2005.
[92] Ragelle H, Danhier F, Préat V, Langer R, Anderson DG. Nanoparticle-based drug delivery systems: a commercial and regulatory outlook as the field matures. Expert Opin Drug Deliv 2016;1−14. Available from: https://doi.org/10.1080/17425247.2016.1244187.
[93] Lamberti MJ, et al. A study on the application and use of artificial intelligence to support drug development. Clin Ther 2019;41:1414−26.
[94] Henstock PV. Artificial intelligence for pharma: time for internal investment. Trends Pharmacol Sci 2019;40:543−6.

CHAPTER 5

Cancer diagnostics and treatment decisions using artificial intelligence

Reza Mirnezami
Department of Surgery and Cancer, Imperial College London, London, United Kingdom

5.1 Background

Recent advances in computational science mean that the capacity of artificial intelligence (AI) to predict disease risk, diagnose illness, and guide therapeutic decision making is now a realistic prospect in the foreseeable future [1]. The use of AI has already shown promise in a wide variety of fields in medicine including cardiology, renal medicine, critical care, and mental health [2–5]. For a variety of reasons, cancer is an area where AI is expected to have a significant impact in the near term, and cancer is in many ways an obvious choice for enhancing the "reputation" of AI-based approaches in healthcare. First, cancer is common and imposes a considerable burden in terms of physical disability, emotional trauma, and economic cost for patients and society as a whole. Thus, almost without exception, across all cancer subtypes, there is a critical need to improve cancer-related outcomes. Second, at the individual level, there is huge patient demand for enhanced cancer diagnostics, prognostication, and therapy; cancers represent a major and indiscriminate cause of morbidity and mortality, and they evoke fear in a way that few other conditions do, because of the perception of cancer as a fatal diagnosis. From the cancer clinician's perspective, there is an equally urgent drive since tumor heterogeneity and the exponential growth in onco-therapeutic options present major challenges in terms of decision making, not least due to sheer data burden. Rather alarmingly, it has been estimated that the modern-day cancer physician would need to spend in excess of 20 hours per day reading, in order to stay up to date with developments in the scientific literature. This challenge is well exemplified by current attempts to unravel the cancer genome—millions of molecular disparities are now coming to light that could, alone or in combination, influence cancer cell survival and progression, but harnessing these discoveries in a meaningful way is impossible without state-of-the-art computing. The impact of AI approaches is also expected to be significant for academia, where cancer research initiatives are likely to be more precisely steered by AI-based data-mining discoveries. This in turn is likely to mean that certain avenues of research in terms of cancer diagnostics and therapy are "shelved" either permanently or temporarily until new developments emerge that suggest otherwise. Furthermore, it is anticipated that for the pharmaceuticals industry, anticancer drug development and the development of companion diagnostics will benefit from an additional layer of premarket delivery validation, through AI-based analyses of actual and simulated clinical outcomes. One can thus appreciate that all stakeholders within the cancer health space stand to make potentially significant gains with the application of AI solutions (Fig. 5.1).

To better understand where the technology is headed, it is first essential to recognize where we are currently and to define the areas of critical unmet need in cancer medicine. Currently, in the case of solid organ tumors, criteria that determine patient prognosis and management relate to a variety of factors, such as histological tumor grade, tumor stage, and the presence/absence of prognostically relevant features such as vascular and/or lymphatic invasion. More recently, molecular phenotyping data have been successfully incorporated into the patient pathway, notable examples including hormone receptor status in breast cancer and KRAS mutational status in Stage IV colorectal cancer [6,7]. As our understanding of cancer development and progression becomes more sophisticated, so data pertaining to a given patient become ever more multidimensional. The ideal computational pipeline under these circumstances would be capable of seamlessly integrating data from multiple heterogeneous sources (molecular/clinicopathological/radiological/exposomal) to arrive at a given answer (risk prediction/cancer diagnosis/cancer prognostication/determination of therapeutic efficacy). Conventional dimensionality reduction statistical approaches of linear regression and logistic regression (LR) have yielded some useful insights into these areas. However, they are now increasingly subject to fundamental limitations imposed by the increasing complexity of biomedical data. In recent years, AI approaches have shown promise in this area as they utilize a variety of statistical, probabilistic, and optimization techniques and parameters to "learn" from past case examples and to detect latent clinically relevant patterns from "noisy"/multidimensional datasets.

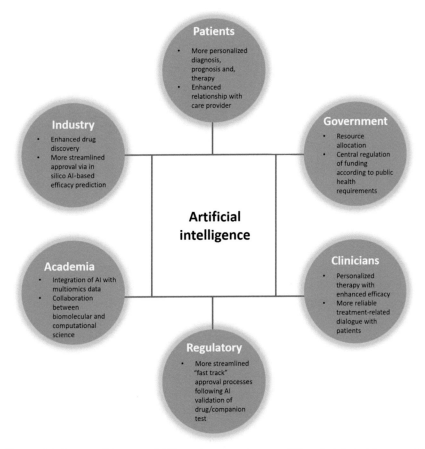

Figure 5.1 Expected impact of AI-based approaches on different stakeholder groups in the cancer health space.

In this chapter, current and emerging trends in the application of AI approaches in all phases of the cancer patient pathway are summarized.

5.2 Artificial intelligence, machine learning, and deep learning in cancer

The term *artificial intelligence* was first coined by computer scientists in 1956 and now represents a wide umbrella term encompassing a growing selection of algorithmic disciplines and subdisciplines [8]. Fig. 5.2 provides a graphical summary of research trends over the past two decades and illustrates the steady growth of the AI sector in cancer research. Presently,

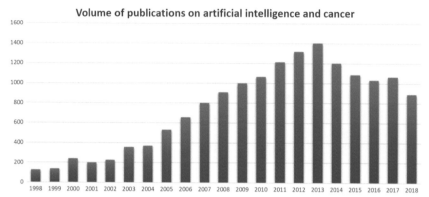

Figure 5.2 Graphical representation of the steady growth of AI-based studies in cancer research.

the terms AI, machine learning (ML), and deep learning (DL) are used somewhat interchangeably in the scientific literature, and to a greater extent in mainstream media. A detailed overview of these computational methods is beyond the scope of this chapter and is provided in Chapter 2, Introduction to Artificial Intelligence. However, it is critical that cancer scientists and cancer physicians have an appreciation of the fundamental differences between these interconnected methods, as they are likely to have impact across different areas. Simplistically, one can consider the key data challenges in cancer medicine as (1) logistical, (2) precision centered, and (3) next generation. It is useful to consider the application of AI, ML, and DL in cancer, across these three domains (Fig. 5.3). From the point of view of *logistics*, there are essential, often labor intensive, human error prone, but relatively basic tasks performed in cancer medicine, where a computational approach that could be trained to perform "as well as" a human counterpart would be highly desirable.

This comes down to time efficiency, cost, and logistics of human endeavor in an era where services are attempting to expand, while simultaneously being subjected to an unprecedented level of pressure in terms of available resources. For example, in February 2019 a National Audit Office review reported that more than 150,000 untested cervical screening samples were discovered in laboratories across England [9]. The report concluded that laboratory staff shortages had resulted in a decline in performance against turnaround time targets. In a situation such as this, an AI platform capable of selecting patients for national cervical screening, based on the predefined criteria, and interpreting smear test results as "normal,"

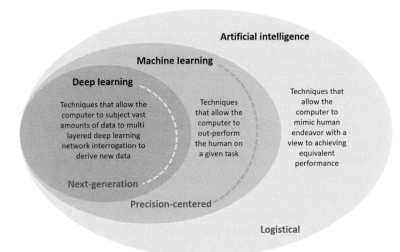

Figure 5.3 The role of AGI, ML, and DL in cancer medicine and how these are likely to address logistical, precision-centered, and next-generation objectives in cancer medicine in the future.

"inadequate," or "abnormal," before algorithmically defining the necessary follow-up strategy for the patient, would be enormously beneficial. Although this would represent a huge undertaking in practical terms, in computational terms this would constitute artificial general intelligence (AGI). From a *precision-centered* perspective, the requirements are slightly different. Here, the aim is to have a computational pipeline that outperforms a person. For example, oncologists and computer scientists in the Netherlands recently reported that a ML approach was capable of achieving better diagnostic performance than a panel of pathologists for the evaluation of breast cancer lymph node metastases [10]. Here, the algorithm is not coming up with anything "new" or "revolutionary," and it has been trained by a human to look for the same tissue morphological features that a pathologist would recognize. However, it is seemingly capable of performing the pattern recognition task more consistently and more reliably than the human counterpart. This is ML applied to cancer medicine. In terms of *next-generation* developments, here the goal with an AI approach is to go beyond what is possible with human interpretation and to thereby generate entirely novel knowledge which would otherwise not be derived. This is where DL methods are showing huge promise. DL is more of a computational search of the unknown, potentially revealing latent links and network connections between seemingly disconnected clouds of information. The algorithm is essentially left to its own devices and mines the data using multiple levels of abstraction to learn connections on a deeper level than is capable with the human mind. The program learns through iteratively processing data and requires large amounts of computational power and a database large enough for it to learn. For example, a study published recently in Cancer Research reported on the use of a convolutional neural network (CNN) to evaluate cancer tissue sections and determine the likely radio-sensitivity of the cancer based on a DL approach. The authors found that the way cancer cells clustered together was found by the algorithm to correspond strongly with whether they were radioresistant or radiosensitive [11]. Simplistically, one can consider the key data challenges in cancer medicine as (1) logistical, (2) precision centered, and (3) next generation. It is useful to consider the application of AI, ML, and DL in cancer, across these three domains (Fig. 5.3).

5.3 Artificial intelligence to determine cancer susceptibility

Clinical vignette: *Aiko is 38 years of age and has a strong family history of colorectal cancer. At the insistence of her family, she agrees to go and see a gastroenterologist following several episodes of fresh rectal bleeding and mild unintentional weight loss. The colonoscopy demonstrates an impassable malignant-looking stricture in the sigmoid colon, which is subsequently confirmed as adenocarcinoma on histology. Her staging CT scan reveals two metastatic deposits in the liver. She undergoes treatment in the form of systemic chemotherapy, right hemi-hepatectomy, and colorectal resection of her primary tumor. She is now under a strict 6 monthly surveillance program as her oncologist tells her that she is still at risk of cancer relapse, given her advanced stage at presentation. She asks her cancer physician whether more could have been done at an earlier stage to identify and quantify her risk of cancer, given the family history.*

The above clinical scenario is a common one in cancer medicine and highlights the critical need for enhanced methods with which to define an individual's risk of cancer. The principles of P4 medicine applied to cancer were first introduced by Hood and Friend in 2011 [12], describing a model that encapsulates the four key facets of precision healthcare—prediction, prevention, personalization, and participation. Prediction of an individual's risk of developing a given cancer has the potential to significantly minimize the physical, psychological, and socio-economic burden imposed by cancer. However, at the present time, there is only very limited understanding of the factors that govern cancer susceptibility. It is estimated that only 5%−10% of cancers are the result of inherited genetic mutations [e.g., breast invasive carcinoma 1 (BRCA 1) and 2 in hereditary breast cancer]. On the other hand, approximately 90% of all cancers are thought to develop as a consequence of the accumulation of acquired somatic mutations, triggered by complex interplay between environmental factors, lifestyle factors, and an individual's genomic cancer likelihood. At the present time, large-scale, prospectively maintained, high-quality data on environmental factors, lifestyle behaviors, and dietary practices are lacking, and this is an area that requires urgent attention. However, rapid developments in whole-genome sequencing in recent years have led to the emergence of publicly available genomic data repositories (e.g., the Cancer Genome Atlas), which now offers the potential to be mined using ML approaches for prediction of cancer genomic susceptibility to different cancer types. This potential is exemplified by a recent study by Kim and Kim, who utilized data from The Cancer Genome Atlas (TCGA) database with the aim of developing ML-based genomic models with which to predict risk across 20 major cancer subtypes [13]. The authors specifically sought to address two objectives with this work. First, to estimate the relative proportion of patients with a specific cancer who acquired the condition primarily through inherited genomic susceptibility. In turn this would, by proxy, lead to the indirect identification of patients developing cancer due to uninherited factors (e.g., environment and/or lifestyle). Second, the authors derived estimates of the rank-ordered probability of an individual patient's likelihood of developing one or more cancer phenotypes. The authors evaluated genotype to phenotype association in 5919 individuals with the cohort including 20 different cancer classes and one "healthy" phenotype, and genomics data were entered into a ML data treatment pipeline using a k-nearest neighbor method. The authors reported that 33%−88% of cancer cases were associated with a demonstrable inherited genomic abnormality, with the remainder primarily attributed, by corollary, to environmental and lifestyle factors. Importantly, the authors also found that at the individual level they were able to determine the likelihood of developing other linked cancers by evaluating commonality in genomic expression patterns across different cancer subtypes. This work represents a significant breakthrough in terms of the application of AI-based approaches to defining cancer susceptibility. The authors furthermore suggest that information regarding an individual's highest susceptibility to 1 of the 21 phenotypes (20 cancer, 1 healthy) may be useful at the patient-specific level to guide clinician and patient on ways with which to minimize that particular cancer risk through environment and lifestyle modifications (where the inherited genomic susceptibility for cancer is predicted "low"), or to proactively monitor for early detection and intervention of the cancer in individuals with "high" predicted genomic susceptibility.

In a 2016 study, Rau et al. focused on evaluating the association between type II diabetes mellitus (T2DM) and risk of liver cancer, utilizing data from the Taiwan National Health Insurance Research Database subjected to AI-based algorithmic interrogation [14]. A total of 2060 patients were included in the final study and patients were grouped according to whether or not they went on to develop liver cancer following an adult diagnosis of T2DM. They incorporated 10 variables into a model of cancer risk (gender, age, alcoholic cirrhosis, nonalcoholic cirrhosis, alcoholic hepatitis, viral hepatitis, other types of chronic hepatitis, alcoholic fatty liver disease, other types of fatty liver disease, and hyperlipidemia) and subsequently constructed artificial neural network (ANN) and LR predictive modeling approaches. The authors found significantly superior classification accuracy using the ANN approach and have now moved on to the next stage, which is construction of a web-based application for ML directed prediction of liver cancer risk in diabetic patients using the described ANN algorithm.

Although these studies illustrate the potential for AI-based approaches in cancer risk stratification, both examples involved analysis of well maintained, prospectively curated databases. The key challenge in real-world clinical environments is that patient data is all too often anything but well maintained and complete, and accessing and collating this data in a meaningful way to extract a quantifiable "risk" of cancer for patients is challenging as a direct consequence. A recent collaborative venture between clinicians at MD Anderson Cancer Center and IBM Watson Health sought to develop an AI application (Oncology Expert Advisor, OEA) capable of providing real-time, real-world patient-specific decision support [15]. This study was primarily borne out of an acknowledgment of the disparate nature of patient health records. The collaborative group, comprised of clinicians and computational scientists, set about creating an AI solution equipped with an advanced natural language processing (NLP) algorithm capable of automatically locating, extracting, and analyzing patient data to compile a composite patient-specific profile, which in turn can be used to drive clinical decision making. The authors found the process of applying NLP to medical records far more challenging than applying AI-based approaches to the evaluation of the published literature or publicly available data repositories. This was due to the fact that medical record documentation frequently comprises fragmented sentence structure, grammatical errors, acronyms, and statements of ambiguous meaning. The developers attempted to overcome these problems by training the OEA to recognize data specifically across nine "concept" modules. Through this approach, the OEA application was found to be capable of locating, searching, and extracting relevant endpoints from medical records, and using this data to create an "integrated patient profile" comprised of multidimensional clinicopathological data. The profile for a given patient, as depicted by OEA, is continually updated, as new data is inputted. The authors assessed the performance of OEA for searching disparate sources within electronic medical records and extracting complex clinical concepts from unstructured text documents varied, reporting F1 scores of 90%—96%. Based on OEA-based patient profile constructs, approved therapies linked to the available supporting evidence were identified with 99.9% recall and 88% precision, and patients were accurately screened for eligibility for clinical trial enrollment (97.9% recall; 96.9% precision). These and other emerging results demonstrate the technical feasibility of an AI-powered solution with which to construct context-specific longitudinal patient profiles with which to propose evidence-based treatment and/or trial options. The authors highlight the critical need for collaboration across clinical and AI domains, and the requirement for clinical expertise throughout all steps of the process, from study inception to design, training, and clinical validation. The developers have yet to test the platform outside of MD Anderson Cancer Center, but undoubtedly this represents an important development and highlights the potential for AI approaches to circumvent some of the fundamental challenges clinicians face with data mining.

5.4 Artificial intelligence for enhanced cancer diagnosis and staging

Clinical vignette: *Penny is a 40-year-old telecommunications executive who has recently been diagnosed with breast cancer. Her surgeon councils her about the treatment options and it is recommended that she undergoes a sentinel lymph node biopsy (SLNB) before deciding whether or not to proceed to more radical surgical axillary nodal clearance. Biopsy tissue acquired at the time of SLNB is sent to the* *histopathologist who subjects the slides to hematoxylin and eosin (H&E) staining, as well as immunohistochemical analysis. The verdict based on pathologist evaluation is that her nodes are free of cancer invasion. However, Penny's cancer physician has also enrolled her to a new trial evaluating the application of deep learning algorithms in detecting lymph node metastases in women with breast cancer. On AI-based analysis, her tissue sections are computationally considered to be cancer-bearing. She decides to undergo the more radical treatment option and currently remains in remission.*

The above case serves to illustrate the challenges facing conventional immunohistochemical evaluation of biopsy samples in cancer medicine, where the stakes are high, and where a verdict one way or another, can be the difference between major surgical intervention and more conservative treatment. The patient in the case example elected, on the advice of her treating physician, to undergo more extensive surgical treatment but felt that this decision was offered credence by the fact that the DL algorithm had identified her lymph node as being cancer-bearing. A study published in JAMA in 2017 by Bejnordi et al., on behalf of the CAMELYON16 Consortium, sought to evaluate this very clinical scenario, by assessing the discriminative accuracy of DL algorithms in detecting lymph node metastases in tissue sections of women with breast cancer, compared with conventional histopathological interpretation [10]. This study is particularly novel since the principal investigators organized a competition (Cancer Metastases in Lymph Nodes Challenge 2016, CAMELYON16) inviting research groups around the world to produce an automated solution for breast cancer metastases detection in sentinel lymph nodes (SLNs). The intention was that once developed, the performance of each algorithm would be compared to that of a panel of 11 breast cancer pathologists. Over a 1-year period, 390 research teams signed up for the challenge, demonstrating the tremendous enthusiasm in computer science for the development of cancer-directed AI solutions. Of these initial declarations of interest, 23 teams submitted 32 algorithms by the competition closing date. In terms of methodology, slides corresponding to SLN tissue from 339 patients were included in the analysis. The top performing algorithm by team HMS and MIT II used GoogLeNet architecture and outperformed all other competition entrants with area under the curve (AUC) of 0.994 (95% CI, 0.983−0.999). This AUC demonstrated statistically significant superiority over mean pathologist performance (AUC of 0.810; 95% CI, 0.738−0.884; $P < .001$). Although this is an early proof of concept study, the findings would appear to suggest that the potential for computationally augmented cancer diagnostics seems indisputable, and moreover, the potential for these approaches to surpass human capabilities seems genuinely possible.

There are also ways where AI approaches can represent not only precise, but also frugally innovative solutions to current cancer challenges. A study published in Nature in 2017 by Esteva et al. acknowledged that skin cancer, representing the most common human malignancy, is primarily diagnosed based on visual evaluation of the lesion(s) in question, commencing with initial clinical examination in the doctor's office, followed potentially by dermatoscopic evaluation, biopsy, and histopathological assessment [16]. The authors developed and implemented a deep CNN in the evaluation of 129,450 clinical images, including 2032 different skin conditions. The authors compared classification accuracy of the CNN with the performance of 21 board certified dermatologists and found the CNN to demonstrate equivalent skin cancer classification accuracy compared with dermatologists. The really telling point here, as the authors point out, is that mobile devices outfitted with deep neural network capabilities would have the potential to greatly extend the "reach" of dermatologists trying to diagnose skin cancers at the early stage, outside of over-subscribed clinics. The potential here to provide a low cost, universally available solution for diagnosing such a common cancer using an AI approach is tremendous.

5.5 Artificial intelligence to predict cancer treatment response

Clinical vignette: *Tomas is a 58-year-old university lecturer from Stockholm who has recently been diagnosed with a locally advanced rectal cancer. Preoperative imaging shows no evidence of distant site metastases, but pelvic MRI reveals that the primary tumor is bulky and extends to the predicted margin of surgical resection. As per conventional practice, it is suggested that Tomas undergoes pelvic radiotherapy, with a view to down-sizing the primary tumor and improving the likelihood of achieving complete surgical tumor clearance, with cancer-free margins of excision. Tomas receives 50.4 Gy of pelvic radiotherapy in divided fractions and then undergoes repeat imaging 10 weeks later to assess tumor response. Unfortunately, his tumor has shown no signs of meaningful regression and now appears to have spread to the liver and lungs. Tomas is bitterly disappointed and wants to know why he was put forward for radiation therapy if it was not going to work?*

Typical questions in radiation oncology include "what is my risk of treatment-related toxicity?", "what is the likelihood of achieving a good local response?," and "will my tumor spread elsewhere while I am being treated?". Unfortunately, the molecular and tumor-specific mechanisms that govern and regulate radiation response are not completely understood at the present time. However, increasingly, patients are wanting to be presented with a factually accurate projection of what radiation therapy will mean for them. AI approaches may represent an attractive way of generating a credible answer to this question and are beginning to demonstrate potential in this area. The processes of target and normal adjacent tissue image segmentation, treatment planning, dose optimization, and toxicity estimation are all areas that could benefit from AI solutions. It is possible in the future, with advances in AI to envisage a system that precisely identifies target and normal adjacent tissue volumes, estimates optimal radiation modality and beam arrangement, and determines a strategy that maximizes local tumor control, while minimizing toxicity. The AI approach should ideally have integrated access to electronic health records comprised of prospectively curated clinicopathological data as well as molecular data. The issue of how we realize the potential of AI-based approaches in the field was the subject of a recent review by Thompson et al. which explores some interesting themes [17]. The authors review the potential for AI as a positively transformative element in radiation oncology and provide a detailed discussion on key challenges in realizing this potential in terms of data availability, data equity, education and training and funding. They suggest that the international radiation oncology community will need to work collaboratively with industry and academia, to ensure optimal utilization of investment, resources, technology, and personnel.

Achieving the goal of precision radiation oncology using AI will be contingent on high-quality research in the field. Toratani et al. published a study in Cancer Research in 2018 where they employed a CNN to evaluate cancer tissue sections and found that this approach was capable of distinguishing cancer regions likely to be radiation sensitive versus radiation resistant, based on the manner in which they were seen to cluster, on computational image segmentation [11]. The ability to use such a technique at the time of diagnosis and treatment planning will potentially allow patients who are predicted to have radiosensitive tumors to be put forward for radiotherapy, while others may need to go directly to surgical resection, as the unnecessary delay to definitive treatment (surgery) while undergoing ineffective local treatment, risks distant cancer spread, as pointed out in the illustrative vignette provided above.

As with radiation therapy, chemotherapy works effectively in a proportion of patients and in the precision oncology era, it is essential to be able to reliably select the right drug for the right patient to maximize anticancer drug efficacy, ideally with minimal toxicity. Drug resistance, a phenomenon that describes tumor evasion of anticancer drug actions, is considered to represent the primary reason for chemotherapeutic treatment failure. The implication here is that a small cohort of cancer cells either acquire (acquired drug resistance) or innately feature (primary drug resistance) the ability to become "immune" to therapy, via multiple, complex drug-resistance mechanisms. A key challenge in drug resistance is the development of methods with which to predict, ideally in advance of treatment initiation, the likelihood of drug effectiveness/resistance. At the present time, monitoring treatment response typically involves metabolic imaging using positron emission tomography. However, this is usually only undertaken 1−2 cycles into the course of treatment, which in recurrent cases risks cancer progression and unnecessary toxicity. Given the unmet need in this area for enhanced response prediction, Liu et al. conducted a study to evaluate the use of single cell mass spectrometry combined with ML to provide a means of ultrafast prediction of treatment response [18]. Single cell MS metabolomic profiling was used to derive rich biochemical information, considered by the authors to be the summative endpoint of upstream genomic/proteomic/transcriptomic activity. Acquired data were then subjected to ML interrogation with the aim of identifying "hidden" metabolic patterns via clustering, regression, and predictive modeling. A variety of algorithms were assessed for predictive accuracy, and the ANN approach demonstrated overall superiority with over 95% predictive accuracy of drug-resistant cancer phenotypes. This work is particularly exciting for several reasons. First, it demonstrates the ability to explore cancer cellular heterogeneity effectively with the single cell MS technique. Second, the combination of minimal sample preparation combined with ML methods that provide rapid results led the authors to conclude that this approach could be used for point-of-care prognostic/therapeutic assays. An area where this would have game-changing potential for the cancer surgeon would be in the field of peritoneal surface malignancy. In operable cases, this is usually treated with a combination of radical cytoreductive surgery, followed by intraperitoneal instillation of hyperthermic chemotherapeutic agents, in liquid form. At the present time, there is no effective means of defining which intraperitoneal chemotherapeutic drug to use for a given patient. A method with which to acquire a sample, intraoperatively, perform single cell MS analysis and combine this with ML to select the most effective drug for a given patient would be highly desirable and shows how high-throughput bioanalytical approaches combined with AI will create the next generation of methods with which to not only select chemotherapeutic regimens, but also to inform the cancer surgeon in real time.

Other previous studies have also demonstrated the ability of AI approaches to predict treatment response. For example, Huang et al. recently looked at ML as a means of predicting individual cancer patient responses to chemotherapy [19]. The authors subjected gene expression profiles of 152 cancer patients to ML interrogation and found that the computational approach permitted highly accurate ($>81\%$) prediction of likely response to either gemcitabine or 5-FU. The authors had earlier published work describing an open access support vector machine (SVM)-based algorithm with which they were able to predict the likely collective response of 273 ovarian cancer patients to seven commonly prescribed chemotherapeutic agents [20]. However, with this more recent study they demonstrated the ability of AI-based approaches to predict response at the individual level, using person–specific gene expression data.

5.6 Artificial intelligence to predict cancer recurrence and survival

Clinical vignette: *Feroz is a 62-year-old retired mechanical engineer who was diagnosed with colon cancer in 2011. His surgical oncologist subjected him to a full raft of investigations before offering surgery, which took the form of a laparoscopic right hemicolectomy. His resection specimen revealed a completely excised Stage I tumor of the ascending colon. His case was discussed at the colon cancer multidisciplinary meeting, and given the favorable pathology, he was not felt to require adjuvant chemotherapy. He underwent yearly CT scanning and follow-up and was told by his physician in 2016, that having completed 5 years of follow-up, and having remained recurrence free, he would be discharged from further routine surveillance as this was not indicated. In May of 2018, Feroz sees his doctor for unintentional weight loss and a CT scan is arranged. This reveals a lesion in the right lung, which is subsequently confirmed on biopsy as a colon cancer metastasis. Feroz and his family are confused; they were told to complete 5 years of follow-up, which they did, and cannot understand why this has happened.*

The vignette above illustrates the critical need to develop more personalized and more precise cancer surveillance pipelines with which to follow up cancer patients. The fundamental problem with conventional cancer surveillance approaches is that they are essentially designed to detect recurrence. Ideally, the objective should be to predict likelihood of recurrence and prevent this from happening if possible with appropriate treatment modifications. At the present time, this is not possible, though there are mounting efforts toward the use of AI-based approaches in predicting the likelihood of cancer recurrence, and these are beginning to show genuine potential.

Gastric cancer is a good case exemplar, as it is a leading cause of cancer-related mortality worldwide, associated with abbreviated 5-year survival, even in treated cases, due to a strong propensity for recurrence. A recent study by Zhang et al. used gastric cancer as a case model and the authors described identification of long noncoding RNAs (lnc-RNAs) involved in gastric cancer development, progression, and patterns of relapse [21]. A composite signature comprised of 11 lnc-RNAs was identified using a ML approach applied to gene expression profiles obtained from the National Center for Biotechnology Information (NCBI), Gene Expression Omnibus (GEO), ArrayExpress, and TCGA data repositories. A weighted correlation network analysis algorithm was constructed to permit ML-based mining of gastric cancer-related molecular data. A "risk-score" was calculated for each patient based on expression patterns of the 11 lnc-RNAs, and patients were classified as "low-risk" or "high-risk," accordingly. The authors demonstrated that ML derived risk scores correlated strongly with overall and recurrence-free survival in gastric cancer patients. They suggest that this technique could offer a means of stratifying patients more individually and planning their treatment and subsequent surveillance in more personalized fashion.

Cancer surveillance is fundamentally intricate. We tend to survey patients for largely arbitrarily defined periods of time following their cancer treatment in completely generic fashion. The problem here is that the inherent longitudinal complexity of what follows once a patient has had cancer treatment means that the "one size fits all" approach in surveillance is severely limited. Ideally, a method for precisely surveying the cancer patient comes down to factors encompassing the past, the present, and the future, and computationally these are difficult to model. There has been interest in the application of next-generation DL approaches in order to longitudinally map the cancer patient "trajectory" post treatment using NLP architecture. These approaches currently remain in the early phases of development but are demonstrating strong potential. It is again important to acknowledge here the unavoidable computational workload that comes with models of this sort that are trained by ever larger and larger datasets. An elegant example of how AI can be successfully applied in cancer surveillance, while simultaneously mitigating problems of computational over-burden is provided by a study published by Qui et al. who looked at distributing the computational workload required for such a task across multiple computing cluster nodes, using a state-of-the-art data parallelism algorithm [22]. A high-performance computing environment was used where a Gradient Descent data parallelism algorithm was used to train a CNN for a NLP algorithm that involved extracting and modeling information from a large-scale cancer database. The authors reported encouraging results and concluded that this approach could represent a feasible means of defining surveillance pathways for cancer patients in the future.

In another study evaluating prediction of oral squamous cell carcinoma (OSCC) recurrence patterns, Exarchos et al. proposed a multiparametric decision support system in order to investigate the basis for OSCC relapse after seemingly total remission in patients undergoing radical treatment [23]. They modeled multiple heterogeneous sources of data (demographic, pathological, imaging, and genomic) in order to predict the likelihood of OSCC recurrence from a cohort of 86 patients, of whom 13 succumbed to loco-regional recurrence. A variety of ML algorithms were used to interrogate the data and among the evaluated variables, an AI generated a composite "panel" of markers consisting of smoking history, tumor thickness, radiological evidence of extraoral tumor spread, number of enlarged lymph glands seen on CT imaging, and mutational status of a panel of genes comprising p53, SOD2, TCAM, and OXCT2 genes, allowed robust discrimination of patients into those with a strong likelihood of tumor recurrence and those predicted to remain disease-free.

Park et al. reported on the use of ML techniques for the analysis of breast cancer recurrence [24]. The proposed algorithm is based on the construction of a graph model depicting recurrence risk which integrates gene expression data with gene network interactions. Based on biological knowledge, the authors selected gene pairs that indicate strong biological interactions. Patient samples were classified into three groups: (1) recurrence, (2) nonrecurrence, and (3) unlabeled samples. Using a combined total of 108,544 gene-encoded protein network interactions, the authors were able to demonstrate that a semisupervised ML method was capable of identifying a number of critical genes related to cancer recurrence.

They also claimed that this approach was found to outperform other existing methods suggested for breast cancer recurrence prediction.

A recent study by Hasnain et al. from the University of Southern California evaluated the role of ML approaches in predicting recurrence patterns and survival outcomes in a cohort of 3503 patients undergoing cystectomy for bladder cancer [25]. The authors constructed initial multivariate models using a series of "base" algorithmic models consisting of SVM, bagged SVM, K-nearest neighbor (KNN), adaptive boosted trees (AdaBoost), random forest (RF), and gradient boosted trees (GBT). These were then used to create more sophisticated "ensemble" models incorporating multiple algorithms simultaneously along different phases of the data treatment pipeline. ML methods developed a meta-classifier comprised of clinical and pathological data that predicted 5-year recurrence with sensitivity and specificity of 70% and 70%, respectively. Using the same approach, 5-year survival was predicted with sensitivity and specificity of 74% and 77%, respectively.

5.7 Artificial intelligence for personalized cancer pharmacotherapy

The potential for AI as a powerful positive disruptor in cancer pharmacotherapy is tremendous. In broad terms, the application(s) are likely to fall under three main headings: (1) *drug selection and toxicity prediction*; (2) *drug pairing*; and (3) *drug repurposing*.

In terms of drug selection, Section 5.5 has highlighted the irrefutable potential for AI-based approaches to determine whether a given patient will respond favorably to a particular anticancer treatment. On the flip side of this point lies toxicity prediction, where the case is equally clear regarding the need for improved prediction of unwanted treatment-related side effects; to put this need into perspective, recently published figures from the National Cancer Institute (NCI) estimate that there are currently 16.9 million cancer survivors in the United States, a figure set to rise to a projected 26 million by 2040 [26]. A significant proportion will receive chemotherapy in the course of treatment. Unfortunately, approximately 50% of these patients will experience drug-related side effects, including cardiovascular, dermatological, gastrointestinal, and neurological [26]. Neurotoxicity is common and chemotherapy-induced peripheral neuropathy, during and after treatment, frequently with permanent and debilitating consequences, has been reported to occur in an estimated

30%—40% of patients receiving chemotherapy. Bloomingdale and Mager recently reported the use of a ML method combined with drug structural analysis to computationally model the likelihood of drug induced neurotoxicity (peripheral neuropathy) from a manually curated library of 95 approved chemotherapies [27]. Molecular descriptors were generated for each of the 95 drugs. MDL MOL files for each drug were obtained from public databases (PubChem, DrugBank, and ChEMBL). MDL MOL files were imported into ADMET Predictor, and 227 molecular descriptors were generated, which included descriptors from the following categories: electrotopological states, topological indices, charge based, functional groups, hydrogen bonding, simple constitutional, and ionization. ANN models were developed for the quantitative prediction of peripheral neuropathy incidence and SVM models were subsequently developed to classify drug compounds into three groups (high, medium, or low) according to their propensity to lead to peripheral neuropathy. Using various algorithmic approaches, they reported sensitivity and specificity of 82%—90% and 87%—97%, respectively. They were able to isolate the number of aromatic rings in the molecular structure of evaluated drugs as a frequent and sensitive factor associated with propensity for neurotoxicity. What was particularly impressive with this study was the fact that the authors then applied the same ML approach to predicting the likely incidence of peripheral neuropathy for 60 small molecular drug compounds currently under clinical trial investigation. This has enormous potential implication for anticancer drug development, as increasingly perceived tumoricidal benefits will need to be weighed up against likely treatment-induced toxicity, and up to this point, it has not been possible to reliably quantify the latter for most drugs.

Drug pairing is another area where AI-based methods are likely to feature strongly moving forward. Cancer is highly complex with a single tumor thought to develop on the order of 100 million coding region mutations, with this molecular diversity believed to be a fundamental cause for drug resistance and treatment failure. Targeting multiple areas of a given tumor's molecular infrastructure requires both a detailed understanding of intratumor biomolecular interactions and also an understanding of how multiple drugs if used in combination will interact with one another. In this regard, the concept of drug pairing in cancer treatment is generating considerable enthusiasm. The fundamental complexity of this endeavor has proved prohibitive until recently, but AI-based approaches are starting to have a big impact here. An elegant example of this was provided recently by Xia et al. who applied a DL module to evaluating and predicting tumor cell-line responses to single-agent therapy versus combinations of drugs selected from 104 FDA-approved anticancer drug compounds, using the recently published NCI ALMANANAC resource [28]. The authors modeled the combined activity of anticancer drugs and noted good model performance ($R^2 = 0.94$) concluding that the methods employed represent a promising means of selecting drug combinations with greater anticancer efficacy compared with single-agent treatment. The DL algorithm considers tumor molecular phenotypic information including gene expression, microRNA expression, and protein expression as well as drug compound structural fingerprints to create the optimal therapeutic "match" for a given tumor cell line and the best performing combination treatment. This approach has the potential to revolutionize the way cancer patients are treated in the future—one can imagine acquiring a biopsy sample from a given tumor, subjecting it to organoid culture and molecular phenotypic interrogation, before using an AI algorithm to predict the most effective treatment and validate this in vitro.

Drug repurposing, or repositioning, represents an area of rapidly growing interest in cancer, with the potential to reduce the cost of new drug development by applying repurposed existing drugs on novel drug–target networks. The "one gene, one drug, one cancer" paradigm is now being challenged and AI based are at the forefront of network-based algorithms now in development that explore drug–drug similarities, drug–target relationships, and gene–gene interactions. An early example of this potential was presented by Iorio et al. who demonstrated that drugs exhibiting chemical structural commonality and similar transcriptional response profiles can potentially be repositioned with one another for use in, at times, highly disparate clinical pharmacological indications [29]. The authors predicted similarity in terms of the drug therapeutic effect by comparing gene expression profiles following drug treatment across multiple cell lines. An alternative way of using AI to repurpose drugs evaluates drugs and targets as part of a wider cancer molecular network. For example, Cheng et al. used a ML approach to predict drug–target relations based on a network of three distinct types of network connectivity: (1) drug–drug structural commonality, (2) target–target similarity, and (3) drug–target/drug–target interaction commonality [30]. The authors performed experimental validation of proposed drug repositioning strategies, for example, demonstrating that simvastatin, a drug used to lower cholesterol, through exertion of an acknowledged inhibitory action on

HMG-CoA reductase, also demonstrated powerful antagonistic effects on estrogen receptor β which is thought to be involved in breast cancer development and progression. On validatory in vitro work, the authors found simvastatin to exert a moderate, novel antiproliferative effect on breast cancer cell lines.

5.8 How will artificial intelligence affect ethical practices and patients?

Fostering public and commercial enthusiasm for AI in cancer medicine is critical to the success of the precision oncology endeavor. From the public perspective, there is interest, but this is also perhaps tempered by a degree of skepticism. One of the issues not discussed in the previous sections relates to the fundamentals of the doctor–patient axis. The rise of AI and the dawn of computationally directed decision making for the cancer patient will fundamentally challenge the age-old relationship between patient and treating physician. In some instances, the patient is likely to perceive this change in dynamic as a beneficial one—for example, the drug carefully selected, via AI, to minimize side effects and maximize treatment response. However, what if these algorithmic approaches are introduced to the cancer multidisciplinary board, as a "virtual" member? This is on the horizon, and recent studies have been seeking to evaluate the potential of "Intelligent Systems Support" for multidisciplinary medical teams [31]. Medicine has at its core a humanistic foundation—decisions on treatment for cancer are reached through consensus opinion of the cancer board at most units, taking into account local experience, local capabilities, and the available evidence. When a team of human physicians, including experts from multiple specialties, arrive at a conclusion (e.g., to offer the patient curative intent surgery, or to accept that the situation is too advanced and instead offer the patient palliative therapy) this news is discussed with the patient and often makes for an emotional and difficult encounter for all parties. But it is one human physician, discussing and determining the right treatment, with a human patient; and the encounter benefits from the centrally "human" nature of this. An AI colleague in the cancer multidisciplinary team (MDT) is probably an inevitably going forward, but it will present an additional layer of complexity to these difficult clinical situations. An AI platform is likely to be able to predict the patient's anticipated gains with treatment in terms of prolonging survival, estimate likely costs to the institution and estimate costs to society as a whole, and will come up with an answer of its own with unparalleled precision. But is precision really the panacea in such a situation? Many patients with cancer when faced with slim prospects for long-term cure still want to give radical treatment a go. It is the only chance they have; and yet AI will dispense with sentiment and instead will suggest proceeding with costly and risky treatment only when it is computationally felt to be feasible/advantageous. If we introduce state-of-the-art AI into the cancer MDT, what happens where there is fundamental disagreement between human board members and AI? There are no right or wrong answers with these questions currently, but what is clear is that AI approaches to cancer medicine will have social and legal implications that will need to be thoroughly explored in the future. From the point of view of industrial buy-in, the world would undoubtedly be better off without cancer, but how would this effect the pharmaceutical industry and is big pharma ready for cancer to be cured? Here, it is important to have an appreciation of spending; in 2015 the estimated spending on cancer medicines was around $107 billion worldwide, according to data from IMS Institute for Healthcare Informatics [32]. This figure is expected to increase further to around $150 billion by the year 2020. Although AI could have a highly disruptive effect on drug development and sales, this need not have a negative impact on the pharmaceutical industry. For example, recent analysis by McKinsey suggests that big pharma companies could generate in excess of $100 billion using AI annually through research and development (R&D) alone [33]. The potential is therefore clear, and all stakeholders stand to make significant gains, provided AI approaches are introduced into mainstream cancer medicine with foresight and careful direction. Consortia made up of international experts in cancer medicine, regulatory organizations, academia, industry, computational scientists, and patient groups will need to steer the cancer AI venture to ensure that it lives up to the promise and has the positive impact that is hoped for.

5.9 Concluding remarks

AI-based approaches comprise a broad range of rapidly expanding and increasingly sophisticated computational methods which have the potential to act as major disruptive influences across all phases of the cancer patient pathway, from prediction of cancer risk, through to next-generation diagnostics, targeted therapy, and personalized cancer surveillance (Fig. 5.4).

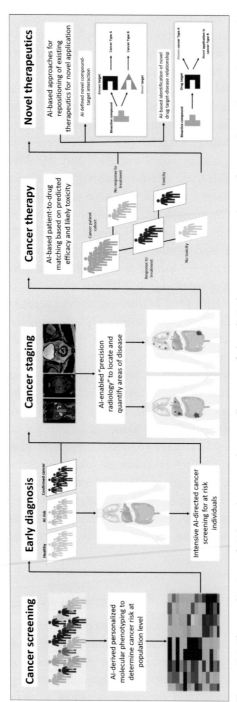

Figure 5.4 The application of AI-based approaches at all stages of the clinical cancer care pathway, from enhanced population-wide screening, to development of individual-specific pharmacotherapeutic regimens.

Although the currently available literature demonstrates the undoubted potential of these techniques, there are vast differences between studies in terms of the specific areas of application and the algorithmic method(s) employed. International consortia comprised of experts in clinical oncology, computational science, and translational research are required to try and define a more unified strategy moving forward and to outline an initial set of realistic deliverables, which can be subjected to large-scale, multinational randomized controlled trial evaluation. In parallel, the same challenges that broadly apply to the successful implementation of precision medicine approaches, apply here: pan-stakeholder collaboration, cost-effectiveness and quality assurance and high-quality research to confirm the benefits of any novel proposed approaches [34].

References

[1] He J, Baxter SL, Xu J, Xu J, Zhou X, Zhang K. The practical implementation of artificial intelligence technologies in medicine. Nat Med 2019;25(1):30−6.

[2] Johnson KW, Torres Soto J, Glicksberg BS, Shameer K, Miotto R, Ali M, et al. Artificial intelligence in cardiology. J Am Coll Cardiol 2018;71(23):2668−79.

[3] Thishya K, Vattam KK, Naushad SM, Raju SB, Kutala VK. Artificial neural network model for predicting the bioavailability of tacrolimus in patients with renal transplantation. PLoS One 2018;13(4):e0191921.

[4] Rush B, Celi LA, Stone DJ. Applying machine learning to continuously monitored physiological data. J Clin Monit Comput 2018;33(5):887−93.

[5] Hatton CM, Paton LW, McMillan D, Cussens J, Gilbody S, Tiffin PA. Predicting persistent depressive symptoms in older adults: a machine learning approach to personalised mental healthcare. J Affect Disord 2019;246:857−60.

[6] Sotiriou C, Pusztai L. Gene-expression signatures in breast cancer. N Engl J Med 2009;360(8):790−800.

[7] Allegra CJ, Jessup JM, Somerfield MR, Hamilton SR, Hammond EH, Hayes DF, et al. American Society of Clinical Oncology provisional clinical opinion: testing for KRAS gene mutations in patients with metastatic colorectal carcinoma to predict response to anti-epidermal growth factor receptor monoclonal antibody therapy. J Clin Oncol 2009;27(12):2091−6.

[8] Patel VL, Shortliffe EH, Stefanelli M, Szolovits P, Berthold MR, Bellazzi R, et al. The coming of age of artificial intelligence in medicine. Artif Intell Med 2009;46 (1):5−17.

[9] National Audit Office. Available from: <https://www.nao.org.uk/report/investigation-into-adult-health-screening/>.

[10] Ehteshami Bejnordi B, Veta M, Johannes van Diest P, van Ginneken B, Karssemeijer N, Litjens G, et al. Diagnostic assessment of deep learning algorithms for detection of lymph node metastases in women with breast cancer. JAMA. 2017;318 (22):2199−210.

[11] Toratani M, Konno M, Asai A, Koseki J, Kawamoto K, Tamari K, et al. A convolutional neural network uses microscopic images to differentiate between mouse and human cell lines and their radioresistant clones. Cancer Res 2018;78(23):6703−7.

[12] Hood L, Friend SH. Predictive, personalized, preventive, participatory (P4) cancer medicine. Nat Rev Clin Oncol 2011;8(3):184−7.

[13] Kim BJ, Kim SH. Prediction of inherited genomic susceptibility to 20 common cancer types by a supervised machine-learning method. Proc Natl Acad Sci USA 2018;115(6):1322−7.

[14] Rau HH, Hsu CY, Lin YA, Atique S, Fuad A, Wei LM, et al. Development of a web-based liver cancer prediction model for type II diabetes patients by using an artificial neural network. Comput Methods Prog Biomed 2016;125:58−65.

[15] Simon G, DiNardo CD, Takahashi K, Cascone T, Powers C, Stevens R, et al. Applying artificial intelligence to address the knowledge gaps in cancer care. Oncologist. 2019;24:772−82.

[16] Esteva A, Kuprel B, Novoa RA, Ko J, Swetter SM, Blau HM, et al. Dermatologist-level classification of skin cancer with deep neural networks. Nature. 2017;542 (7639):115−18.

[17] Thompson RF, Valdes G, Fuller CD, Carpenter CM, Morin O, Aneja S, et al. Artificial intelligence in radiation oncology: a specialty-wide disruptive transformation? Radiother Oncol 2018;129(3):421−6.

[18] Liu R, Zhang G, Yang Z. Towards rapid prediction of drug-resistant cancer cell phenotypes: single cell mass spectrometry combined with machine learning. Chem Commun 2019;55(5):616−19.

[19] Huang C, Clayton EA, Matyunina LV, McDonald LD, Benigno BB, Vannberg F, et al. Machine learning predicts individual cancer patient responses to therapeutic drugs with high accuracy. Sci Rep 2018;8(1):16444.

[20] Huang C, Mezencev R, McDonald JF, Vannberg F. Open source machine-learning algorithms for the prediction of optimal cancer drug therapies. PLoS One 2017;12 (10):e0186906.

[21] Zhang Y, Li H, Zhang W, Che Y, Bai W, Huang G. LASSO based CoxPH model identifies an 11 lncRNA signature for prognosis prediction in gastric cancer. Mol Med Rep 2018;18(6):5579−93.

[22] Qiu JX, Yoon HJ, Srivastava K, Watson TP, Blair Christian J, Ramanathan A, et al. Scalable deep text comprehension for cancer surveillance on high-performance computing. BMC Bioinforma 2018;19(Suppl. 18):488.

[23] Exarchos KP, Goletsis Y, Fotiadis DI. Multiparametric decision support system for the prediction of oral cancer reoccurrence. IEEE Trans Inf Technol Biomed 2012;16 (6):1127−34.

[24] Park C, Ahn J, Kim H, Park S. Integrative gene network construction to analyze cancer recurrence using semi-supervised learning. PLoS One 2014;9(1):e86309.

[25] Hasnain Z, Mason J, Gill K, Miranda G, Gill IS, Kuhn P, et al. Machine learning models for predicting post-cystectomy recurrence and survival in bladder cancer patients. PLoS One 2019;14(2):e0210976.

[26] Bluethmann SM, Mariotto AB, Rowland JH. Anticipating the "Silver Tsunami": prevalence trajectories and comorbidity burden among older cancer survivors in the United States. Cancer Epidemiol Biomarkers Prev 2016;25(7):1029−36.

[27] Bloomingdale P, Mager DE. Machine learning models for the prediction of chemotherapy-induced peripheral neuropathy. Pharm Res 2019;36(2):35.

[28] Xia F, Shukla M, Brettin T, Garcia-Cardona C, Cohn J, Allen JE, et al. Predicting tumor cell line response to drug pairs with deep learning. BMC Bioinforma 2018;19 (Suppl. 18):486.

[29] Iorio F, Bosotti R, Scacheri E, Belcastro V, Mithbaokar P, Ferriero R, et al. Discovery of drug mode of action and drug repositioning from transcriptional responses. Proc Natl Acad Sci USA 2010;107(33):14621−6.

[30] Cheng F, Liu C, Jiang J, Lu W, Li W, Liu G, et al. Prediction of drug-target interactions and drug repositioning via network-based inference. PLoS Comput Biol 2012;8(5):e1002503.
[31] Buzaev IV, Plechev VV, Nikolaeva IE, Galimova RM. Artificial intelligence: Neural network model as the multidisciplinary team member in clinical decision support to avoid medical mistakes. Chronic Dis Transl Med 2016;2(3):166−72.
[32] Informatics IIfH. Available from: <https://www.iqvia.com/-/media/iqvia/pdfs/institute-reports/global-medicines-use-in-2020>.
[33] McKinsey & Company Digital in R&D: The $100 billion opportunity.
[34] Mirnezami R, Nicholson J, Darzi A. Preparing for precision medicine. N Engl J Med 2012;366(6):489−91.

CHAPTER 6

Artificial intelligence for medical imaging

Khanhvi Tran[1], Johan Peter Bøtker[2], Arash Aframian[3] and Kaveh Memarzadeh[4],*

[1]Sonohaler, Copenhagen, Denmark
[2]Department of Pharmacy, University of Copenhagen, Copenhagen, Denmark
[3]NHS, London, United Kingdom
[4]ChemoMetec, Lillerød, Denmark
Corresponding author

6.1 Introduction

An image may be defined as a two-dimensional function $f(x,y)$ where x and y represent the spatial coordinates and the function (f) represents the amplitude at any given pair of coordinates (x,y). The amplitude (f) is also often referred to as the gray level or the intensity of that point in the image. A digital image is thus composed of a finite number of these x, y elements where all of them have an exact location and value. These picture elements are often referred to as pixels. 3D images may subsequently be defined by a three-dimensional function (x,y,z) and the individual elements are often referred to as voxels. The image quality is an important parameter in the field of radiology and the term spatial resolution refers to the capability of differentiating two neighboring objects in the image. Another important resolution property is the temporal resolution, which refers to the quality of the imaging with respect to time.

DICOM (Digital Imaging and Communications in Medicine) is the standard protocol for managing and communicating the medical image information and related data. DICOM is a registered trademark of the National Electrical Manufacturers Association. The standard is widely used within radiology, cardiology, oncology, obstetrics, and dentistry. The DICOM files convey images and information to systems capable of receiving patient data and images in the DICOM format. A DICOM file consists of site of origin, patient identification, the image itself, and attributes of the image such as pixel size. Importantly, DICOM files insure that patient data and picture data cannot be separated. Consequently, the DICOM image is always linked to the patient.

There is no consensus on where simple image processing ends and where a more advanced image analysis and computer vision begins. Occasionally, a definition is made by characterizing image processing as an operation where both the input and the output of an operation is an image. This is perhaps a bit too limiting definition boundary as a trivial task of computing, for example, the mean image intensity would then not be considered as an image processing operation. In the other end of the scale, computer vision exists, where the goal is to emulate human perception and device actions based on the inputs. Computer vision is a branch of artificial intelligence (AI) and AI uses human intelligence as an inspiration. There is no threshold boundary between ordinary image processing, computer vision, and AI. However, it might be useful to characterize low-level processes as ordinary operations such as noise reduction, image sharpening, and contrast enhancement. Mid-level processes may involve object detection within the image (image segmentation) and subsequent classification of those objects. High-level processes may involve an ensemble of classified objects where the conclusions can be compared to what would be obtained by a human assessment.

6.2 Outputs of artificial intelligence in radiology/medical imaging

6.2.1 Preprocessing

Image operations such as image enhancement, image normalization, and noise removal are often performed in a preprocessing step. Image enhancement is a process where the image is adjusted so that the image is more suitable for either visualization or further analysis. Specific examples include removal of noise from the image, sharpening of the image, image intensity adjustments, or facilitating easier object detection (image segmentation). A variety of methods exist for obtaining such goals.

Filtering with morphological operators may be used as a method for removing small objects in the image and to correct nonuniform background illumination. The filtering may be performed using a structuring element of a customizable size and shape. A larger structuring element will remove larger objects than a smaller structuring element. In the case of correcting nonuniform background illumination, all objects in the image are removed and the background intensity can subsequently be assessed and subtracted from the original image. Median filtering may be used for, for example, noise removal in images. The median filtering approach works by assigning the output pixel value to the median value of the neighboring input pixels. This approach removes outliers in the pixel values, as they would be far from the median value neighboring pixel values.

Histogram equalization, also known as image normalization, may be used for performing contrast adjustments to the image so that the intensity values of the image span the entire intensity range. Furthermore, this larger span of the intensity values allows for sharper differences between dark and bright regions.

In the field of image acquisition, it is also possible to encounter periodic noise in the images which typically originate from electrical and/or electromechanical interference that affect the image acquisition. To remove the periodic noise, it is necessary to determine its parameters. These are typically assessed by analyzing the Fourier spectrum of the image. This periodic noise produces frequency spikes that can be easily detected and when sufficiently pronounced, automated analysis can be invoked to ease the burden on determining the input parameters.

6.2.2 Segmentation

Image segmentation is an important step in many areas of the medical imaging field. This can be characterized as a process where objects or regions of interest are subdivided within the image. The subdivision or segmentation should hence be stopped when region of interest or object has been isolated. Segmentation of images is one of the most demanding tasks and its accuracy is fundamental for subsequent analysis failures or successes.

Segmentation algorithms are often based on either image intensity value similarities or image intensity value discontinuities. In the first case, segmentation is based on subdividing image into objects or regions of interest which appear to have similar intensity values based on a set of predefined criteria. In the second case, the segmentation is based on subdividing image by following abrupt changes in the intensities. Within segmentation of images using intensity discontinuities, there exists three basic types of approaches: point, line, and edge detection. In respect to the point and the line detection, these techniques involve the detection of isolated points or straight lines in a given image, respectively, and are of course of importance in image segmentation in general. However, edge detection is by enlarging the most utilized method for detecting intensity value discontinuities. This is because an edge in this case can be described as an arbitrarily curved line and especially in medical imaging it is often so that we are not interested in detecting either points or straight lines. For edge detection first derivatives are used as they have properties that necessitates that they are zero in regions of the image with constant intensity and nonzero and numerically correlated to the degree of intensity change in regions with variable intensities.

For segmentation of images using intensity value similarities, thresholding is often used. Basic thresholding enjoys widespread use due to its simplicity of implementation and intuitive properties. In the most basic setting, a threshold value can be determined by trial and error and subsequently ending up with a value that fits the purpose as judged by the user. Alternatively, a threshold value can be assessed by the visual inspection of a histogram of the image. These two proposed methods are of course highly user dependent, so it is often advisable to utilize an algorithm that automatically chooses a threshold value based on the image data in question.

6.2.3 Object detection

Object detection using bounding boxes is often used in conjunction with deep learning neural network algorithms and has been implemented in medical images to detect anatomical structures in computed tomography (CT) scans [1]. This technique is often utilized for identifying one or multiple regions of interest in an image. It could be utilized for the automated detection of the gallstones in patients while also locating the position of the liver and spine in the image (Fig. 6.1) [2].

6.3 Using artificial intelligence in radiology and overcoming its hurdles

With the current availability of big data, enhanced computer power, and new algorithms, many challenges have become manageable. However, there are still hurdles when using AI in radiology. One of the biggest hurdles is the need for a vast amount of high-quality labeled datasets that have a satisfactory training model, a balanced dataset that is representative of all data. In comparison to general databases, the volume of medical imaging datasets is still immensely lacking with only hundreds to thousands of images per dataset compared to millions of images, for example, in ImageNet [3,4]. The amount of data from patients with the same disease is scarce, even more so with rare diseases, and high-quality labeled data from

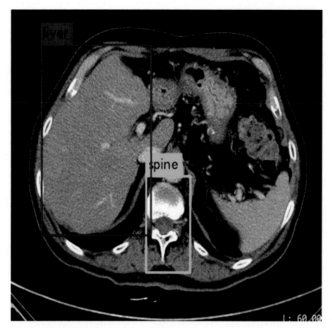

Figure 6.1 A correctly recognized negative sample. *Reprinted from Pang S, et al. A novel YOLOv3-arch model for identifying cholelithiasis and classifying gallstones on CT images. PLoS One 2019;14(6):e0217647. doi:10.1371/journal.pone.0217647.*

experts are limited and expensive [5−7]. Consequently, an underfit or often overfit model is trained and incorrectly classifies or fails to detect critical information. In radiology, it is especially important to avoid false negative results, as it can lead to severe repercussions for the patient and the physician, for example, if the model does not detect a benign lesion [6,8].

When models have many parameters and not enough training data to balance, the problem of overfitting arises. To compensate for the lack of training data or just limited labeled data, different methods can be applied. In some cases, datasets with fewer than 100 patients have yielded adequate performance [6]. In the following, we focus on how to expand the training set by transforming current data, generating synthetic data, regularizing the model, and leveraging external labeled datasets.

6.3.1 Not enough training data (data augmentation)

In traditional data augmentation, the images' spatial structure, appearance, and quality gets manipulated but preserves the labels to become new images that in turn enlarges the dataset.

- *Spatial* transformations alter the image through rotation, flips, scaling, and deformation, which mimics different orientations during scanning, variability of shapes, and motion in tissue.
- *Appearance* augmentation transforms the statistical characteristics, such as brightness and contrast in the image intensities, mimicking different scanning protocols and device vendors.
- *Image quality* is manipulated through blurriness, sharpness, and noise level. The former two are the reverse of each other and commonly caused by MR/ultrasound motion artifacts and resolution.

The images from traditional augmentations can be highly correlated, have insufficient focus on rare conditions, and therefore have a limiting impact. It is also computationally expensive on large 3D datasets [9,10].

Synthetic data augmentation can be categorized by same-domain and cross-domain image synthesis, respectively, generating labeled data in the target domain, or transferring labeled data from a different domain to the target domain.

The latter category is similar to domain adaptation and is touched upon in detail further down. Various methods for generating synthetic images with the characteristics of the given dataset are proposed by using a class of neural networks called generative adversarial networks (GANs); these are generative models that create new data. Utilizing GANs, alternate versions or characteristics of the images can be generated to enlarge the dataset for further training [10,11].

6.3.2 Unbalanced training data (data weighing)

To combat overfitting, regularization of methods can be used. This is the means by which the weights between two connecting layers are adjusted to a more effective range. The traditional method is weight regularization. Weights of the model are kept small to make it simpler and more robust. An alternative method is the dropout method, where the weights are randomly set to zero. The results will clearly show which weights affect performance, and after enough iterations, only the important weights are kept [8,10,12].

6.3.3 Not representative training data (transfer learning, domain adaptation)

In addition to requiring a huge amount of data, the training data must be representative of the model that it will "meet" in the future, otherwise the model may not generalize to the new data. Consequently, a model trained on data from one or more hospitals may perform worse on data from different hospitals. To increase a model's generalization capability, the learning algorithm must include *inductive bias*—a set of assumptions about the true distribution of data to predict the results of input that it has not yet seen. To achieve this, the model can either train on all data available, which is very inefficient, or it can transfer "knowledge" from an abundant dataset to the current model to generate a more representative data [13—15].

6.3.3.1 Transfer learning

Humans can abstract knowledge from one source domain and apply the knowledge to learn a new target domain. Compared to a person with no musical experience, a guitar player can transfer their previously learned music knowledge (source domain) to faster and more effectively learn the task of playing piano (target domain). Likewise, a model that can classify ImageNet images most likely performs better on a medical image dataset, as opposed to a model with random weights [16]. In transfer learning, also called "fine-tuning" or "pre-training," the weights of a pretrained network with abundance of data (usually ImageNet for 2D images) are fine-tuned to adapt to a new target domain (Fig. 6.2).

Transfer learning allows for transfer of knowledge between domains, tasks, or distributions and is grouped according to what is transferred (instance transfer, feature-representation transfer, parameter transfer,

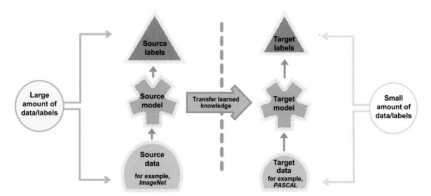

Figure 6.2 A depiction of the flow of transfer learning (inspired by Transfer Learning Explained by integrate.ai).

relational-knowledge transfer) and where the transfer takes place [15,17,18]:

- *Inductive*: Different tasks are transferred through the same domain. The source domain is used to improve the target domain by leveraging the knowledge of the source task to choose or adjust the target task's inductive bias. If the source domain contains labeled data then source and target tasks are learned simultaneously and are called multitask learning, whereas no labeled data available uses self-learning.
- *Unsupervised*: The tasks are different, domains are similar, but with no labeled data available.
- *Transductive*: The tasks are similar and domains are different. Usually with the source domain containing a large collection of labeled data, whereas the target domain has none. This is also referred to as domain adaptation.

6.3.3.2 Domain adaptation

Images trained and tested in the same domain, that is, the same protocol, magnetic resonance imaging (MRI) scanner, resolution and image contrast, are most likely to face the problem of Domain Shift when the model is used in real-world cases: the data distribution only represents learnings from the training set, making the model less generalizable [15,19].

Domain adaptation is a form of transfer learning, where the task of source and target domain is the same (i.e., the domains share class labels), but the source and target domain, data representation or distribution are different.

When labels are only available in the source domain, the images are translated from the source to the target domain. This is called *Unsupervised Domain Adaptation (UDA)*. An example of UDA is translating MRI images containing spleen masks to CT spleen images for the task of spleen segmentation [20]. While keeping the anatomical structure, and with that the segmentation masks, the source images are translated to the characteristics of the target domain. Predictions of the target images are then made by using the synthesized images and source masks to train a segmentation network [10,13,21,22].

If a small set of target examples are labeled, the domain adaptation is considered semisupervised, while both domains containing labeled data are categorized as supervised domain adaptation. In this case, a shared representation is learned between the two domain to predict data labels regardless of input domain [10,21].

6.3.3.2.1 Black box—algorithm explanation

As new algorithms evolve to make these predictions, how do we make sure that they are accurate, when the deep neural network is a black box. Therefore a new field is emerging to combat this issue, called *Explainable AI* that looks to increase interpretability, visualize features, and measure sensitivity by other means. Another way to increase trustworthiness is by computing uncertainty estimates while making predictions, which can be done with Bayesian Deep Learning [13,23].

6.3.3.2.2 Implementation/integration

Building an accurate prediction algorithm can only get so far, if the radiologist cannot or will not use it, which builds upon another point in trustworthiness. Developing a system for clinical use should be in collaboration with the end-user to make sure the entire workflow will also work in a practical setting and build trust in the systems [5,13]. The developers should also look at privacy concerns, and building for regulatory, ethical, and legal use [13,23,24].

6.4 X-rays and artificial intelligence in medical imaging— case 1 (Zebra medical vision)

6.4.1 X-rays and their role in medicine

The discovery of X-ray imaging revolutionized the field of medicine. This discovery was an intrinsic part of our modern society, manufacturing, airports, and important discoveries such as the DNA. X-ray imaging was discovered in the year 1896. Initially, X-rays were used for entertainment purposes and individuals would use "bone portrait studios" to acquire a radiograph of their hands, head, legs, or chest and use these images as objects of beauty to be admired by all. During this period, handheld fluoroscopes (Fig. 6.3) were invented so that individuals would wear a headpiece (much like the Virtual Reality headsets of today) and will lift one hand in front of the screen and were able to see their own skeletons. Of course, there is now a good reason to know why these gadgets are not safe to use. During the same period, the lifesaving potential of this technology was also identified by a few scientists and physicians who noticed that bone fractures could be visualized when exposed to X-ray.

Figure 6.3 A handheld fluoroscope, which is oddly similar to some of the "Virtual Reality" headsets of today.

6.4.2 X-ray discovery

The individual responsible for discovering the X-ray was a German physicist and engineer named Wilhelm Conrad Roentgen. Roentgen was a physics professor at the University of Würzburg, Germany. Near the time of his discovery he was working on a device referred to as a "Crookes tube," an experimental electrical discharge tube that was thought to produce different types of rays. On November 8, 1895, Roentgen setup his Crookes tube in his office to follow his ambition of discovering new "invisible" rays. He wrapped the glowing tube in a black paper to prevent its visible glow and with the lights off in the room, Roentgen noticed a bright green glow on a primitive fluorescent screen (a piece of cardboard with a few barium platinocyanide crystal on it) laying atop of a table in the room, it was curious that the bright glow on the cardboard would disappear when the experimental setup was not functional. Roentgen did not stop there. He further experimented by placing various materials (including his own hand) between the cardboard and the tube and observed that some materials like paper would not block rays but metallic objects and his own hand would cast an outline on the fluorescent screen. Over the next few weeks, Roentgen allegedly refined the method and called the newly discovered rays "X" because of its unknown nature. During the same period, he took the photograph of the first X-ray of his wife hand that can be seen in Fig. 6.4.

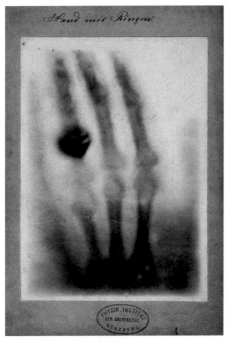

Figure 6.4 The first ever X-ray image of a hand.

6.4.3 Chest X-rays

Short after its discovery and because of its unique potential, X-ray was continuously used and until very recently all images were handled physically by healthcare professionals. However, with the recent advances in computer technology, most X-ray images are digitized and with millions of patients examined on a daily basis, the data produced is simply vast.

Chest X-rays (CXRs) are essential in the field of radiology because various life-changing diagnoses are made using them. These are cheap and can be performed quickly, even in the most underdeveloped regions in the world. However, there is often a huge workload associated with the obtained images. Interpretations of these are often hard, and human error can be common due to anatomical complexities such as the interfaces between lesions and bone. Hence, AI-based algorithms can be implemented to assist and guide clinical experts in their decision-making process. Furthermore, lack of radiology experts for diagnostics purposes is a rising concern. In the United Kingdom alone, between 2012 and 2017, the workload demand for radiologists has increased by 30% while the percentage of expert clinicians has risen by only 15% [25]. Therefore implementation of smart systems can bring significant benefits and enhance workflow optimization for clinicians and health workers [26].

6.4.4 RadBot-CXR—clinical findings using deep learning

One cannot ignore the influence and innovative nature of small companies that are increasingly emerging within the field of radiology. Today's world allows for vast fields of digital "fertile ground" for keen and enthusiastic entrepreneurs and inventors to experiment and play with ideas. Here, we will briefly highlight the achievement of an innovative company named, Zebra medical vision (ZMV), a software start-up that is "empowering radiologists" with its revolutionary AI and helps health providers manage the ever-increasing workload without compromising quality. According to ZMV, this can be achieved by utilizing hundreds of thousands of old and new scans to create a software that analyzes present and future data with human level accuracy, thus providing professionals with the assistance that they require.

As stated previously, image analysis and diagnostic radiology are in high demand. Today, often other (nonexpert) clinicians and radiographic technicians are called upon to provide a preliminary interpretation of images for patients. This is an effort to decrease the waiting interval for patients but can lead to a reduction in diagnostic accuracy. Even among expert radiologists, serious clinical errors are made [27]. There are thus some rather challenging diagnostic decisions that have to be made and because of the long working hours, workload, and the volume of examinations, conditions such as early lung cancer are often missed in 19%—54% of cases [28] and minor errors can be made around 30% of the time [29].

Shortage of expertise for a thorough, accurate, and fast analysis of images has prompted ZMV to employ convolutional neural networks (CNNs) to tackle the problem. RadBot-CXR has been developed and validated to an expert level automatic interpretation system for the detection of four categories that relates to seven distinct radiographic findings on CXR: alveolar consolidation, lung mass, atelectasis, pleural effusion, hilar prominence, diffuse pulmonary edema, and cardiomegaly. The four main categories are general opacity, cardiomegaly, hilar prominence, and pulmonary edema, where the software output reports will verify the existence or the nonexistence of each of these categories. These categories provide an in-depth insight into the patient's diagnosis. Furthermore, the data obtained from this CNN has shown that the level of agreement between the RadBot-CXR system and the radiologists is slightly higher than the interobserver agreement among a team of three expert radiologists [30]. Additionally, these results show a significant improvement of the radiologists' performance following unblinding of the algorithm results. ZMV claims that the significance of such software comes into play where there is a shortage of experienced clinical staff and a high demand for expert level diagnosis.

6.4.5 Detecting osteoporosis using artificial intelligence

In recent years, medical technology and effective interventions have led to an aging population. This means that while we live longer, we are increasingly prone to new types of conditions. Musculoskeletal (MSK) conditions are high on this list and have been projected to grow and potentially add to mortality rates [31]. Osteoporosis is a common MSK condition that leads to a reduction of bone mass, leading to an increased risk of bone fracture in later years of life [32]. This condition is worse when it affects the vertebrae, leading to vertebral compression fractures (VCFs), a phenomenon that leads to a significant weakening of the spine and often manifesting itself in the elderly with a characteristic "rounded back" and a "bent look." Despite its incapacitating nature and the fact that it increases the chances of later hip fracture, conventional diagnostic measures for identifying this condition by utilizing imaging systems like CT have been less than satisfactory.

By creating a solid algorithm that can detect VCF in X-rays obtained from chest and abdomen, the researchers and the ZMV team were able to obtain segments of patches of the vertebral column. These patches were binary classified, put together, and trained using a CNN and then run on recurrent neural networks (RNNs). RNNs are a powerful deep learning tool that are often used to model sequences. Together with CNNs, RNNs can predict the probability of the presence of VCFs within the CT scan images. ZMV claims that trained RNNs resulted in 83.9% sensitivity, 93.8% specificity, and 89.1% accuracy [33].

With such promising outcomes, deep learning is on its way to become an integral feature for the future of medical image analysis.

6.5 Ultrasound and artificial intelligence in medical imaging—case 2 (Butterfly iQ)

6.5.1 Ultrasound and its role in medicine

Medical ultrasound (sonography) is a medical technology that uses high-frequency soundwaves to allow its users to observe events inside the human body. A transducer is often used to produce and direct soundwaves to the region of interest and the results are often observed live on an integrated monitor.

The uses of ultrasound for diagnostic purposes are often either anatomical or functional. Anatomical ultrasound is often used to produce images of individual structures inside the body, while functional ultrasound builds on this and measures the movement and speed of blood or the softness or hardness of tissue. This allows clinicians to see changes of function within a body structure or organ. This technology is now widely used as a therapeutic to treat a variety of medical conditions. High-intensity ultrasonic soundwaves have the capability to modify and destroy unwanted tissue such as tumors. These soundwaves can also potentially deliver drugs and can also disrupt unwanted blood clots that otherwise could lead to catastrophic health outcomes.

Because of its diverse, safe, and relatively low cost, ultrasound is increasingly used in diagnostic and therapeutic medicine (Fig. 6.5).

6.5.2 The Butterfly iQ

The Butterfly iQ is an innovative pocket-sized ultrasound device, developed by the Butterfly Network. It utilizes capacitive micromachined ultrasound transducer (CMUT)/complementary metal oxide semiconductor-based parts in the ultrasonic probe. It differentiates itself from all other competitors by utilizing silicon chip CMUTs instead of using conventional piezoelectric crystal-based transducers. The CMUT has a much wider bandwidth compared to the piezoelectric transducers that are tuned to oscillate at specific frequencies. A CMUT can thus be used to detect and emit many different frequencies and as a result a CMUT can be utilized for whole-body imaging. A major future area of application for the Butterfly iQ is interfacing with AI. The company has announced that this is a direction it will investigate by feeding the uploaded images from the users to a deep learning algorithm. The AI aim is to enable the software to provide guidance to the user both during image interpretation and during image acquisition. Such AI capabilities could dramatically expand the

Figure 6.5 A handheld Butterfly iQ device. Similar devices to the iQ can revolutionize the medical field by allowing an easy to use and accessible imaging anywhere. *Reproduced with permission from https://www.butterflynetwork.com/.*

application of the device into different realms of uses. In one setting, the company envisions that in low-resource settings AI-guided interpretations and guidance will be valuable. In another setting, the low cost of the device combined with possible AI-guided capabilities may entail that it can be offered as a personal medical device [34].

The deep learning algorithms of the Butterfly network are used to power their computer vision applications. By training these on enough images related to a medical condition, their program learns how to scan the correct regions of the patient's body and can distinguish normal from abnormal tissue. Currently, most of the image processing and analysis is taking place in the cloud and not on the device or local networks.

The ease of using devices like Butterfly iQ and the low cost of obtaining each image makes it attractive and fast to collect a large number of images for analysis. It is envisioned that this device would be available at retailers including pharmacies or at clinics and the images captured with the device can be sent to a doctor for further analysis, similar to telehealth applications.

The connected and automated nature of the device would also enable a diagnosis to be delivered almost instantaneously instead of weeks of delay and waiting.

6.6 Application of artificial intelligence in medical imaging—case 3 (Arterys)

Cardiovascular disease is a class of diseases that encompasses both the heart and blood vessels. Therefore an important factor that is often missed is the blood flow. Additionally, the relevant measurements are performed manually, which are time consuming and are carried out with little automation.

The founders of Arterys, a medical imaging platform with the first FDA approval for use of AI in a clinical setting, combines deep learning, 4D flow (a technique allowing the direct examination of the flow and the function of the heart using MRI), and cloud computing to aid radiologists in their daily work [35] (Fig. 6.6).

In a single, short scan, compared to a conventional scan, 4D flow acquires volumetric anatomical, functional, and flow information during the entire cardiac cycle. This is while the deep learning model classifies

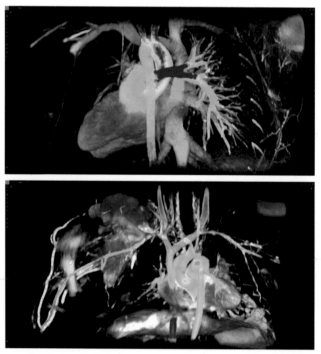

Figure 6.6 4D flow can be used to look at blood flow and heart function at the same time. Using this technique, a comprehensive evaluation of the heart can be obtained in 10 min. This will allow the clinicians to quantify the best therapeutic approach for patients. *www.arterys.com.*

and segments the heart as accurately as manual measurements by experts and increases the quality and image resolution in less time. With these two technologies combined, files that usually took days to process in a hospital setting can now take minutes [36]. Arterys uses cloud computing to run data processing and augmentation at a much higher rate, while gathering images from different hospitals, scanner models, and patients, which increases the model's generalization capability and thereby its predictions. This functionality also puts the question of patient data privacy and security into play. To tackle this issue, Arterys has incorporated a system called Protected Health Information (PHI) service that removes identity-related data from images before uploading to the cloud. Therefore, when an authorized medical staff requires the identity of the data for patient evaluation, PHI rebuilds the image with the patient identity data from the secure hospital server [37]. Thus the stripped and raw image is uploaded, converted into DICOM data, reconstructed, corrected, and the real-time interactive 3D postprocessing tools are applied. Subsequently, flow visualization, quantitative analysis, and statistical analysis are performed in a short period of 12 minutes in the cloud [36].

With utilization of deep learning, it is specifically classification and segmentation that are improved by saving time and making the process more efficient. Segmentation of the heart's ventricles is one of the most time-consuming tasks in cardiac MRI. According to experienced clinicians, it takes approximately 1 hour to segment the left and right ventricles, ~ 30 minutes per ventricle [38]. Arterys model automatically segments the inside and outside of the ventricles as accurately as expert annotators in just 10 seconds [39]. The most suited layers are then identified for the deep learning model to yield the most accurate results in an appropriate time frame for inference, as more layers would make the model more accurate but can take longer to process [37].

The Arterys model also decreases the average total examination time from 71 to 52 minutes. This includes a significant reduction of anesthesia used in pediatric patients and eliminating breaths held during MRI examinations. As compared to the conventional manual task, the AI model can process tremendous volumes of new imaging data. Additionally, new and relevant clinical insights are extracted from the complex and nonlinear relationships of the data points in the deep learning models. The platform was developed together with the end-users to integrate the system into their daily workflow, capturing their inputs and implementing change to improve the flow of work [37].

Currently, Arterys is aiming to predict treatment response and the risk of disease progression on images from clinical trials, as well as adding data from lab results to individualize patient care by building a more comprehensive predictive model [37]. The model has also been extrapolated to analyze the lungs, liver, and CXR, while developing neuro and prostate platforms, creating a healthcare AI suite [40,41].

6.7 Perspectives

While the future for AI in radiology looks very promising, for radiologists things would be somewhat more bleak according to Geoffrey Hinton credited as a "leading figure in deep learning." He is quoted in the The New Yorker as saying that "I think that if you work as a radiologist you are like Wile E. Coyote in the cartoon, you're already over the edge of the cliff, but you haven't yet looked down. There's no ground underneath . . . It's just completely obvious that in five years deep learning is going to do better than radiologists, it might be ten years . . . I said this at a hospital. It did not go down too well . . . They should stop training radiologists now" somewhat similar thoughts were presented by the well-known computer scientist and statistician, Andrew Ng. These sentiments are also (to some degree) supported within the profession itself—the past president of the Radiological Society of North America has [predicted] that "within ten years, no medical imaging study will be reviewed by a radiologist until it has been pre-analyzed by a machine." He further explains that computers are not taking over quite just yet; the machines are there to support and not to replace. While all the above statements can be true, it is important to note that various limitations in the current application of AI in radiology exist and that they are still very much task specific (narrow). It is interesting to note that FDA-approved algorithms can detect that no wrist fractures are present within an X-ray radiograph, but fail to identify that there has been a proximal row carpectomy. That is to say that an entire row of bones in the wrist has been surgically removed. This is because the software is only designed to look for fractures, so while the function is robust, it has its limitations.

Finally, we hypothesize that hybridization of the field of radiology is the likely outcome in the near future. This entails an understanding that within this field, AI will increasingly augment the needs of the healthcare workers, reduce the burden on clinicians, integrate all relevant patient data, and will ultimately lead to a better patient experience. Often, these processes are faced with dilemmas that require human intervention and therefore a "collective intelligence" of both human and machine can lead to a more efficient and cohesive workflow.

References

[1] Onieva Onieva J, González Serrano G, Young TP, Washko GR, Ledesma Carbayo MJ, San José Estépar R. Multiorgan structures detection using deep convolutional neural networks. Progress in biomedical optics and imaging—Proceedings of SPIE, 10574. 2018.

[2] Pang S, et al. A novel YOLOv3-arch model for identifying cholelithiasis and classifying gallstones on CT images. PLoS One 2019;14(6):e0217647.

[3] Hussain Z, Gimenez F, Yi D, Rubin D. Differential data augmentation techniques for medical imaging classification tasks. AMIA Annu Symp Proc 2017;2017:979—84.

[4] Li L-J, et al., ImageNet: a Large-Scale Hierarchical Image Database, Characterization of natural fibers, View project, Display Wall immersive imagery research, 2009.

[5] Miotto R, Wang F, Wang S, Jiang X, Dudley JT. Deep learning for healthcare: review, opportunities and challenges. Brief Bioinform 2017;19(6):1236—46.

[6] Ker J, Wang L, Rao J, Lim T. Deep learning applications in medical image analysis. IEEE Access 2017;6:9375—9.

[7] Mazurowski MA, Habas PA, Zurada JM, Lo JY, Baker JA, Tourassi GD. Training neural network classifiers for medical decision making: the effects of imbalanced datasets on classification performance. Neural Netw 2008;21:427—36.

[8] Erickson BJ, Korfiatis P, Akkus Z, Kline TL. Machine learning for medical imaging. Radiographics 2017;37(2):505—15.

[9] Zhang L, et al., When unseen domain generalization is unnecessary? Rethinking data augmentation; 2019.

[10] Tajbakhsh N, Jeyaseelan L, Li Q, Chiang J, Wu Z, Ding X. Embracing imperfect datasets: a review of deep learning solutions for medical image segmentation; 2019.

[11] Bowles C, et al., GAN augmentation: augmenting training data using generative adversarial networks; 2018.

[12] Tripathi N, Jadeja A. A survey of regularization methods for deep neural network. Int J Comput Sci Mob Comput 2014;3:429—36.

[13] Lundervold AS, Lundervold A. An overview of deep learning in medical imaging focusing on MRI. Z Med Phys 2019;29(2):102—27.

[14] Zech JR, et al. Variable generalization performance of a deep learning model to detect pneumonia in chest radiographs: a cross-sectional study. PLoS Med 2018;15:e1002683.

[15] Torrey L, Shavlik J. Transfer learning. Handbook of research on machine learning applications. IGI Global; 2009.

[16] Shorten C, Khoshgoftaar TM. A survey on image data augmentation for deep learning. J Big Data 2019;6:60.

[17] Lu J, Behbood V, Hao P, Zuo H, Xue S, Zhang G. Transfer learning using computational intelligence: a survey. Knowl Based Syst 2015;80:14—23.

[18] Pan SJ, Yang Q. A survey on transfer learning. IEEE Trans Knowl Data Eng 2010;22(10):1345—59.

[19] Kushibar K, et al. Supervised domain adaptation for automatic sub-cortical brain structure segmentation with minimal user interaction. Sci Rep 2019;9.

[20] Huo Y, Xu Z, Bao S, Assad A, Abramson RG, Landman BA. Adversarial synthesis learning enables segmentation without target modality ground truth. Proceedings—International symposium on biomedical imaging, 2018. 2018. p. 1217—20.

[21] Csurka G. Domain adaptation for visual applications: a comprehensive survey. Domain adaptation in computer vision applications. Springer; 2017.

[22] Kouw WM, Loog M. A review of domain adaptation without target labels. IEEE Trans Pattern Anal Mach Intell 2019;.

[23] Lee JG, et al. Deep learning in medical imaging: general overview. Korean J Radiol 2017;18(4):570–84.

[24] Chartrand G, et al. Deep learning: a primer for radiologists. Radiographics 2017;37 (7):2113–31.

[25] Board of the Faculty of Clinical Radiology. Clinical Radiology UK Workforce Census Report 2018, R Coll Radiol, April 2018.

[26] Kao E-F, Liu G-C, Lee L-Y, Tsai H-Y, Jaw T-S. Computer-aided detection system for chest radiography: reducing report turnaround times of examinations with abnormalities. Acta Radiol 2015;56(6):696–701.

[27] Robinson PJ, Wilson D, Coral A, Murphy A, Verow P. Variation between experienced observers in the interpretation of accident and emergency radiographs. Br J Radiol 1999;72(856):323–30.

[28] Hanna TN, Lamoureux C, Krupinski EA, Weber S, Johnson JO. Effect of shift, schedule, and volume on interpretive accuracy: a retrospective analysis of 2.9 million radiologic examinations. Radiology 2018;287:205–12.

[29] Bruno MA, Walker EA, Abujudeh HH. Understanding and confronting our mistakes: the epidemiology of error in radiology and strategies for error reduction. Radiographics 2015;35(6):1668–76.

[30] Brestel C, Cohen-sfaty M. RadBot-CXR: classification of four clinical finding categories in chest X-ray using deep learning. Medical Imaging with Deep Learning (MIDL 2018). 2018. p. 1–9.

[31] Kiadaliri AA, Englund M. Mortality with musculoskeletal disorders as underlying cause in Sweden 1997–2013: a time trend aggregate level study. BMC Musculoskelet Disord 2016;17(1):163.

[32] Liu W, Yang L-H, Kong X-C, An L-K, Wang R. Meta-analysis of osteoporosis: fracture risks, medication and treatment. Minerva Med 2015;106(4):203–14.

[33] Bar A, Wolf L, Bergman Amitai O, Toledano E, Elnekave E. Compression fractures detection on CT. Medical imaging 2017: computer-aided diagnosis. 2017.

[34] Joyce YLB, Jiajun XM, Flemming FP, Ji-Bin LMF. CMUT/CMOS-based Butterfly iQ—a portable personal sonoscope. Adv Ultrasound Diagnosis Ther 2019;3:115.

[35] Marr B. First FDA approval for clinical cloud-based deep learning in healthcare. Forbes. 2017.

[36] Arterys. Transforming cardiac MR: advances in AI, 4D flow and cloud computing; 2018.

[37] Natarajan P, Frenzel JC, Smaltz DH, Leibowitz C. Arterys: deep learning for medical imaging. Demystifying big data and machine learning for healthcare. CRC Press; 2018. p. 169–73.

[38] Lieman-Sifry J, Le M, Lau F, Sall S, Golden D, FastVentricle: cardiac segmentation with ENet.

[39] Suinesiaputra A, et al. Quantification of LV function and mass by cardiovascular magnetic resonance: multi-center variability and consensus contours. J Cardiovasc Magn Reson 2015;17(1):63.

[40] Platform - Arterys.

[41] Arterys FDA clearance for liver AI and lung AI lesion spotting software | Medgadget.

CHAPTER 7

Medical devices and artificial intelligence

Arash Aframian[1], Farhad Iranpour[2] and Justin Cobb[3]
[1]Trauma and Orthopaedics, Imperial College London, United Kingdom
[2]Trauma and Orthopedic surgeon, Imperial College London, London, United Kingdom
[3]Professor and the Chair of Orthopedics at Imperial College London, London, United Kingdom

7.1 Introduction

From diagnoses to guiding treatments, the rapid growth of medical technology into a multitrillion dollar industry has inevitably helped to accelerate the development of AI in medical technologies which has enabled a near endless multitude of possibilities. There is near infinite potential for artificial intelligence (AI) in medical devices, with current technologies having the potential for improvement using AI. This chapter presents a history of AI in medical devices, some examples of how it is currently being used and how this may change over time along with limitations that are currently faced. The increasing use of medical devices worldwide has led to the need for new regulations in both Europe and worldwide and these too are discussed. A clear division must be made between physical hardware devices and software or virtual devices, in part because they are different and also because the regulatory requirements vary for different medical devices. However, this is beginning to change, with the latest regulations reflecting the increasing importance of medical software, allocating equal importance in the regulation of software as medical devices where it will influence patient care decision-making, as would have traditionally been limited to hardware devices only. Expectations must be tempered, however, as it is an emerging field and the ultimate responsibility of patients' welfare must be protected.

7.2 The development of artificial intelligence in medical devices

Many would credit the early history of AI to Alan Turing as far back as the 1950s. Since then, much work has been done in developing systems which can "think," "speak," and "behave" in a human fashion. As with many

Artificial Intelligence in Healthcare
DOI: https://doi.org/10.1016/B978-0-12-818438-7.00007-1
163

developments, much of the early investment came from the defense sector with projects at leading technology universities in the 1960s and it was some decades before autonomous cars and then chess-playing computer programs made this commercially available in the 1980s and 1990s [1]. The rising costs of healthcare means that not only this market is gaining economic value with an increasing percentage of GDP (gross domestic product) but it is doing so at a faster rate than the national economic growth in the United States [2]. This increasing market value has led to a number of different applications of AI that now exists in the form of both hardware and software. In software solutions, the mathematical principles of machine learning are applied into algorithms and this becomes the backbone of the development. These algorithms can then be transferred to be processed in hardware where use "in the field" or "on the fly" is required and this does not require the highest levels of processing power—in those instances, data is usually transferred to a mainframe of HPCs (high powered computers) and the results transferred back. The field is developing and growing at an exponential rate with new applications of concepts being developed regularly and the existing ones being improved with ever increasing datasets.

During the "Transforming medicine through AI-enabled healthcare pathways" talk in The Alan Turing Institute, UK, the development of a large dataset of patients treated with cancers and their data over several time points was described. The granularity of this data has allowed AI to begin to suggest possible best management solutions on a patient-specific basis, rather than simply by disease state alone. Each decision is based not just on disease but patient-specific factors such as age, comorbidities, microscopic and radiological findings, medications, and development of the disease over time. As it was put, our understanding has changed from seeing "the patient that is the needle in the haystack, to us all being different so that we are now essentially a haystack of needles." Finding the best treatment for a patient may now be possible by finding other groups of patients who are matched not only by diagnosis but multiple covariant factors of age, gender, medical history, and so on. While we might be at an early stage of data accumulation, the recurrent neural networks being used mean that it will continue to get stronger every day, being able to provide guidance to oncologists and surgeons about how they might best manage patients.

7.2.1 Activity tracking devices

The various devices now used in healthcare are able to capture huge volumes of data for analysis and the emergence of the Internet of Things

(IoT) has enabled devices to be linked to one another, allowing control of all manner of things around the home and environment. Diseases that were traditionally difficult for clinicians to manage because they relied heavily on patient compliance, can now be managed using AI technology. While AI can be used to detect which kind of cardiac pacemaker a patient has when they have a chest x-ray radiograph in an emergency [3], its use in everyday life is perhaps where most patients look forward to seeing the benefits of the latest advances. In the United States alone, failure of patients adhering to medication doses prescribed has been attributed to 10% of hospital admissions and as many as 125,000 deaths per annum due to lack of compliance with medications [4]. The AI could help detect anomalies by linking users' implanted devices to a central database— Cardiologs (Paris, France) has obtained clearance in both Europe and the United States to use deep learning on a database of over one million heart tracings to enable it to learn to detect a number of arrhythmias. Traditionally, diabetic patients who did not comply with manually and regularly measuring and managing their own levels of sugar in their blood—whether because they did not like needles, did not understand the importance or make it a priority and so their health suffered. Now, this technology has progressed, allowing patients to control their blood sugar with implanted insulin pumps and gives them the ability to monitor their blood sugar on smartphone devices, even granting access to others who can help them to monitor this. The constant streams of data allow the use of AI to help control the blood sugars in keeping with meals and sleep—bolstered by the ability to tie into other devices which monitor activity.

Activity trackers in every device are now commonplace and they generate huge amounts of data which are then used to develop new tools—a single major brand managed to collect sleep data equivalent to six billion nights [5] in a year and this in turn can be used to create AI apps [6], pillows [7], mattresses [8,9], or headphones and even compose lullaby music based on machine learning [10]. These new technologies also help to perform studies which were previously restricted to inpatient study from specialist centers, for example, for sleep studies, which can now be done using smart technologies in the patients' own home (Fig. 7.1). It has also been demonstrated that adults whose breathing at night compromises their airway (obstructive sleep apnea, OSA) are shown to have higher degradation in areas of the brain, which are known to be linked to dementia [11]. A principal components analysis determined that of the factors

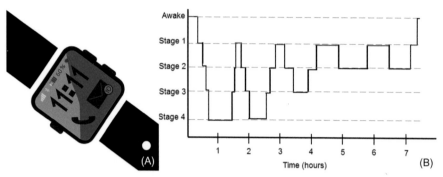

Figure 7.1 Personalized sleep with AI powered Bryte bed (A) and sleep monitoring with Fitbit smart watches (B, C).

considered, the two most significant were the level of deoxygenation and the degree of sleep disturbance. Given that we know sleep is essential to life and bodily repair and that there are AI technologies for the detection and treatment of OSA, it is possible that using AI to help with OSA may in fact be helping to delay the onset of dementia in this group of patients.

The use of AI in devices is not limited to external machines alone, with great advances being made in wearable devices—while much work is done in the software world, the data from machine learning algorithms can be used to phase out and replace the hardware with software solutions. Predictions made about wearables and their future two decades ago [12] have been fulfilled and the technology has continued to develop at a fast pace as hardware capabilities have improved and these wearables have become smaller and portable enough to become part of everyday life for many [13], with an aim to stem spiraling healthcare costs. Some proposals even claim to be adaptive and personalized to offer individual users the ability to create, in effect, their own AI personal trainer [14] or help to track recovery from medical illness such as a stroke [15].

The concepts of AI can be combined with existing technology such as electrodes to measure or stimulate brain activity (electromyography, EMG) or cameras to assess the type of hand grip required to clasp an object [16] and allows better control of robotics such as prosthetic limbs. Deep brain stimulation has grown in use as the benefits in recalcitrant patients are seen and the processes to offer the procedure become more widely available—it is used for movement disorders and is also being investigated for psychiatric disorders [17]. It relies on accurate electrode placement within the brain and effective programming but the number of variables involved in programming the stimulator, makes it complex and time-consuming, so artificial neural networks are being used in research to improve the process. The Wyss Center for Bio and Neuroengineering (Switzerland) is pioneering a new model in neurotechnology translation to accelerate the development of novel concepts to help people with nervous system disorders such as tinnitus. Neurofeedback is the process of using a real-time display of brain activity to learn self-regulation. If the clinical trial is successful it may lead to the development of implantable devices which can detect and transfer the electrical activity of the auditory cortex to a screen available to the patient. This can enable visualization of brain activity in real time and possibly self-regulation of tinnitus.

7.2.2 Implants, bionics and robotics devices

The availability of technological solutions to improve medical care and patient treatments have led to the development of a new field of research (Biomechatronics), which specializes in the mimicry of human motion. To achieve this, machines need to learn and then reproduce human motion. There are now a handful of laboratories in the world specializing in such research [18]. Such centers are working on developing tools and technologies for interfacing; bridging the human-technology divide, in some cases, surgery is even undertaken to redirect nerve function to new areas where the brain may better cocontrol the prosthetic limb [19,20]. AI is being used by bioengineers to create human machine interfaces which enable prosthetic limbs to feel like and extension of the body [21]. Ottobock (Nader Holding Gmbh & Co KG, Duderstadt, Germany) acquired the BeBionic hand from Steeper [22], which allows amputees to regain a number of hand functions and has specific programming for different kinds of finger grips, including the ability to use AI to sense slipping objects and increase pressure to maintain hold of the object. The company has also developed sensors which can learn from the activation of groups of muscles in sequence to allow amputees to carry out actions which would usually require users to manually coactivate different muscles [23] (Fig. 7.2).

The US Food & Drug administration have released draft guidance on implanted Brain-Computer Interfaces [24] as the technology moves from external AI in the prosthetics such as those previously offered by Össur (Reykjavik, Iceland) which help users to walk better, with a gait which more closely resembles a "normal" one, to their new mind-bionic

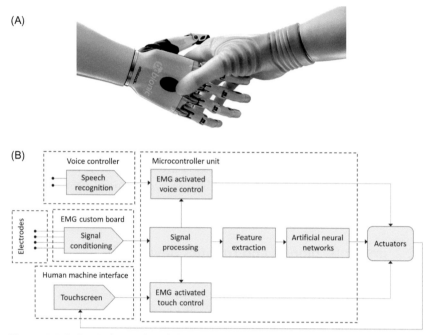

Figure 7.2 BeBionic bionic hand showing hand posture (A) and example of a block diagram for an electromyography activated bionic hand showing different operating modes and neural network classifier (B). *From: https://www.ottobock.com/de/newsroom/media/medieninformationen-hand-prothesen/*

offering which offers an interface between the human and the machine. The pairing of a hardware (camera) with a software algorithm has also allowed a bionic hand to adapt the kind of grip it uses to clasp different objects that the camera has been trained to recognize [16]. The bionic limbs are also being developed to provide feedback to users, allowing them to have a perception of sensation—in the hand, sensors offer kinesthetic feedback to amputees so that they can do more than just provide movement but sensation too, through which they can improve motor control [25].

The compliance of the stump tissues and limb sockets remains a major issue in the management of prosthetics and now AI is being utilized to maintain good contact and fit for prosthetic limbs [26]. Furthermore, AI can be utilized to maintain fit as body temperature changes and tissues swell or contract [27]. The growth of "biomechatronics" is leading to advances not only in how limb prostheses fit, but also in how they function. The traditional prostheses simply served as artificial body parts and historic examples have been found as far back as ancient Egypt. Today, contemporary designs use muscle activation in the residual limb stump with the latest models employing signals directly from the brain. Using load cell sensors in the fingertips, the robotic division at John Hopkins is developing technology that allows the users to "feel" when they interact with objects through their prostheses. Combining this with the EMG technology would allow users to closely replicate the functions they lack with traditional technology.

Prostheses were traditionally constructed from solid materials in a single process. However, the increasing need to prototype and produce systems rapidly has led to the growth of smaller scale and limited production run manufacturing, as it does not require entire production facilities to be reprogrammed. The American Society for Testing and Materials International Committee has introduced the term additive manufacturing (AM) for techniques that produce physical objects from three-dimensional digital data via the "process of joining materials to make objects from 3D (three-dimensional) model data, usually layer upon layer, as opposed to subtractive manufacturing methodologies" [28]. AM also known as 3D printing can use materials such as powders, plastics, ceramics, metals, liquids, or living cells. AI has extensive applications in AM with machine learning being used for auto-segmentation and testing in the prefabrication stage. Medical applications of 3D printing in medicine include: surgical planning [29,30], prosthetic design and manufacturing [31–33], medical education and training [34,35], organ printing [36,37], drug delivery [38,39], and medical research [36].

Robotics are not just limited to patients but can now also be utilized by their clinicians. The increasing fidelity of appendages and improvements in sensor and camera technologies mean that it is now possible to practice operations ahead of performing them. This can be done both in simulator laboratories using more expensive six-degrees-of-freedom robotic arms and with less expensive simulators, even in the comfort of one's own home using either virtual or augmented reality. Our own teams and MSK Lab colleagues at Imperial College, London have published validated examples of these [40–44]. Both surgical and anesthetic skills have been trained into AI systems, although previous attempts to automate the complex practice of anesthesia [45] have been largely unsuccessful. The advent of developments in machine learning and AI "may however usher in a new area of automation" in anesthesia [46,47] (Fig. 7.3).

The use of AI technology is not just restricted to medical needs within the limits of a healthcare setting alone and has also been used to enhance

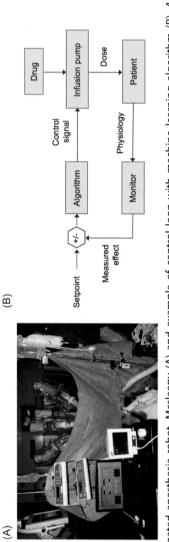

Figure 7.3 Automated anesthesia robot, Mcsleepy (A) and example of control loop with machine learning algorithm (B). *Adapted from https://www.clinmedjournals.org/articles/ijaa/international-journal-of-anesthetics-and-anesthesiology-ijaa-6-098.php?jid = ijaa and https:// www.popsci.com/technology/article/2010-10/worlds-first-all-robot-surgery-performed-montreal/.*

sporting performance. The combination of machine learning with activity data has enabled the creation of headphones that help select music based on the user's preferences [48] or even to help coach them through a run [49]. This remains a developing field despite having been described a quarter of a century ago [50] and uses vary from performance analysis [51] of many sports—from running to cricket bowling techniques [52]—all being amenable to AI assessment [53]. Nerve and muscle function and recovery after injury have also been studied using wearables and a supervised learning approach to track functional recovery after a stroke [54].

7.3 Limitations of artificial intelligence in medical devices

One of the major difficulties with the use of new technologies is the learning curve. A good example for this is the struggle faced when developing the algorithm which was eventually adopted into the Apple watch to detect electrocardiograms [55]. The original dataset was well structured yet too refined and hence limited the ability of the machine learning engine to function fully; the eventual solution was to provide a larger, unrestricted dataset and let the mathematics work. The difficulty lies in how to use AI technologies in areas where decisions are critical compared to just recognizing faces from pictures or translating languages [56]. Medical records can be used to predict the postoperative mortality risks for patients in a hospital setting based on intraoperative data "but are not (yet) superior to existing methods" [57]. The limits are not just pure data but also in technical tasks. There is already a published example where a robotic arm was controlled using a real-time video screen and a computer joystick for intubation with a 91% success rate [58]. In time, this technology could be trained with image recognition and perhaps the operator all but replaced. However, while there is a potential benefit from integration of such systems, with faster procedure times [59], the learning curve required must also be considered, though new users may find learning new technologies easier than learning to use traditional methods [60].

Concerns over potential safety issues are a major limiting factor in the success and adoption of novel technologies. AI can be used to highlight and identify patients whose observations or laboratory results raise medical concern [61] as well as potential likely diagnoses and risks of in-hospital mortality and unplanned readmissions [62]. To safely integrate such technologies takes time and while some of these technologies date back to the 1980s [63], much work remains until they are safe, reliable, and socially accepted.

The ever-growing list of AI functions and applications also means that there is a developing need for regulatory oversight, with both ethical and moral considerations. These raise such questions as to who should be responsible for making the decisions about how the AI should behave? If and when machine learning technologies become commonplace in healthcare is unclear, with one estimation being that this will take a further two decades [64]. Listening capable AI tools have also been used to track sleep stages without the need for wearable sensors, using radiofrequency waves [65], and listening to sleep patterns and sounds. The growth of these "hearable" devices will mean that the AI technology can capture conversations and help to plan user's lives. However, this raises moral and ethical questions about what the devices listen to and the decisions they make for users [66].

7.4 The future frontiers of artificial intelligence in medical devices

The future of machine learning is secure and AI in medical devices has become integral in daily practice. Robots have already been used by surgeons as tools in surgical procedures but have even been taught how to perform basic autonomous tasks such as suturing [67]. Indeed, it has been shown that the robotic suturing is comparable to or even superior to humans [68] and it is even possible that in time, the assessment of surgical skills will move from being objective to being automated with machine learning robots [69]. In both developed and developing countries, surgeons are operating remotely using robots [70,71] and the concept of using augmented reality in surgery is becoming increasingly adopted [72]. The datasets captured may enable the application of image recognition technology in future and automated labeling for teaching purposes too, though some are skeptical if and when the true promise of AI in medicine will bear fruit [73].

The growth of "day case" and "ambulatory" surgery centers initially in the United States and Canada but now also across the world, demonstrates the realization of a vision to enable enhanced recovery. Patients are able to return home more quickly, reducing risks of an inpatient stay and helping them to return safely to their normal environment at a reduced cost [74–76]. However, determining who are the most suited for such procedures can help manage expectations for patients, therapists, and surgeons as well as helping hospitals to plan resource utilization [76].

Adherence with physiotherapy exercises are known to be poor and using AI technology to monitor and promote these can increase compliance and improve outcomes. Efforts to resolve these questions are underway [77] where the role of AI in determining injury and pathology, and eventually tying this into pathoanatomy relies on a good understanding of the data required. It is a unique scenario where the solutions offered now also serve as a gateway to the future.

By understanding "big data" and AI, it is possible to ascertain which data needs to be collected in a way which will be useful in years to come, with rich datasets structured in ways that allow the machine learning concepts to be applied and AI engines built and refined (Fig. 7.4). A lack of understanding of the concepts discussed in this chapter and in this book could cost a company many years in research and development which could have been saved with the right person being involved early on, with seemingly trivial details being the cornerstone of future AI technology development. Companies who seek to develop medical AI, need to involve a team of experts who understand and have an appreciation for scientific, ethical, legal and clinical applications and combine these in ways to deliver the best results for patients and healthcare workers:

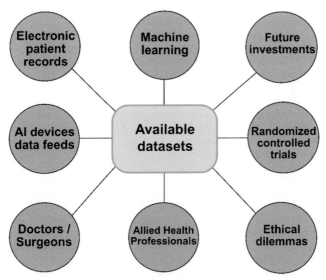

Figure 7.4 Application of AI to the relevant datasets can bring immense benefits for multiple disciplines within the healthcare ecosystem.

References

[1] Londhe VY, Bhasin B. Artificial intelligence and its potential in oncology. Drug Discov Today 2019;24:228−32.

[2] Bhardwaj R, Nambiar AR, Dutta D. A study of machine learning in healthcare. Proc Int Comput Softw Appl Conf 2017;2:236−41.

[3] Howard JP, et al. Cardiac rhythm device identification using neural networks. JACC Clin Electrophysiol 2019;5:576−86.

[4] Viswanathan M, et al. Interventions to improve adherence to self-administered medications for chronic diseases in the United States. Ann Intern Med 2012;157:785.

[5] Yahoo Finance. What Fitbit discovered from 6 billion nights of sleep data, <https://uk.finance.yahoo.com/news/exclusive-fitbits-6-billion-nights-sleep-data-reveals-us-110058417.html> [accessed 27.10.19].

[6] Sleeptrackers. SleepScore raises the bar with clever new app, <https://sleeptrackers.io/sleepscore-releases-free-app/> [accessed 27.10.19].

[7] Liu L, Pottim KR, Kuo SM. Ear field adaptive noise control for snoring: an real-time experimental approach. IEEE/CAA J Autom Sin 2019;6:158−66.

[8] HEKA. HEKA mattresses, <http://47.89.189.225/blog.html> [accessed 28.10.19].

[9] Bryte. Intelligent Sleep Platform—BRYTE™, <https://bryte.com/pages/intelligent-sleep-platform> [accessed 29.10.19].

[10] Telegraph. Jukedeck has taught a computer program to write songs in seconds, <https://www.telegraph.co.uk/technology/2016/05/21/jukedeck-has-taught-a-computer-program-to-write-songs-in-seconds/> [accessed 27.10.19].

[11] Cross NE, et al. Structural brain correlates of obstructive sleep apnoea in older adults at risk for dementia. Eur Respir J 2018;52. Available from: https://doi.org/10.1183/13993003.00740-2018 pii:1800740.

[12] Pentland AP. Wearable intelligence. Sci Am 1998;1:90−5.

[13] Downey T. Enhancing the quality of life through wearable technology. IEEE Eng Med Biol Mag 2003;41−8. Available from: https://doi.org/10.1007/978-1-84628-863-0_4.

[14] Buttussi F, Chittaro L. MOPET: a context-aware and user-adaptive wearable system for fitness training. Artif Intell Med 2008;42:153−63.

[15] Kundu R, Mahmoodi T. Mining acute stroke patients' data using supervised machine learning. Mathematical Aspects of Computer and Information Sciences (MACIS 2017). Lecture notes in computer science (including subseries lecture notes in artificial intelligence and lecture notes in bioinformatics), vol. 10693. Springer; 2017. p. 364−77.

[16] Ghazaei G, Alameer A, Degenaar P, Morgan G, Nazarpour K. Deep learning-based artificial vision for grasp classification in myoelectric hands. J Neural Eng 2017;14.

[17] Hell F, Palleis C, Mehrkens JH, Koeglsperger T, Bötzel K. Deep brain stimulation programming 2.0: future perspectives for target identification and adaptive closed loop stimulation. Front Neurol 2019;10:1−11.

[18] Biomechatronics—Wikipedia, <https://en.wikipedia.org/wiki/Biomechatronics> [accessed 29.12.19].

[19] Cheesborough JE, Smith LH, Kuiken TA, Dumanian GA. Targeted muscle reinnervation and advanced prosthetic arms. Semin Plast Surg 2015;29:62−72.

[20] Kuiken TA, Barlow AK, Hargrove LJ, Dumanian GA. Targeted muscle reinnervation for the upper and lower extremity. Tech Orthop 2017;32:109−16.

[21] Alshamsi H, Jaffar S, Li M. Development of a local prosthetic limb using artificial intelligence. Int J Innov Res Comput Commun Eng 2016;4:15708−16.

[22] Georg H, et al. Ottobock acquires BeBionic from Steeper; 2017.

[23] Ottobock SE & Co. KGaA. Myo Plus pattern recognition, <https://www.ottobock.com/en/press/media-information/media-information-myo-plus/> [accessed 27.10.19].

Medical devices and artificial intelligence 175

[24] U.S. Food and Drug Administration (FDA). Implanted brain-computer interface (BCI) devices for patients with paralysis or amputation—non-clinical testing and clinical considerations: draft guidance for industry and Food and Drug Administration staff; 2019.
[25] Marasco PD, et al. Illusory movement perception improves motor control for prosthetic hands. Sci Transl Med 2018;10:1–13.
[26] Paternò L, Ibrahimi M, Gruppioni E, Menciassi A, Ricotti L. Sockets for limb prostheses: a review of existing technologies and open challenges. IEEE Trans Biomed Eng 2018;65:1996–2010.
[27] Gupta S, Loh KJ, Pedtke A. Sensing and actuation technologies for smart socket prostheses. Biomed Eng Lett 2019;10:103–18. Available from: https://doi.org/10.1007/s13534-019-00137-5.
[28] ASTM International. Standard terminology for additive manufacturing technologies. Rapid Manuf Assoc 2013;1–3. Available from: https://doi.org/10.1520/F2792-12A.2.
[29] Vodiskar J, Kütting M, Steinseifer U, Vazquez-Jimenez JF, Sonntag SJ. Using 3D physical modeling to plan surgical corrections of complex congenital heart defects. Thorac Cardiovasc Surg 2017;65:031–5.
[30] Hermsen JL, et al. Scan, plan, print, practice, perform: development and use of a patient-specific 3-dimensional printed model in adult cardiac surgery. J Thorac Cardiovasc Surg 2017;153:132–40.
[31] Ho CMB, Ng SH, Yoon YJ. A review on 3D printed bioimplants. Int J Precis Eng Manuf 2015;16:1035–46.
[32] Elahinia MH, Hashemi M, Tabesh M, Bhaduri SB. Manufacturing and processing of NiTi implants: a review. Prog Mater Sci 2012;57:911–46.
[33] Suaste-Gómez E, Rodríguez-Roldán G, Reyes-Cruz H, Terán-Jiménez O. Developing an ear prosthesis fabricated in polyvinylidene fluoride by a 3D printer with sensory intrinsic properties of pressure and temperature. Sensors (Switzerland) 2016;16:1–11.
[34] Ploch CC, Mansi CSSA, Jayamohan J, Kuhl E. Using 3D printing to create personalized brain models for neurosurgical training and preoperative planning. World Neurosurg 2016;90:668–74.
[35] Andolfi C, Plana A, Kania P, Banerjee PP, Small S. Usefulness of three-dimensional modeling in surgical planning, resident training, and patient education. J Laparoendosc Adv Surg Tech 2017;27:512–15.
[36] Lode A, et al. Green bioprinting: Fabrication of photosynthetic algae-laden hydrogel scaffolds for biotechnological and medical applications. Eng Life Sci 2015;15:177–83.
[37] Ollé-Vila A, Duran-Nebreda S, Conde-Pueyo N, Montañez R, Solé R. A morphospace for synthetic organs and organoids: the possible and the actual. Integr Biol 2016;8:485–503.
[38] Goyanes A, Det-Amornrat U, Wang J, Basit AW, Gaisford S. 3D scanning and 3D printing as innovative technologies for fabricating personalized topical drug delivery systems. J Control Release 2016;234:41–8.
[39] Khaled SA, Burley JC, Alexander MR, Roberts CJ. Desktop 3D printing of controlled release pharmaceutical bilayer tablets. Int J Pharm 2014;461:105–11.
[40] Sugand K, Mawkin M, Gupte C. Validating Touch Surgery™: a cognitive task simulation and rehearsal app for intramedullary femoral nailing. Injury 2015;46:2212–16.
[41] Akhtar K, et al. Training safer orthopedic surgeons. Acta Orthop 2015;86:616–21.
[42] Sugand K, Wescott RA, Carrington R, Hart A, Van Duren BH. Teaching basic trauma: validating FluoroSim, a digital fluoroscopic simulator for guide-wire insertion in hip surgery. Acta Orthop 2018;89:380–5.
[43] Logishetty K, et al. Can an augmented reality headset improve accuracy of acetabular cup orientation in simulated THA? A randomized trial. Clin Orthop Relat Res 2019;477:1190–9.

[44] Liu H, Auvinet E, Giles J, Rodriguez y Baena F. Augmented reality based navigation for computer assisted hip resurfacing: a proof of concept study. Ann Biomed Eng 2018;46:1595−605.

[45] Lee H-C, Ryu H-G, Chung E-J, Jung C-W. Prediction of bispectral index during target-controlled infusion of propofol and remifentanil. Anesthesiology 2018;128:492−501.

[46] Alexander JC, Joshi GP. Anesthesiology, automation, and artificial intelligence. Bayl Univ Med Cent Proc 2018;31:117−19.

[47] Connor CW. Artificial intelligence and machine learning in anesthesiology. Anesthesiology 2019;131:1346−59.

[48] Kickstarter. Vinci—first smart headphones with artificial intelligence by Inspero Inc., <https://www.kickstarter.com/projects/inspero/vinci-first-smart-3d-headphones-that-understand-yo> [accessed 27.10.19].

[49] Forbes. Blade headphones from Soul Electronics use artificial intelligence to coach you while you run, <https://www.forbes.com/sites/suziedundas/2019/05/28/soul-electronics-blade-headphones-uses-artificial-intelligence-to-coach-you-while-you-run/> [accessed 26.10.19].

[50] Lapham AC, Bartlett RM. The use of artificial intelligence in the analysis of sports performance: a review of applications in human gait analysis and future directions for sports biomechanics. J Sports Sci 1995;13:229−37.

[51] Bartlett R. Performance analysis: can bringing together biomechanics and notational analysis benefit coaches? Int J Perform Anal Sport 2001;1:122−6.

[52] Mukherjee S. A.I. versus M.D. The New Yorker; 2017.

[53] Bartlett R. Artificial Intelligence in sports biomechanics: new dawn or false hope? J Sport Sci Med 2006;5:474−9.

[54] Parnandi A, Wade E, Matarić M. Motor function assessment using wearable inertial sensors. In: 2010 Annual International Conference of the IEEE Engineering In Medicine and Biology. EMBC'10. IEEE; 2010. p. 86−9. Available from: https://doi.org/10.1109/IEMBS.2010.5626156.

[55] Topol E. Deep medicine: how artificial intelligence can make healthcare human again; 2019.

[56] Mathis MR, Kheterpal S, Najarian K. Artificial intelligence for anesthesia: what the practicing clinician needs to know more than black magic for the art of the dark. Anesthesiology 2018;129:619−22.

[57] Lee CK, Hofer I, Gabel E, Baldi P, Cannesson M. Development and validation of a deep neural network model for prediction of postoperative in-hospital mortality. Anesthesiology 2018;129:649−62.

[58] Hemmerling TM, et al. First robotic tracheal intubations in humans using the Kepler intubation system. Br J Anaesth 2012;108:1011−16.

[59] Göpel T, Härtl F, Schneider A, Buss M, Feussner H. Automation of a suturing device for minimally invasive surgery. Surg Endosc 2011;25:2100−4.

[60] Leeds SG, et al. Learning curve associated with an automated laparoscopic suturing device compared with laparoscopic suturing. Surg Innov 2017;24:109−14.

[61] Maheshwari K, Ruetzler K, Saugel B. Perioperative intelligence: applications of artificial intelligence in perioperative medicine. J Clin Monit Comput 2019;1−4. Available from: https://doi.org/10.1007/s10877-019-00379-9.

[62] Rajkomar A, et al. Scalable and accurate deep learning with electronic health records. NPJ Digit Med 2018;1:1−10.

[63] Lanfranco AR, Castellanos AE, Desai JP, Meyers WC. Robotic surgery: a current perspective. Ann Surg 2004;239:14−21.

[64] Sayburn A. Will the machines take over surgery? Bull R Coll Surg Engl 2017;99:88−90.

[65] Zhao M, Yue S, Katabi D, Jaakkola TS, Bianchi MT. Learning sleep stages from radio signals: a conditional adversarial architecture. In: 34th International Conference on Machine Learning (ICML 2017), vol. 8. 2017. p. 6205−14.

[66] Medium. How hearables could change everything, <https://medium.com/s/story/hearing-is-believing-the-future-of-hearables-in-media-705ce0891750> [accessed 27.10.19].

[67] Schulman J, Gupta A, Venkatesan S, Tayson-Frederick M, Abbeel P. A case study of trajectory transfer through non-rigid registration for a simplified suturing scenario. In: IEEE International Conference on Intelligent Robots and Systems. IEEE; 2013. p. 4111−7. Available from: <https://doi.org/10.1109/IROS.2013.6696945>.

[68] Shademan A, et al. Supervised autonomous robotic soft tissue surgery. Sci Transl Med 2016;8.

[69] Fard MJ, et al. Machine learning approach for skill evaluation in robotic-assisted surgery. Lect Notes Eng Comput Sci 2016;2225:433−7.

[70] Eveleth, R. The surgeon who operates from 400km away. BBC Future, <https://www.bbc.com/future/article/20140516-i-operate-on-people-400km-away>; 2014 [accessed 15.09.19].

[71] Patel TM, Shah SC, Pancholy SB. Long distance tele-robotic-assisted percutaneous coronary intervention: a report of first-in-human experience. EClinicalMedicine 2019;14:53−8.

[72] Vávra P, et al. Recent development of augmented reality in surgery: a review. J Healthc Eng 2017;2017.

[73] Insel TR. AI and the new medicine. Nature 2019;567:172−3.

[74] Larsen JR, et al. Feasibility of day-case total hip arthroplasty: a single-centre observational study. Hip Int 2017;27:60−5.

[75] Richards M, et al. An evaluation of the safety and effectiveness of total hip arthroplasty as an outpatient procedure: a matched-cohort analysis. J Arthroplasty 2018;33:3206−10.

[76] Masaracchio M, Hanney WJ, Liu X, Kolber M, Kirker K. Timing of rehabilitation on length of stay and cost in patients with hip or knee joint arthroplasty: a systematic review with meta-analysis. PLoS One 2017;12:e0178295. Available from: https://doi.org/10.1371/journal.pone.0178295.

[77] On the Mend. <https://www.onthemend.com/> [accessed 01.01.20].

CHAPTER 8

Artificial intelligence assisted surgery

Elan Witkowski and Thomas Ward
Department of Surgery, Massachusetts General Hospital/Harvard Medical School, Boston, MA, United States

8.1 Introduction

The practice of surgery involves physical manipulation of tissue to treat disease. Over many centuries, surgical outcomes have improved with increasing human knowledge and the development of novel tools. Artificial intelligence (AI) has been defined as the study of algorithms that give machines the ability to reason and perform functions such as problem solving, object and word recognition, inference of world states, and decision-making [1]. Surgery is therefore a natural and yet complex application for AI technologies. Operations require surgeons to synthesize data from multiple sources to make decisions, identify anatomy, and carry out physical tasks in rapidly changing scenarios. Outside of the operating room, components of surgical care include diagnosis, preoperative assessment, postoperative care, assessment of outcomes, and training of surgeons (Fig. 8.1). AI promises to improve the quality and efficiency of perioperative care, improve surgical decision-making, augment the physical capabilities of human surgeons, and offer many exciting opportunities for future investigation—though not without potential pitfalls and challenges. In this chapter, past, present, and future applications of AI in surgery are reviewed.

8.2 Preoperative

Sun Tzu's declaration that "Every battle is won or lost before it is fought" rings especially true with surgery. Surgeons lay the operation's true foundation before the case itself with adequate preoperative patient selection, assessment, and optimization. Surgery is a controlled insult on the human physiology, and therefore the patient must have sufficient reserve to

179

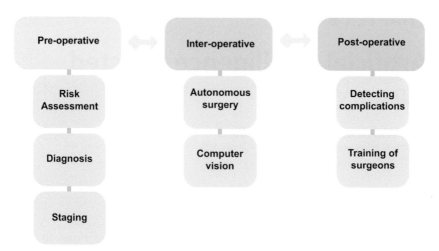

Figure 8.1 Applications of artificial intelligence in surgery and perioperative care.

tolerate the insult. The surgeon must also have adequate preoperative planning to correctly diagnose and assess the extent of the disease they will address intraoperatively. Lastly, the progress of medicine, in particular, neoadjuvant chemoradiotherapy prior to surgical resection of cancer, may obviate the need for curative resection and surgeons will need to identify these patients and prevent unnecessary operations.

8.2.1 Preoperative risk assessment

Before proceeding with an operation, the surgeon must determine if the patient can tolerate the procedure. Of paramount importance is adequate cardiac function. The American College of Cardiology and American Heart Association publish guidelines to guide physicians with risk-stratification and preoperative optimization [2]. These guidelines stratify patients into low-, medium-, and high-risk categories which correlate to probability of a perioperative major adverse cardiac event (MACE) of death or myocardial infarction. To evaluate a patient preoperatively, first physicians use risk-assessment calculators to estimate perioperative risk of MACE. These calculators include the Revised Cardiac Risk Index (RCRI) developed in the 1990s and the more recent Gupta Myocardial Infarction or Cardiac Arrest (MICA) risk model [3,4] After establishing that a patient is at elevated risk ($>1\%$), they then use subjective assessment of a patient's metabolic equivalents (METs) functional capacity. As long as the patient has adequate functional capacity, they are deemed to not require further expensive preoperative cardiac testing aimed at identifying intervenable coronary disease that could be optimized prior to surgery.

Recent studies call into question this algorithm [5,6]. The POISE trial showed 8000 patients who underwent noncardiac surgeries had a MACE rate of 6.9% [5]. The majority who had a MACE had either one or two points on the RCRI scale which should have correlated to 1.0%−2.4% MACE rate. Clearly, the scale underperformed. More recently, a group looked at patients' subjective assessment of their METs functional capacity. Patients' subjective assessments were only 19% sensitive to predict what METs they could obtain during formal cardiopulmonary exercise testing. This underestimation led to a significant proportion of high-risk patients placed incorrectly into a low-risk category. Obviously, preoperative cardiac evaluation methods need further refinement.

The American Society of Anesthesiologists (ASA) has also proposed a classification system to assess perioperative risk [7]. It look beyond cardiac status alone and rate patients on a scale from I to V based upon their overall physical status. Examining over 6000 patients, higher classes correlated to increased risk odds ratio (OR) of postoperative complications: ASA class III and IV with OR 2.2 and 4.2, respectively. Perhaps, the ASA scale's primary limitation is the inherit subjectivity (e.g., mild vs severe systemic disease vs unlikely to survive 24 hours). One survey of 304 board-certified anesthesiologists gave them ten patients to classify according to the ASA scale. Only six of the patients received consistent ratings [8]. Again, current methods of risk assessment are found lacking.

Clearly, current methods of preoperative risk assessment need improvement. Their shortcomings stem from one of two issues. The first issue is subjectivity. This subjectivity limits consistent application of models such as the ASA classification and METs functional capacity and can lead to striking underestimation of risk. The second issue is the assumption of models that risk factors are additive and consistently related. Often times, these models assess risk factors in a binary fashion (present vs absent). In addition to the cardiac risk calculators previously named, one such calculator is the American College of Surgeons National Surgical Quality Improvement Program (ACS-NSQIP) calculator. This calculator uses a generalized linear mixed model to predict risk for multiple complications [9].

Medicine is obviously more nuanced, which reflects how these models consistently underperform. Application of machine learning (ML) methods could remove subjectivity and may allow for a better incorporation of variables in a nonlinear way that better reflects clinical reality. As ML models have gained popularity, many analyses and risk calculators that were previously created with traditional regression modeling techniques have been reconceived (using the same or similar data sets) with neural networks and other algorithmic approaches. A group from Massachusetts General Hospital recently created a ML-based score for emergency surgery patients [10]. Their algorithm (POTTER) used optimal classification trees to create a comprehensive decision tree to predict both mortality and morbidity [11]. Their score outperformed the ASA classification ACS-NSQIP calculator for both morbidity and mortality prediction, with a c-statistic of 0.9162 for over 382,960 patients. The strength of the POTTER approach was in its ability to more appropriately draw on data from representative patients to achieve risk prediction. Being one of the first ML-based calculators of its kind, the future should bring further scores and therefore refinement in our ability to stratify patients at risk for undergoing surgery and ultimately improve decision-making and counseling, assess appropriateness, and optimally prepare them for surgery—examples.

8.2.2 Preoperative diagnosis

Early diagnosis and detection of malignant lesions are essential. Almost universally, earlier detection leads to improved outcomes. One example is melanoma. Excised melanomas with a thickness under 1 millimeter (mm) have a 95% 5-year survival rate. More locally advanced melanomas with a thickness over 4 mm only have a 45% 5-year survival rate [12]. Therefore there has been a push for earlier and improved detection of melanomas. Local tools such as dermoscopy initially served as an extension of the examiner's eye when looking for malignant skin lesions. Dermoscopy uses a combination of a magnifying device with cross-polarized light to make the initial few layers of the skin translucent which affords analysis of deeper skin structures and improved diagnosis. The use of dermoscopy leads to an OR of 9.0–15.6 for detection of melanoma [13].

One trial sought to go beyond extending physicians' ability to visualize a lesion by extending a physician's judgment with ML. Research has previously shown that physicians are willing to change their clinical decisions a quarter of the time when presented with conflicting output from a clinical decision-support system [14]. The study enrolled expert clinicians (dermatologists) and nonexpert clinicians to examine 511 patients who had a

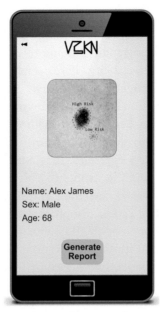

Figure 8.2 Example of a smart phone with a neural-network decision-support tool to analyze skin lesion images.

combined 3827 pigmented skin lesions [15]. The nonexpert physicians were allowed to use a neural-network decision-support tool to analyze skin lesion images they acquired with dermoscopy to increase their accuracy to the level of an expert clinician. The nonexperts, with the decision-support system's assistance, were actually able to exceed the specificity of an expert dermatologist (82% vs 72%).

Subsequently, a deep convolutional neural network was trained on 129,450 clinical images to classify images malignant versus benign categories [16]. Not only did it analyze dermoscopy images but also regular photos. When tested against 21 dermatologists, the system achieved superior sensitivity and specificity. The authors further added that the images could be acquired with only smartphones. Through a smartphone alone, the public could then have the ability to diagnosis malignant skin lesions (Fig. 8.2). This would provide almost universal access to specialist-level diagnostic care at a low cost through the power of AI and ML.

8.2.3 Preoperative staging

Once diagnosed, malignancies next need staging to determine resectability. The American Joint Committee on Cancer publishes a manual that is the gold standard and stages of tumors based on Tumor size/depth, Nodal status, and Metastases (TNM) [17]. Clinicians use history, physical exam, laboratory values, and radiographic imaging to determine a tumor's clinical stage and therefore resectability. If resectable, then the patient traditionally would go straight to surgery and then postoperative chemotherapy and radiation therapy (known as "adjuvant therapy"). More-and-more, there has been a move with certain tumors to perform chemoradiotherapy prior to surgical resection (neoadjuvant therapy) with improved survival and cure for patients. Examples include both esophageal and rectal cancer [18–20]. Interestingly, with the switch of chemoradiotherapy treatment prior to resection, surgeons found some resected specimens with no residual cancer. Esophageal and rectal cancer both had pathological complete responses (pCRs) with no residual tumor around 30% of the time [19,21]. With this observation, groups looked at the previously unimaginable: in patients with a complete clinical response opting to "watch and wait" rather than removing resectable disease [21,22]. Avoiding surgery limits the high morbidity that can accompany both rectal and esophageal resections. Initial results are promising, with overall 5-year survival and disease-free survival rates in rectal cancer of 100% and 92%, respectively, in one study [21] and no difference in 3-year overall survival in another [22]. In the latter study, 26% of patients were spared the morbidity of a permanent colostomy.

Of course, a complete clinical response does not necessarily correlate to a pCR which can only be known if the tumor is resected. As expected, local rectal cancer recurrences occurred in 34% of patients from assumed residual disease after chemoradiotherapy [22]. Interestingly, even in patients without complete clinical response, 7% had pCR in one cohort [21]. Groups have therefore looked at additional factors to predict pCR and limit the number of recurrences in patients initially manage nonoperatively. One group looked at pretreatment levels of carcinoembryonic antigen (CEA), a colorectal cancer serum tumor marker, and found that elevated pretreatment CEA was associated with a lower rate of pCR [23]. Another group performed multivariate analysis across over 300 rectal cancer patients and found that sex, age, body mass index (BMI), radiation dose, tumor differentiation, clinical stage, and CEA were not associated with pCR but that delaying surgery more than 8 weeks after chemoradiotherapy was associated with pCR [24]. With these confounding results, other researchers looked at other possible predictors including fluorodeoxyglucose-positron emission tomography (FDG PET) scans, which detect metabolically active tumors with radiolabeled glucose molecules. They found that the lower standardized uptake value posttreatment and pretherapy correlated with pCR [25]. Macomber et al. then applied ML techniques to various FDG PET features (mean uptake, max uptake, total lesion glycolysis, and volume) of esophageal cancers to predict pCR [26]. Their methods included k nearest neighbors, decision trees, support vector machines, and Naïve Bayes decision analysis (Fig. 8.3). The k nearest neighbors had the highest accuracy at 0.74. Future methods will work to incorporate not only imaging, but a variety of other clinical data with ML techniques to improve our ability to predict pCR. This improved prediction, surgery may transition from the first-line treatment to a rescue option; and with this transition, we may spare more-and-more patients the serious morbidity associated with these large operations.

8.3 Intraoperative

Some of the most exciting and promising applications of AI technology occurs in the operating room. An estimated 234 million surgeries happen across the globe annually and there exists a significant room for improvement: up to 20% of surgical patients experience complications [27,28]. AI technology (including computer vision and computer-assisted or eventually autonomous surgery) offers promising solutions to assist surgeons in the operating room with the hope of further decreasing complications.

8.3.1 Autonomous surgery

Bowel resection and subsequent reconnection (anastomosis) is one of the fundamental tasks for a general surgeon. Gastrointestinal anastomoses, across the board, leak 4.8% of the time. With each leak came a 10-times increased risk of return to the operating room, almost tripled length of hospital stay, tripled 30-day mortality, and doubled long-term mortality [29]. Reviewing 21,902 patients with a colon cancer resection and colorectal anastomoses, a meta-analysis also shockingly found an OR of 2 for local recurrence, OR 1.38 for distant recurrence, and OR 1.75 for long-term mortality [30]. As for quality of life, a colorectal anastomotic leak leads to more frequent stools, worse control of stools, and worse overall mental and physical component scores [31]. Surgeons traditionally sutured bowel anastomoses. However, leaks plagued the surgical community and innovators sought to ameliorate the problem with technology.

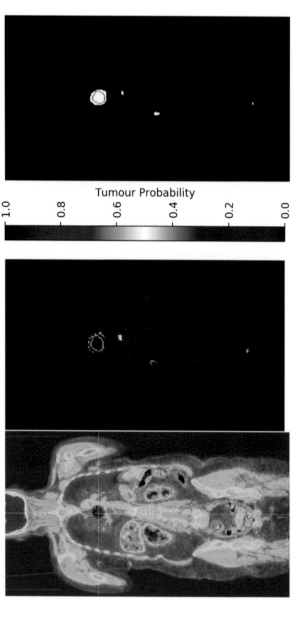

Figure 8.3 Application of deep learning for detection of esophageal lesions in positron emission tomographic-computed tomographic scans. *From I. Ackerley, R. Smith, J. Scuffham, M. Halling-Brown, E. Lewis, E. Spezi, V. Prakash, K. Wells, Using deep learning to detect oesophageal lesions in PET-CT. In Medical Imaging 2019: Biomedical Applications in Molecular, Structural, and Functional Imaging (Vol. 10953, p. 109530S), 2019, International Society for Optics and Photonics.*

Ravitch and Steichen popularized stapled (ST) anastomoses in the United States after earlier efforts by Russian surgeons [32]. The reliability and speed of uniformly applied staples versus a hand-sewn (HS) anastomosis theoretically promised faster and safer anastomoses. Unfortunately, this promise failed to bear fruit: most trials and meta-analyses show equivalence in outcomes between ST versus HS anastomoses [33,34]. Emergency general surgery is one area that may show an edge to HS anastomoses. One retrospective study showed almost double failure rate for ST compared to HS anastomoses [35]. The authors hypothesized that these relatively sicker patients had more likely bowel wall edema and size mismatch which a uniformly applied staple load could not properly address while a surgeon could customize for the discordance with each individual stitch. A later multicenter prospective trial enrolled patients and found that surgeons given this theoretical advantage preferred HS anastomosis for sicker patients (higher lactates, lower albumin, more vasopressor utilization, higher BMIs). Their leak rates were identical ($\sim$10%) despite the patients with HS anastomoses being significantly sicker [36]. Therefore HS anastomoses offer the potential promise for safer anastomoses in sicker patients, but still require an expert hand. With leak rates still approaching 5%—10%, the surgical community has room for improvement.

Autonomous surgery offers hope for safe surgery, particularly for gastrointestinal anastomoses. Researchers at Children's National Hospital developed a vision-guided robotics system that can perform autonomous suturing [37,38]. The Smart Tissue Anastomosis Robot (STAR), later named Smart Tissue Autonomous Robot, is a robotic arm with a suturing tool based on the Endo360° suturing device by EndoEvolution, LLC mounted to a light-weight robot. In its first iteration, it came with two modes: a manual mode and an automatic mode. The manual mode meant a surgeon selected each location for stitch placement. The automatic mode had a surgeon outline a linear incision and the STAR would take bites at a predefined interval. Comparing the STAR systems versus laparoscopic (both plain and robotic-assisted) surgery, the STAR system could finish a 5.5 cm incision closure in manual and automatic modes in 64.51 and 70.93 seconds, respectively. The laparoscopic surgeon took 9.34 minutes and the robotic-assisted surgeon took 5.71 minutes [37]. Not only was the STAR system more efficient, but it was more precise; stitches placed by surgeons had double the standard deviation for distance variation compared to the robot (Fig. 8.4). In addition, due to being able to

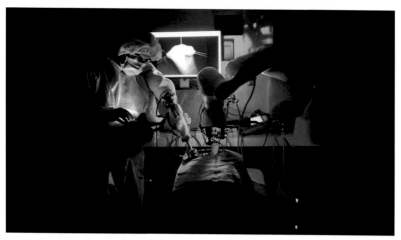

Figure 8.4 Image of smart tissue autonomous robot during preparation for intestinal anastomosis procedure. *From E. Svoboda, Your robot surgeon will see you now, Nature 573 (Sept), 2019, S110–S111.*

set defined tension for each suture bite, there was an order of magnitude less deflection of the stitches when they were pulled on tension compared to the much looser surgeon placed stitches (0.25 vs 2 mm).

Minimally invasive surgery, like autonomous surgery, aspires for safer and improved surgical practices. It was popularized as laparoscopic surgery matured in the 1980s, where a surgeon insufflates the abdomen then uses instruments through small incisions [39]. With the smaller incisions and decreased intraabdominal trauma, laparoscopic surgery compared to open comes with reduced postoperative pain, decreased wound infection rate, and quicker time to resumption of usual activities [40]. Laparoscopic surgery continues to present numerous technical challenges, however, with limited degrees of freedom for instrument articulation, diminished tactile feedback, and limited 2D vision, which makes technically challenging procedures more difficult except for the master laparoscopic surgeons. Recently, computer-assisted laparoscopic surgery offers augmented capabilities to address some of these issues. The da Vinci robotic system (Intuitive Surgical Inc) offers a "master-slave" configuration with the surgeon at a bedside console using inputs to control the robotic arms inside the patient [41]. Its display gives a three-dimensional image with instruments that have increased degrees of freedom to address the prior listed shortcomings of laparoscopic surgery. Despite these theoretical advantages, the use of robotic-assisted surgery has not created significant outcome differences. Short-term results are comparable in large reviews of colorectal, gynecologic, and urologic surgery [42–44]. The robotic-assisted surgery also negatively brought with increased cost ($2000 per case in the gynecologic operations) and length of procedure time. Despite the promise of laparoscopic and robotic-assisted surgery, they so far failed to deliver concrete improvements in patient outcomes.

Patients' surgical outcomes still rely on a human's technical ability which may inherently limit the impact of laparoscopy and robotic-assisted surgery. The creators of STAR built upon their prior work with a particular focus on bowel anastomoses [38]. Prior iterations could only suture on simple planar surfaces. To address the more realistic elastic surfaces one encounters throughout surgery (bowel, airways, blood vessels), the creators used both a 3D visual tracking system and a plenotopic 3D surface reconstruction system to plan suture placement. First, a surgeon placed a few near-infrared fluorescent imaging markers onto the corners of the bowel to give STAR the outlines of the planned anastomosis. Then, the plenotopic camera created a 3D point cloud in-between the markers to map the remaining bowel edges. The STAR then placed stitches based on an algorithm for optimal suture spacing and tension. Compared to open, laparoscopic, and robotic-assisted techniques, the STAR created equivalent if not superior anastomoses. With the consistent stitch placement and ideal tension application on the suture, the STAR anastomoses could withstand nearly double leak pressure compared to the other anastomoses. This technology offers great promise as it could enable any surgeon to instantly have the ability to perform a technically excellent anastomosis. The device adjusts tension with each stitch, therefore giving it unparalleled ability to adjust for varying tissue conditions. As discussed previously, HS anastomoses showed an edge over ST ones in patients with thick edematous bowels likely due to surgeons adjusting their tension and stitch placement for the tissue. Staplers do not adjust. The STAR therefore gives the benefit of an automated anastomosis that can adjust with each stitch and form the ideal connection to even the most novice of surgeons.

Though exciting, truly autonomous surgical systems still are several years from clinical application. In the interim, the current surgical landscape still has much room for improvement [38]. The STAR takes 50 minutes to perform a bowel anastomosis compared to just 8 minutes for an open surgeon. It also currently operates in a "supervisory autonomous mode," which allows a surgeon to make minor adjustments prior to each stitch placement and is necessary for 42% of stitches. Obviously in its infancy, the technology will continue to grow. Already the creators expanded the system to semiautonomously resect tumors with electrocautery [45]. Then once viable the technology will need to go through the clinical approval process. While we wait as others develop an autonomous future, AI and ML techniques for computer vision offer a more immediate impact in the intraoperative phase of surgical care as an expansion of a surgeon's collective knowledge and pattern recognition capability.

8.3.2 Computer vision

What ultimately separates a good from an excellent surgeon is not technical ability but intraoperative judgment. Skills such as suturing, knot tying, and stapling form surgical fundamentals necessary for performance of an operation. To dissect out vital structures and reach the stage to use these fundamentals, however, requires a thoughtful surgeon who uses pattern recognition to progress an operation. Surgical trainees and junior surgeons often lack that refined ability to see planes and quickly but safely operate. Birkmeyer et al. showed with an analysis of almost half-a-million patients that for more complicated procedures (esophagectomy, lung resection, aortic valve replacement, abdominal aortic aneurysm repair, pancreatic resection, coronary artery bypass grafting, and cystectomy) that a surgeon's case volume had an inverse relationship to patient mortality. The OR for operative death between a low-volume and high-volume surgeon varied from 1.24 for lung resections to 3.61 for pancreatic resections [46]. Similarly, 20 bariatric surgeons in Michigan submitted videos of their complex laparoscopic bariatric surgeries. Independent surgeons then rated each video. Surgeons in the top quartile had lower rates of reoperation, readmission, emergency department visits, surgical complications, and medical complications [47]. Performing an operation frequently likely gave those surgeons a better appreciation for the nuances of the case to more safely and expediently perform them and minimize postoperative complications.

Currently, the only way to improve is to do more surgery. For pancreatic resections, surgeons become dramatically more proficient after 60 cases, with reduced blood loss, operative time, and rate of removed specimen's positive cancer margins [48]. In laparoscopic colonic resections, Tekkis et al. showed that completion of a case without need to convert to open also correlated to experience, with about 60 cases (55 for right-sided resection, 62 cases for left-sided resections) needed for competency

[49]. While surgeons acquire experience, they subject patients to objectively worse outcomes. Fortunately, AI and ML offer a potential solution: computer vision.

Computer vision gives a machine understanding of images and videos. This chapter mentioned some examples of computer vision earlier, with diagnosis of dermatologic processes and STAR's soft-tissue mapping capabilities [16,38]. In the 1990s and 2000s, computer vision had few applications due to the limited capabilities of earlier AI approaches. Convolutional neural networks (ConvNets) led to a revolution. ConvNets were designed to process multiple arrays, such as data representation of images which have multiple arrays of images' pixel values. They, in fact, ultimately take on a networked structure quite similar to the visual cortex pathway in primates [50]. Krizhevsky et al. applied ConvNets with great success on the ImageNet challenge, an image recognition test across 1000 image categories, to obtain an error rate half that of all other competitors [51]. This advance, combined with increased processing ability of computers, led to an explosion of computer vision applications throughout the industry, including recent advances for intraoperative real-time guidance.

Early applications of computer vision to intraoperative recognition of surgical phases and structures were limited. As expected, trying to identify a constantly changing operative field brings challenges. Some groups identified that instruments were easily identifiable and tried to use the appearance of certain instruments (e.g., the clips used in gallbladder surgery to ligate the ducts and arteries) to predict the current stage of an operation [52,53]. A typical operation typically uses instruments at multiple stages and some surgeons use different instruments at different times, often in quite unique ways, which limited this approach. Other groups, similar to the STAR approach described earlier, used near-infrared markers (NIFs). At a case's start, surgeons would apply biodegradable inert markers at key locations in the operative field to allow the system to then identify structures and predict an operation's course—essentially augmented reality [54]. The surgeon's need for placement at an operation's start and the inherent limitations of a constantly changing operative field with instruments and organs intermittently concealing these markers restricted NIFs' ultimate ceiling.

More recently, a group developed a method that used a standard high-definition video of laparoscopic surgery to identify surgical phases in real time that did not rely on instrument detection nor surgeon-placed markers [55]. One issue with processing live video is that 1 minute of data contains 25 times the amount of data in a high-resolution computer tomography scan image [56]. To solve this, images were made into a representative histogram for analysis based off visual cues determined from discussions with surgeons. These cues included color, position, shape, and texture. Next, the histograms were processed into a compressed amount of data called a k-segment coreset stream which stripped out all information not useful to determine real-time operative phase identification. These compressive steps reduced the size of the data by over 90% with no accuracy loss [55]. With a useable amount of information, they could then apply support vector machine and hidden Markov models to classify and predict surgical phases. After training the system on only 10 hours of video, they could, with 92.8% accuracy, predict the stage of a laparoscopic sleeve gastrectomy in real time using just a standard laparoscopic surgery high-definition video feed.

Real-time operative computer vision analysis promises exciting possibilities. Upon review of surgical videos, we already can reliably detect negative intraoperative events. A Canadian group reviewed 54 recordings of bariatric laparoscopic surgery. In 70% of the recordings, they detected near-miss events (events that if not detected and addressed would lead to adverse outcomes and injury). Two-thirds of the events needed additional interventions such as further suturing or attempts at hemostasis. Fortunately, in these select videos, the surgeon detected all events intraoperatively [57]. Computer vision techniques could detect these events in real time and alert the surgeon, hopefully with even earlier detection than the surgeon whose focus may be on a separate area of the visual field. Computer vision could possibly even identify before a surgeon performs a near-miss event and alert them to prevent it from happening in the first place. In this sense, any surgeon, no matter how novice, would have the collective surgical experience from every surgeon whose videos trained the machine. Each operation will have a "surgical fingerprint" that once deviated from, the machine can alert the surgeon and help guide them back to a safer course. No longer would patients' outcomes suffer as the surgeon climbed the previously described learning curves. Instead, they would have the real-time decision support of multiple senior surgeons to guide them through their operations. As we continue to face a shortage of surgeons, particularly in rural communities, general surgeons will need to safely perform a myriad of operations, for which many they will not have had the experience or volume to have yet surmounted the learning curve [58]. Decision support from computer vision will level the playing field and shrink healthcare inequality to allow for safe delivery of care to patients, no matter their access to high-volume centers.

8.4 Postoperative

8.4.1 Detecting complications

"When the abdomen is open you control it, when closed it controls you."
Moshe Schein [59].

Much to the chagrin of surgeons, intraoperative excellence only partly determines an operation's ultimate success. Recent literature discusses that the majority of perioperative mortality differences originates from "failure to rescue," where institutions are unable to prevent mortality after an adverse complication [60]. Complication rates across the top and bottom 20 percentile institutions over 270,000 patients are nearly identical. Despite this, the mortality rate is 2.5 times higher for a bottom-tier institution. The difference is even more pronounced for pancreatic surgery with a 13-times higher mortality rate [61]. Some of the treatment gap occurs from delayed recognition of operative complications. There currently exist a number of scoring systems to try and identify patients at higher risk for inpatient complications. The National Early Warning Score (NEWS) from the United Kingdom incorporates clinical parameters to try and predict events in the next 24 hours including cardiac arrest, unanticipated ICU admission, and death. NEWS is only partly effective though, with area under receiver operating characteristic curve (AUROC) values for those respective complications of 0.72, 0.56, and 0.94 [62].

The linear nature of input variables limits current systems such as NEWS. NEWS, for example, only uses seven parameters, a patient's vital signs and level of consciousness, to make predictions about generalized morbidity and mortality. These limited inputs neglect the remaining wealth of information in the electronic medical record (EMR) including notes, laboratory values, and radiology tests. AI is not limited like these linear and logistic regression models, and the fine-grained nature of AI, particularly with nature language processing (NLP), allows for improved and more-refined predictions. NLP allows a computer to understand human language and therefore perform large-scaled analysis of EMR data [63]. Groups have already developed accurate deep learning of EMR data with AUROC values for mortality and various morbidities outperforming traditional logistic regression models [64,65]. These studies however only looked at patient data from an entire admission rather than snapshots early in the admission to predict later hospital-course complications. Rajkomar et al. advanced this work to develop deep learning models that incorporated as many as 175,639 data points per patient and could predict hospital-stay mortality and even discharge diagnoses after only 24 hours of admission [66].

NLP and deep learning offer promise for earlier detection of postoperative complications and hopefully a resultant decrease in "failure to rescue" (Fig. 8.5). To predict postoperative complications, often the answer lies in examination of the operation itself. Ideally, one could read (or use as an input for ML) the narrative operative report from the surgeon to have a clear sense of each step of the operation. The quality though of an operative report varies drastically from surgeon to surgeon. Almost a third of complications during a laparoscopic cholecystectomy are not dictated into the operative note [67]. Similarly, when surgeons examined systematic video recordings versus narrative operative notes, they found that the notes only adequately described 52.5% of the steps [68]. Accurate assessment of potential intraoperative difficulties leading to complications therefore lies upon analysis of intraoperative video and instrument manipulation.

Hung et al. developed objective metrics to rate surgeons from in robotic-assisted radical prostatectomy. They used surgical manipulation

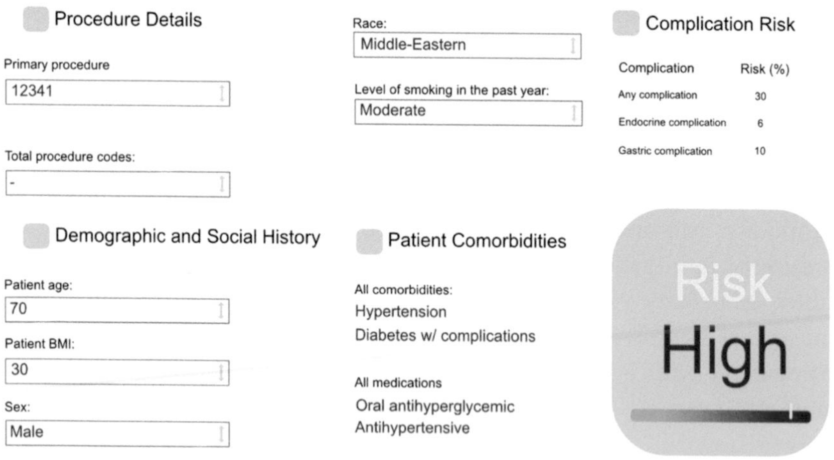

Figure 8.5 Example of surgery classification system for assessing postoperative outcomes including relevant scoring and process functions.

data from the da Vinci Surgical system over 100 cases for ten expert and ten novice surgeons. The data included task completion time, instrument path length, and dominant instrument moving velocity. Through analysis, they were able to objectively classify surgeons into expert and novice categories [69]. Using this objective scale, they then developed a ML algorithm that based on automated performance metrics predicted perioperative outcomes, in particular, prolonged hospital stay over 2 days with 87.2% accuracy [70]. With further development, they hope to broaden predictions to a wider variety of perioperative complications.

Computer vision also offers promise to predict postoperative complications. For example, laparoscopic cholecystectomy, one of the most common surgical procedures, is often an outpatient procedure with few complications. It does, however, have one dreaded complication: common bile duct injury. This injury occurs at a rate of 0.4%—0.6% cases. In one series, surgeons only recognized this complication 32.4% of the time, leading to an increased risk of uncontrolled sepsis and mortality [71]. Furthermore, in another series, 96% of bile duct injuries were found to have occurred as a result of errors in human visual perception [72]. Computer vision offers a chance in real time to prevent such an injury, but also could analyze videos postoperatively to detect the injury for earlier treatment and for future education to prevent the surgeon from making the same mistake again.

8.4.2 Training and certification of surgeons

Continuing education curriculum and recertification processes for surgeons are undergoing significant changes. Historically, to keep one's board-certification active, the American Board of Medical Specialties (ABMS) required board-certified physicians to take a four-part assessment every 10 years. Physicians called this practice into question. The test was "one-size-fits-all" and not tailored to one's clinical practice, so a clinician with a subspecialized practice, such as Breast surgeon, would need to know all of general surgery. The tests additional were quite costly, in terms of both time and cost. Lastly, they only forced a refreshment of knowledge every 10 years [73]. To address the latter concern, the ABMS has transitioned toward promoting lifelong learning with a continuous certification system. The American Board of Surgery (ABS), for example, transitioned the recertification process to a continuous certification which entails a 2-week online assessment process performed every-other-year

[74]. This change unfortunately fails to address that surgery not only requires clinical knowledge, but operative ability, which is not assessed at any point during the recertification process.

Greenberg et al. recognize the importance of operative ability in clinical practice. They note the need for deliberate practice, performance reflection, and intentional adjustments for a surgeon to continue to grow operative skills. Coaches could optimize this growth just as they do in sports. To address the incompleteness of surgical recertification, they advocate that the ABS incorporate active coaching into the certification process. This coaching would provide personalized coaching to a surgeon's practice and also assess and more importantly hopefully improve their surgical technique and outcomes [75]. Obvious limitations include that there are currently almost 23,000 general surgeons across the country [58]. Such a one-on-one coaching experience would be difficult to scale such a population. Computer vision could address this issue with surgical fingerprints. To maintain certification, a surgeon could submit surgical fingerprints obtained from their operative videos. As long as they fall within a level of acceptable technique, their operative ability would then pass recertification [76]. Extrapolating this further, when new techniques (e.g., laparoscopic surgery, robotic-assisted surgery, natural orifice transluminal endoscopic surgery) are invented, surgeons could perform first in a simulated setting then graduate to observed-practice one their surgical fingerprint is acceptable. The computer vision could then in real time guide the surgeons in the learning processes then once they reach acceptable fingerprints, they would become "certified" to perform that procedure independently.

8.5 Conclusion

AI technology offers innumerable promises to advance surgery. In the preoperative phase, it helps surgeons diagnose, adequately stage, and properly risk-stratify patients well-beyond our past capabilities. Intraoperatively, it offers the promise of flawless execution of technical skills through autonomous surgery and with computer vision the augmentation of a surgeon's decision and pattern-recognition abilities with the collective surgical consciousness. Postoperatively, it promises faster recognition of complications to decrease "failure-to-rescue" and integration into the recertification process to more adequately train and evaluate surgeons. Even though we are in the infancy of AI in surgery, it has already achieved startling advancements and promises numerous more in the future. We look forward to the coming years to see how far we can advance surgery and increase the safety of patients everywhere.

References

[1] Hashimoto DA, Rosman G, Rus D, Meireles OR. Artificial intelligence in surgery: promises and perils. Ann Surg 2018;268(1):70−6. Available from: https://doi.org/10.1097/SLA.0000000000002693.

[2] Fleisher LA, Fleischmann KE, Auerbach AD, et al. 2014 ACC/AHA guideline on perioperative cardiovascular evaluation and management of patients undergoing non-cardiac surgery: executive summary: a report of the American College of Cardiology/American Heart Association Task Force on Practice Guidelines. Circulation 2014;130(24):2215−45. Available from: https://doi.org/10.1161/CIR.0000000000000105.

[3] Lee TH, Marcantonio ER, Mangione CM, et al. Derivation and prospective validation of a simple index for prediction of cardiac risk of major noncardiac surgery. Circulation 1999;100(10):1043−9.

[4] Gupta PK, Gupta H, Sundaram A, et al. Development and validation of a risk calculator for prediction of cardiac risk after surgery. Circulation 2011;124(4):381−7. Available from: https://doi.org/10.1161/CIRCULATIONAHA.110.015701.

[5] POISE Study Group, Devereaux PJ, Yang H, et al. Effects of extended-release metoprolol succinate in patients undergoing non-cardiac surgery (POISE trial): a randomised controlled trial. Lancet Lond Engl 2008;371(9627):1839−47. Available from: https://doi.org/10.1016/S0140-6736(08)60601-7.

[6] Wijeysundera DN, Pearse RM, Shulman MA, et al. Assessment of functional capacity before major non-cardiac surgery: an international, prospective cohort study. Lancet Lond Engl 2018;391(10140):2631−40. Available from: https://doi.org/10.1016/S0140-6736(18)31131-0.

[7] Wolters U, Wolf T, Stützer H, Schröder T. ASA classification and perioperative variables as predictors of postoperative outcome. Br J Anaesth 1996;77(2):217−22. Available from: https://doi.org/10.1093/bja/77.2.217.

[8] Owens WD, Felts JA, Spitznagel EL. ASA physical status classifications: a study of consistency of ratings. Anesthesiology 1978;49(4):239−43.

[9] Bilimoria KY, Liu Y, Paruch JL, et al. Development and evaluation of the universal ACS NSQIP surgical risk calculator: a decision aid and informed consent tool for patients and surgeons. J Am Coll Surg 2013;217(5):833−842.e1−3. Available from: https://doi.org/10.1016/j.jamcollsurg.2013.07.385.

[10] Bertsimas D, Dunn J, Velmahos GC, Kaafarani HMA. Surgical risk is not linear: derivation and validation of a novel, user-friendly, and machine-learning-based Predictive OpTimal Trees in Emergency Surgery Risk (POTTER) Calculator. Ann Surg 2018;268(4):574−83. Available from: https://doi.org/10.1097/SLA.0000000000002956.

[11] Bertsimas D, Dunn J. Optimal classification trees. Mach Learn 2017;106(7):1039−82. Available from: https://doi.org/10.1007/s10994-017-5633-9.

[12] Balch CM, Buzaid AC, Soong SJ, et al. Final version of the American Joint Committee on Cancer staging system for cutaneous melanoma. J Clin Oncol 2001;19(16):3635−48. Available from: https://doi.org/10.1200/JCO.2001.19.16.3635.

[13] Vestergaard ME, Macaskill P, Holt PE, Menzies SW. Dermoscopy compared with naked eye examination for the diagnosis of primary melanoma: a meta-analysis of studies performed in a clinical setting. Br J Dermatol 2008;159(3):669−76. Available from: https://doi.org/10.1111/j.1365-2133.2008.08713.x.

[14] Dreiseitl S, Binder M. Do physicians value decision support? A look at the effect of decision support systems on physician opinion. Artif Intell Med 2005;33(1):25−30. Available from: https://doi.org/10.1016/j.artmed.2004.07.007.

[15] Dreiseitl S, Binder M, Hable K, Kittler H. Computer versus human diagnosis of melanoma: evaluation of the feasibility of an automated diagnostic system in a prospective clinical trial. Melanoma Res 2009;19(3):180−4. Available from: https://doi.org/10.1097/CMR.0b013e32832a1e41.

[16] Esteva A, Kuprel B, Novoa RA, et al. Dermatologist-level classification of skin cancer with deep neural networks. Nat Lond 2017;542(7639):115−118G 10.1038/nature21056.

[17] Brierley J, Gospodarowicz MK, Wittekind C, editors. TNM classification of malignant tumours. 8th ed. Chichester, West Sussex, UK ; Hoboken, NJ: John Wiley & Sons, Inc; 2017.

[18] Benson AB, Venook AP, Al-Hawary MM, et al. Rectal Cancer, Version 2.2018, NCCN clinical practice guidelines in oncology. J Natl Compr Cancer Netw 2018;16(7):874−901. Available from: https://doi.org/10.6004/jnccn.2018.0061.

[19] van Hagen P, Hulshof MCCM, van Lanschot JJB, et al. Preoperative chemoradiotherapy for esophageal or junctional cancer. N Engl J Med 2012;366(22):2074−84. Available from: https://doi.org/10.1056/NEJMoa1112088.

[20] Shapiro J, van Lanschot JJB, Hulshof MCCM, et al. Neoadjuvant chemoradiotherapy plus surgery versus surgery alone for oesophageal or junctional cancer (CROSS): long-term results of a randomised controlled trial. Lancet Oncol 2015;16(9):1090−8. Available from: https://doi.org/10.1016/S1470-2045(15)00040-6.

[21] Habr-Gama A, Perez RO, Nadalin W, et al. Operative versus nonoperative treatment for stage 0 distal rectal cancer following chemoradiation therapy. Ann Surg 2004;240(4):711−18. Available from: https://doi.org/10.1097/01.sla.0000141194.27992.32.

[22] Renehan AG, Malcomson L, Emsley R, et al. Watch-and-wait approach versus surgical resection after chemoradiotherapy for patients with rectal cancer (the OnCoRe project): a propensity-score matched cohort analysis. Lancet Oncol 2016;17(2):174−83. Available from: https://doi.org/10.1016/S1470-2045(15)00467-2.

[23] Probst CP, Becerra AZ, Tejani MA, et al. Watch and wait? - Elevated pretreatment CEA is associated with decreased pathological complete response in rectal cancer. Gastroenterology 2015;148(4, Suppl. 1):S-1105. Available from: https://doi.org/10.1016/S0016-5085(15)33766-5.

[24] Kalady MF, de Campos-Lobato LF, Stocchi L, et al. Predictive factors of pathologic complete response after neoadjuvant chemoradiation for rectal cancer. Trans Meet Am Surg Assoc 2009;127:213−20. Available from: https://doi.org/10.1097/SLA.0b013e3181b91e63.

[25] Shanmugan S, Arrangoiz R, Nitzkorski JR, et al. Predicting pathologic response to neoadjuvant chemoradiotherapy in locally advanced rectal cancer using FDG PET/CT. J Clin Oncol 2011;29(4_Suppl). Available from: https://doi.org/10.1200/jco.2011.29.4_suppl.505 505−505.

[26] Macomber MW, Samareh A, Chaovalitwongse WA, et al. Prediction of pathologic complete response to neoadjuvant chemoradiation in the treatment of esophageal cancer using machine learning. Int J Radiat Oncol 2016;96(2, Suppl.):E699. Available from: https://doi.org/10.1016/j.ijrobp.2016.06.2379.

[27] Weiser TG, Regenbogen SE, Thompson KD, et al. An estimation of the global volume of surgery: a modelling strategy based on available data. Lancet 2008;372 (9633):139—44. Available from: https://doi.org/10.1016/S0140-6736(08)60878-8.

[28] Healey MA, Shackford SR, Osler TM, Rogers FB, Burns E. Complications in surgical patients. Arch Surg 2002;137(5):611—18. Available from: https://doi.org/10.1001/archsurg.137.5.611.

[29] Turrentine FE, Denlinger CE, Simpson VB, et al. Morbidity, mortality, cost, and survival estimates of gastrointestinal anastomotic leaks. J Am Coll Surg 2015;220 (2):195—206. Available from: https://doi.org/10.1016/j.jamcollsurg.2014.11.002.

[30] Mirnezami A, Mirnezami R, Chandrakumaran K, Sasapu K, Sagar P, Finan P. Increased local recurrence and reduced survival from colorectal cancer following anastomotic leak: systematic review and meta-analysis. Ann Surg 2011;253(5):890—9. Available from: https://doi.org/10.1097/SLA.0b013e3182128929.

[31] Ashburn JH, Stocchi L, Kiran RP, Dietz DW, Remzi FH. Consequences of anastomotic leak after restorative proctectomy for cancer: effect on long-term function and quality of life. Dis Colon Rectum 2013;56(3):275—80. Available from: https://doi.org/10.1097/DCR.0b013e318277e8a5.

[32] Ravitch MM, Steichen FM. A stapling instrument for end-to-end inverting anastomoses in the gastrointestinal tract. Ann Surg 1979;189(6):791—7.

[33] Neutzling CB, Lustosa SA, Proenca IM, da Silva EM, Matos D. Stapled versus hand-sewn methods for colorectal anastomosis surgery. Cochrane Database Syst Rev 2012; (2). Available from: https://doi.org/10.1002/14651858.CD003144.pub2.

[34] Honda M, Kuriyama A, Noma H, Nunobe S, Furukawa TA. Hand-sewn versus mechanical esophagogastric anastomosis after esophagectomy: a systematic review and meta-analysis. Ann Surg 2013;257(2):238—48. Available from: https://doi.org/10.1097/SLA.0b013e31826d4723.

[35] Farrah JP, Lauer CW, Bray MS, et al. Stapled versus hand-sewn anastomoses in emergency general surgery: a retrospective review of outcomes in a unique patient population. J Trauma Acute Care Surg 2013;74(5):1187—92. Available from: https://doi.org/10.1097/TA.0b013e31828cc9c4 discussion 1192-1194.

[36] Bruns BR, Morris DS, Zielinski M, et al. Stapled versus hand-sewn: a prospective emergency surgery study. An American Association for the Surgery of Trauma multi-institutional study. J Trauma Acute Care Surg 2017;82(3):435—43. Available from: https://doi.org/10.1097/TA.0000000000001354.

[37] Leonard S, Wu KL, Kim Y, Krieger A, Kim PCW. Smart Tissue Anastomosis Robot (STAR): a vision-guided robotics system for laparoscopic suturing. IEEE Trans Biomed Eng 2014;61(4):1305—17. Available from: https://doi.org/10.1109/TBME.2014.2302385.

[38] Shademan A, Decker RS, Opfermann JD, Leonard S, Krieger A, Kim PCW. Supervised autonomous robotic soft tissue surgery. Sci Transl Med 2016;8(337). Available from: https://doi.org/10.1126/scitranslmed.aad9398 337ra64.

[39] Jones SB, Jones DB. Surgical aspects and future developments of laparoscopy. Anesthesiol Clin N Am 2001;19(1):107—24.

[40] Jaschinski T, Mosch CG, Eikermann M, Neugebauer EA, Sauerland S. Laparoscopic versus open surgery for suspected appendicitis. Cochrane Database Syst Rev 2018; (11). Available from: https://doi.org/10.1002/14651858.CD001546.pub4.

[41] Lanfranco AR, Castellanos AE, Desai JP, Meyers WC. Robotic surgery. Ann Surg 2004;239(1):14—21. Available from: https://doi.org/10.1097/01.sla.0000103020.19595.7d.

[42] Kim CW, Kim CH, Baik SH. Outcomes of robotic-assisted colorectal surgery compared with laparoscopic and open surgery: a systematic review. J Gastrointest Surg 2014;18(4):816—30. Available from: https://doi.org/10.1007/s11605-014-2469-5.

[43] Wright JD, Ananth CV, Lewin SN, et al. Robotically assisted vs laparoscopic hysterectomy among women with benign gynecologic disease. JAMA. 2013;309 (7):689−98. Available from: https://doi.org/10.1001/jama.2013.186.

[44] Jeong IG, Khandwala YS, Kim JH, et al. Association of robotic-assisted vs laparoscopic radical nephrectomy with perioperative outcomes and health care costs, 2003 to 2015. JAMA. 2017;318(16):1561−8. Available from: https://doi.org/10.1001/jama.2017.14586.

[45] Opfermann JD, Leonard S, Decker RS, et al. Semi-autonomous electrosurgery for tumor resection using a multi-degree of freedom electrosurgical tool and visual servoing. In: 2017 IEEE/RSJ International Conference on Intelligent Robots and Systems (IROS). IEEE; 2017. p. 3653−60. Available from: <https://doi.org/10.1109/IROS.2017.8206210>.

[46] Birkmeyer JD, Stukel TA, Siewers AE, Goodney PP, Wennberg DE, Lucas FL. Surgeon volume and operative mortality in the United States. N Engl J Med 2003;349(22):2117−27. Available from: https://doi.org/10.1056/NEJMsa035205.

[47] Birkmeyer JD, Finks JF, O'Reilly A, et al. Surgical skill and complication rates after bariatric surgery. N Engl J Med 2013;369(15):1434−42. Available from: https://doi.org/10.1056/NEJMsa1300625.

[48] Tseng JF, Pisters PWT, Lee JE, et al. The learning curve in pancreatic surgery. Surgery. 2007;141(4):456−63. Available from: https://doi.org/10.1016/j.surg.2006.09.013.

[49] Tekkis PP, Senagore AJ, Delaney CP, Fazio VW. Evaluation of the learning curve in laparoscopic colorectal surgery. Ann Surg 2005;242(1):83−91. Available from: https://doi.org/10.1097/01.sla.0000167857.14690.68.

[50] LeCun Y, Bengio Y, Hinton G. Deep learning. Nature 2015;521(7553):436−44. Available from: https://doi.org/10.1038/nature14539.

[51] Krizhevsky A, Sutskever I, Hinton GE. ImageNet classification with deep convolutional neural networks. Adv Neural Inform Process Syst 2012;25:1097−105.

[52] Bouget D, Benenson R, Omran M, Riffaud L, Schiele B, Jannin P. Detecting surgical tools by modelling local appearance and global shape. IEEE Trans Med Imaging 2015;34(12):2603−17. Available from: https://doi.org/10.1109/TMI.2015.2450831.

[53] Padoy N, Blum T, Feussner H, Marie-Odile B, Navab N. On-line recognition of surgical activity for monitoring in the operating room. In: Proceedings of the National Conference on Artificial Intelligence, vol. 3; 2008. p. 1718−24.

[54] Wild E, Teber D, Schmid D, et al. Robust augmented reality guidance with fluorescent markers in laparoscopic surgery. Int J Comput Assist Radiol Surg 2016;11 (6):899−907. Available from: https://doi.org/10.1007/s11548-016-1385-4.

[55] Volkov M, Hashimoto DA, Rosman G, Meireles OR, Rus D. Machine learning and coresets for automated real-time video segmentation of laparoscopic and robot-assisted surgery. In: 2017 IEEE International Conference on Robotics and Automation (ICRA). IEEE. 2017. p. 754−9. Available from: <https://doi.org/10.1109/ICRA.2017.7989093>.

[56] Natarajan P, Frenzel JC, Smaltz DH. Demystifying big data and machine learning for healthcare. Boca Raton, FL: CRC Press Taylor & Francis Group; 2017.

[57] Bonrath EM, Gordon LE, Grantcharov TP. Characterising 'near miss' events in complex laparoscopic surgery through video analysis. BMJ Qual Saf 2015;24(8):516−21. Available from: https://doi.org/10.1136/bmjqs-2014-003816.

[58] Ellison EC, Pawlik TM, Way DP, Satiani B, Williams TE. Ten-year reassessment of the shortage of general surgeons: increases in graduation numbers of general surgery residents are insufficient to meet the future demand for general surgeons. Surgery 2018;164(4):726−32. Available from: https://doi.org/10.1016/j.surg.2018.04.042.

[59] Schein M, Rogers PN, editors. Schein's common sense emergency abdominal surgery: 21 tables. 2nd ed. Berlin: Springer; 2005.

[60] Silber JH, Williams SV, Krakauer H, Schwartz JS. Hospital and patient characteristics associated with death after surgery. A study of adverse occurrence and failure to rescue. Med Care 1992;30(7):615−29.

[61] Ghaferi AA, Birkmeyer JD, Dimick JB. Complications, failure to rescue, and mortality with major inpatient surgery in medicare patients. Ann Surg 2009;250 (6):1029−34. Available from: https://doi.org/10.1097/SLA.0b013e3181bef697.

[62] Smith GB, Prytherch DR, Meredith P, Schmidt PE, Featherstone PI. The ability of the National Early Warning Score (NEWS) to discriminate patients at risk of early cardiac arrest, unanticipated intensive care unit admission, and death. Resuscitation 2013;84(4):465−70. Available from: https://doi.org/10.1016/j.resuscitation.2012.12.016.

[63] Cambria E, White B. Jumping NLP curves: a review of natural language processing research [review article]. IEEE Comput Intell Mag 2014;9(2):48−57. Available from: https://doi.org/10.1109/MCI.2014.2307227.

[64] Murff HJ, FitzHenry F, Matheny ME, et al. Automated identification of postoperative complications within an electronic medical record using natural language processing. JAMA 2011;306(8):848−55. Available from: https://doi.org/10.1001/jama.2011.1204.

[65] FitzHenry F, Murff HJ, Matheny ME, et al. Exploring the frontier of electronic health record surveillance: the case of post-operative complications. Med Care 2013;51(6):509−16. Available from: https://doi.org/10.1097/MLR.0b013e31828d1210.

[66] Rajkomar A, Oren E, Chen K, et al. Scalable and accurate deep learning with electronic health records. NPJ Digit Med 2018;1(1):18. Available from: https://doi.org/10.1038/s41746-018-0029-1.

[67] Wauben LSGL, van Grevenstein WMU, Goossens RHM, van der Meulen FH, Lange JF. Operative notes do not reflect reality in laparoscopic cholecystectomy. BJS 2011;98(10):1431−6. Available from: https://doi.org/10.1002/bjs.7576.

[68] van de Graaf FW, Lange MM, Spakman JI, et al. Comparison of systematic video documentation with narrative operative report in colorectal cancer surgery. JAMA Surg 2019;154:381−9. Available from: https://doi.org/10.1001/jamasurg.2018.5246.

[69] Hung AJ, Chen J, Jarc A, Hatcher D, Djaladat H, Gill IS. Development and validation of objective performance metrics for robot-assisted radical prostatectomy: a pilot study. J Urol 2018;199(1):296−304. Available from: https://doi.org/10.1016/j.juro.2017.07.081.

[70] Hung AJ, Chen J, Gill IS. Automated performance metrics and machine learning algorithms to measure surgeon performance and anticipate clinical outcomes in robotic surgery. JAMA Surg 2018;153(8):770. Available from: https://doi.org/10.1001/jamasurg.2018.1512.

[71] Sicklick JK, Camp MS, Lillemoe KD, et al. Surgical management of bile duct injuries sustained during laparoscopic cholecystectomy. Ann Surg 2005;241(5):786−95. Available from: https://doi.org/10.1097/01.sla.0000161029.27410.71.

[72] Way LW, Stewart L, Gantert W, et al. Causes and prevention of laparoscopic bile duct injuries: analysis of 252 cases from a human factors and cognitive psychology perspective. Ann Surg 2003;237(4):460−9. Available from: https://doi.org/10.1097/01.SLA.0000060680.92690.E9.

[73] Weinberger SE. Can maintenance of certification pass the test? JAMA 2019;321:641−2. Available from: https://doi.org/10.1001/jama.2019.0084.

[74] American Board of Surgery. Continuous certification, <http://www.absurgery.org/default.jsp?exam-moc> [accessed 17.02.19].

[75] Greenberg CC, Ghousseini HN, Pavuluri Quamme SR, Beasley HL, Wiegmann DA. Surgical coaching for individual performance improvement. Ann Surg 2015;261 (1):32−4. Available from: https://doi.org/10.1097/SLA.0000000000000776.
[76] Iglehart JK, Baron RB. Ensuring physicians' competence—is maintenance of certification the answer? N. Engl. J. Med. 2012;367(26):2543−9. Available from: https://doi.org/10.1056/NEJMhpr1211043.

Further reading

Iglehart JK, Baron RB. Ensuring physicians' competence—is maintenance of certification the answer? N Engl J Med 2012;367(26):2543−9. Available from: https://doi.org/10.1056/NEJMhpr1211043.

CHAPTER 9

Remote patient monitoring using artificial intelligence

Zineb Jeddi[1,2] and Adam Bohr[3]
[1]TICLab, International University of Rabat (UIR), Rabat, Morocco
[2]National Institute of Statistics and Applied Economics (INSEA), Rabat, Morocco
[3]Sonohaler, Copenhagen, Denmark

9.1 Introduction to remote patient monitoring

Today, chronic illnesses such as asthma, cancer, cardiovascular, diabetes, and mental health—related conditions have replaced infectious diseases as the top global cause of death and morbidity [1,2]. The number of people living with multiple chronic conditions is likely to increase because of the aging population and leading unhealthy lifestyles. The key challenge offered by chronic conditions is the absence of an immediate treatment. Individuals suffering from chronic conditions find themselves obliged to manage their illness continuously to prevent acute symptoms. This translates into frequent regular diet monitoring, physical activity monitoring, and medication management. Therefore, there is an increasing interest and need in patient monitoring solutions as a key gateway to facilitate chronic disease management, and especially improve self-management [3,4].

Remote patient monitoring is an emerging field of healthcare, which concerns the management of health/illness with the goal of treating or diagnosing illness using information technology and telecommunication tools. It uses these tools to collect health data at the individual user's home or during an everyday activity and stores or transmits this data to the healthcare professionals for evaluation and recommendation [5]. There is an overlap between the terminologies remote patient monitoring, telemedicine, telehealth, and mobile health, all of which cover various aspects of monitoring patients outside of the hospital setting using information technology. Remote patient monitoring serves to meet the interest to study signals and behavior of people with chronic conditions who are at risk of acute symptoms. This provides the patient with improvement of treatment and reassurance of monitoring sudden health attacks and reduces hospital visits. There are many incentives for performing remote monitoring including real-time and continuous monitoring of symptoms, early detection or prevention of illness, reduced healthcare costs, increased information on health conditions, and finally a greater opportunity for service and emergency medical care [6]. Because of an increasing aging population, a rise in demands, limited resources, and the prospects of healthy aging, cost reduction is the main driver for implementing remote health monitoring [7].

A remote monitoring system generally consists of a data acquisition system, data processing system, and monitor/terminal for the end user and a communication network. The data acquisition system consists of different sensors or devices such as a smartphone with in-built sensors and wireless transmission. The data processing system receives and transmits data and the terminal can be a computer at the hospital or a database on the device or smartphone. Often there would be a communication network that connects the user and data processing system with a healthcare professional [8]. In some cases, data is continuously collected and used by the patient and occasionally presented to a healthcare professional, whereas in other cases data is continuously transmitted to the healthcare professional for assessment. These are all dependent on the condition of the patient, the recorded information, and the complexity of the data. The patient can either be prompted to visit the hospital, perform first-aid, take medication, or make contact with a healthcare professional. The current remote monitoring systems vary largely with regard to their level of technology, communication path, and integration of sensors (Fig. 9.1).

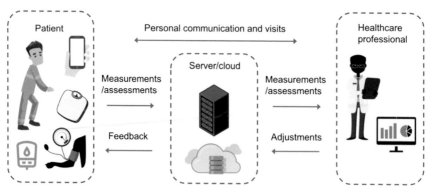

Figure 9.1 Example of remote patient monitoring architecture.

9.2 Deploying patient monitoring

Remote patient monitoring systems record physiological data from patients based on sensors and other data input sources. The most common data collected from current systems include heart rate, blood pressure, respiration rate and air flow, blood oxygen volume, electrocardiogram (ECG), electroencephalogram (EEG), electromyogram, body/skin temperature, and blood glucose levels. Further, these can be supplemented with activity data such as steps walked and calories burnt, sleep data, location/environment, and weight of the patient. Most conventional monitoring setups collect data using sensors that are placed on the body such as electrodes attached onto the skin whereas some can be more invasive such as implanted sensors for monitoring the brain [9]. With the continuous developments in electronics and wireless technology, the sensors and monitoring devices are becoming smaller and increasingly wireless and many recent devices are powered by devices such as a smartphone or tablet.

9.2.1 Patient monitoring in healthcare today

There are many health conditions (e.g., diabetes, cardiovascular disease, and neurological conditions) where remote patient monitoring can help prevent hospitalization, reduce mortality, and provide improved treatment and service.

Diabetes is a prevalent chronic condition that requires strict management to maintain control of glucose levels and reduce the risk of complications. Blood glucose levels in diabetics can vary a lot throughout the day and is important to measure multiple times per day to ensure being in the right range together with insulin treatment. Traditional glucose sensors are based on electrochemical methods where blood glucose is analyzed based on a small blood sample collected from a finger-prick, an inconvenient invasive method [10]. The emergence of continuous glucose monitoring (CGM) has been a particular success story for improving the lives as well as health outcomes of patients with type 1 and type 2 diabetes. These CGMs record blood glucose levels throughout the day and report trends for retrospective evaluation and hypo/hyperglycemia detection. The main benefit of CGMs is the reduced time that the patient spends in hypoglycemia due to continuous monitoring. They are also associated with improvements in hemoglobin A1c (HbA1c), a key diabetes indicator [11]. When used together with an insulin pump, CGMs are especially useful for controlling glucose levels as they enable both more frequent monitoring and adjustments (Fig. 9.2). Such closed-loop systems are currently only offered by a few manufacturers including Medtronic and Dexcom (with Tandem) and are not available for long-term monitoring [12]. Further, the clinical acceptance and implementation of CGM systems has been relatively modest, despite the substantial benefits they provide.

Heart failure is a prevalent condition affecting more than 26 million people worldwide and is associated with an increased rate of hospitalization and mortality [13]. For people with heart failure, lack of monitoring following episodes of treatment is likely to increase the risk of readmission as well as morbidity. Remote patient monitoring offers a cost-effective solution compared with in-person visits by healthcare professionals. The monitoring typically consists of a combination of measurements including ECG, blood pressure, heart rate, and weight combined with telephone-based monitoring and consultations. Some of these are based on telemonitoring whereas others are based on videoconferencing or mobile phone—based monitoring and interventions [14]. There are numerous clinical studies reporting the outcome of such monitoring programs and the conclusions vary regarding their positive outcome for patients. However, there seems to be some consensus that some type of home monitoring has an effect on lowering heart failure hospital readmission and mortality [15].

Remote patient monitoring is also associated with the treatment of neurological disorders such as Parkinson's, epilepsy, multiple sclerosis, and dementia. Some of the clinical challenges with these disorders are to measure the gains and losses of daily functioning over time and assessing the effect of medication on patient's symptoms and function. For many neurological conditions, measurement of physical activity is important, this is

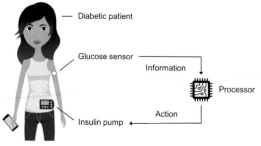

Figure 9.2 Illustration of closed-loop glycemic management system using glucose sensor and insulin pump for insulin delivery.

because of the link between physical inactivity and various morbidities. Remote monitoring provides a means of assessing physical activity outside of the clinic and can be used in the evaluation of predictors, outcomes, and decision support. Physical activity monitors may be used to predict fall risk and assess missteps in patients with Parkinson's using accelerometry devices [16]. Such devices including the commercially available Actigraph and Omron devices measure the movements of the arm or leg while reaching after an object or walking. They assess parameters such as gait stance, swing time, and speed of movement. These systems provide feedback to users and clinicians, allowing them to better assess the situation and offer better rehabilitation and therapeutic management solutions [17] (Fig. 9.3).

9.2.2 Using remote patient monitoring to improve clinical outcomes

There are numerous applications of remote patient monitoring where improved clinical outcomes have been demonstrated based on clinical studies and pilot programs. At the same time, there are many studies reporting negligible improvement in clinical outcome, even for the same type of monitoring tool or disease. Due to the large difference between its applications and different disease indications, it is difficult to speak generally about the effectiveness of remote patient monitoring systems. Most monitoring systems target a single technology and a single disease,

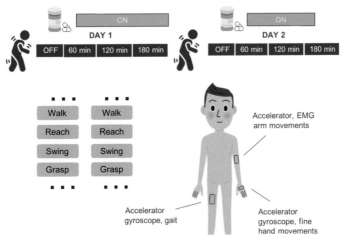

Figure 9.3 Example of data collection and sensor setup for physical activity monitoring of patients with Parkinson's.

indicating the segmented nature of the remote patient monitoring area. So far, most discussions on clinical outcomes have focused on outcomes in terms of reduced morbidity, mortality, or biometric markers such as HbA1c for diabetes. On the other hand, several relevant indicators of clinical effects are overlooked, including adherence. Adherence refers to the degree to which a patient follows medical advice and prescriptions. This encompasses the patient's adherence to medication as well as the use of other advised methods of care. To ensure proper control and achieve improved health outcomes, adherence should be recognized as a central parameter in the implementation of patient monitoring [18].

Monitoring of adherence using remote monitoring systems, including mHealth apps, has grown in the past decade and now constitutes one of the main applications for recent apps and services. Adherence promoting tools include everything from SMS-based reminders to monitor systems using sensors with integrated medical support. A study by Hamine et al. compiled adherence and patient outcome data from 107 clinical studies where both adherence and conventional patient outcome data were monitored and found that 56% of the studies had a significant effect on adherence whereas 40% of the studies had a significant effect on clinical outcomes [19]. It was also shown that SMS reminders are the most widely and successfully used tool to facilitate adherence and that improvements in adherence is highly operator dependent, relying to a large degree on active patient engagement and information exchange.

Active patient engagement is an important factor for achieving high user retention and as a result improved adherence and clinical outcomes. Patients are often not motivated to improve adherence as this requires extra effort and the results may not be evident immediately. Getting people to adopt a new habit is not easy, especially if it does not contain entertainment value or provides an immediate reward. Yet, adoption is a key factor for these adherence tools if they are to move beyond clinical tests and pilot studies and have a meaningful impact on society.

So far adoption has been low for most devices and systems there can be several reasons for this [20]. Monitoring systems are mainly placed on the body and are often large and require a specific placement on body to provide accurate measurements. Connectivity issues are also common with many monitoring systems resulting from delays, data loss, and network communication causing poor user experience. Accuracy and reliability of clinical decision support may also be a concern for clinicians using the monitoring system if it was developed and tested mainly under fixed conditions and using simulations. Finally, user engagement and user inter-action are typically lacking. Users often end up feeling like active data collectors with very limited access to all the data they collect, limiting their acceptance and uptake [21,22]. To obtain higher adoption, these monitoring systems must be improved by focusing more on end-user experience and ensuring that the product provides value to the user. The acceptance and adoption of any product in the healthcare sector depends to a large degree on user awareness as well as acceptance by both the patient and the clinician.

9.3 The role of artificial intelligence in remote patient monitoring

9.3.1 Harnessing the power of consumer technology

The widespread use of artificial intelligence (AI) in remote monitoring has emerged as a result of the rapid growth in the adoption of technologies in society. According to the Pew Research Center in the United States, the last statistics from 2018 indicate that 81% of people in the United States are smartphone owners, where more than 62% of these use their smartphone for health or medical-related searches according to a study conducted in 2015 [23]. Furthermore, in 2018 another study on smartphones ownership and the health apps use among vulnerable categories in United States showed that 39% of participants report the use of a mobile app to track their health condition [24]. In this context, significant advancements in AI, Big Data, and machine learning (ML) are taking place to meet the consumers' demands. At a community level, the consumers' needs in terms of healthcare technologies arise from several factors, related mainly to the demographic growth and population aging, the increasing healthcare costs, the healthcare disparities, and the increasing prevalence of many chronic diseases such as asthma, diabetes, dementia, and Parkinson's [25]. This translates into regular diet monitoring, physical activity, and medication management. Therefore there is an increasing interest in the future of AI solutions for remote patient monitoring, as a key gateway to facilitate chronic disease management and the improvement of self-management. There is also a growing concern about increasing workforce shortages as in the US healthcare system [26]. According to the AAMC Consumer Survey of Health Care Access 2010, these shortfalls are expected to impact mostly the underserved consumers living in rural and inner-city areas [27]. Furthermore, the current US healthcare workforce is aging and leading to an increased retirement rate [28]. The role of AI in remote patient monitoring has therefore been enhanced to develop advanced health monitoring technologies that would help individuals dealing with the rising costs of healthcare services.

The technologies of interest focus mainly on helping stabilize the patient, minimizing the duration of hospitalizations, and reducing its costs, accomplishing the treatment and diagnosis under the physician's supervision and monitoring the recovery process while the patients are not at the hospital. Moreover, there is evidence that individuals with low income have poorer health status, poorer access to health services, and are more likely to be careless about their health and live mostly in unhealthy environments [25]. These disparities keep on existing for several sociocultural, socioeconomic, behavioral, environmental, and healthcare system factors. Therefore considerable levels of mistrust and challenges exist among these individuals and healthcare providers whenever the system is unresponsive or poorly responsive to their concerns [4].

9.3.2 Sensors, smartphones, apps, and devices

The integration and the use of AI and the Internet of things (IoT) for sensors, mobile apps, social media, and location-tracking technology can make early detection, treatment, and better self-management possible for patients with chronic diseases [6] (Fig. 9.4). Remote healthcare sensing technologies fall into three different categories: passive sensing, active sensing, and functional assessment. For passive sensing, smartphones are the most common. The built-in sensors of a smartphone (accelerometer, gyroscope, magnetometer) enable physics-based abilities, such as geographic localization and detecting the number of steps the person takes per day. They can also provide information about atmospheric pressure, ambient light, voice, and touchscreen pressure. In addition to these capabilities, the built-in camera allows other creative uses of these sensors that can transform the smartphone into a fall detector [29], a spirometer [30] (by sensing the acoustic signal captured by the microphone), or a heart-rate sensor [31].

Wearable devices such as wrist sensors are common nowadays and have many of the same sensors as smartphones. They can be used to detect motion such as those associated with smoking [32] and seizure activity [33]. In 2017, 17% of adults in the United States used a wearable device such as a smart watch or a wrist-worn fitness band [34]. Wrist

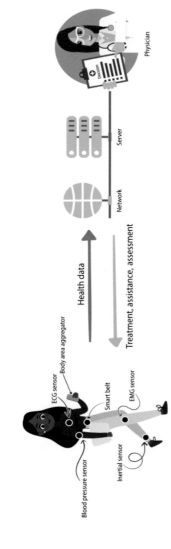

Figure 9.4 Wearable technology in healthcare.

sensors often comprise photoplethysmographic sensors that detect changes in reflected light caused by changes in microvascular blood flow with each heartbeat. This provides measures of heart rhythm parameters [35]. Wearable sensors are also used by clinicians, for instance, wearable patches can measure muscle activity and posture [36], radiofrequency sensors placed over clothes can detect pulmonary function [37], and others can assess the recovery process of several neurological disorders [38]. For medication adherence, there is an emerging technology that consists of a miniature sensor implanted in a pill, which emits signals to a wearable patch when it enters the stomach [39]. While passive sensing only collects observable data, active sensing involves the patient monitoring process. Active sensing, thus, allows a subjective assessment of the health state of a patient using questionnaires to describe their experience during some specific intervals of time. The reported data by the patient is equally important as the measured data and it is necessary to include in the AI models for accurate assessment of the results.

Functional assessments combine both passive and active sensing thereby including the patient's more subjective reporting as well as the more objective data measurement by the device sensor. It requires patients to perform standard tasks using the mobile health technology to assess functional performance. Many solutions are proposed such as the estimation of the 6-minute walk test performance using the smartphone motion sensors [40], assessing the parkinsonian voice vibrations using the smartphone microphone [41], and evaluating the cognitive functions such as memory and reaction time, through mobile applications [42]. Most of the mobile applications to assess the cognitive function are brain training apps that are based on mental stimulation. Many available mental health solutions have been reported to have a significant impact on improving the mental health of their users through enhancing their cognitive functions such as speed, memory, concentration, and problem-solving. One of the first apps in the market were Lumosity and Elevate by Apple, these apps use fun mini games that are designed by researchers to improve neuroplasticity of the brain. The integration of AI in these apps can provide the users with a friendly monitoring interface to evaluate their cognitive function in time. Another major application of AI in cognitive/mental health monitoring consists in Chatbot coaches. Mobile health apps such as WoeBot, Youper, Wysa, Replika, and Unmind [43] provide conversational AI Chatbots that offer day-to-day emotional support and mental health tracking through smartphones. The AI algorithms used by the

Chatbots are trained and tested using tools from Cognitive Behavioral Therapy elaborated by researchers in the psychology/mental health field.

9.3.3 Natural language processing

The volume and the complexity of patient-generated medical data are increasing exponentially. These vast quantities of data contain relevant information that could present a great help in medical decision-making. Unfortunately, the use of these data is restrained due to its unstructured format (e.g., free-text clinical reports) and also no human workforce is capable of handling the large volumes of data because most are unstructured (unorganized stored data). However, a computer program using AI can read all the data and save relevant information for further use [44]. These AI algorithms are based on natural language processing (NLP) and can extract meaning, sentiment, or the intent of a text written by a person. NLP algorithms allow computers to process human language which facilitates clinical research by a rapid extraction of relevant information and automate the analysis of free-text documents (Fig. 9.5). The existing methods are mainly based on statistical, ML, and deep learning (DL) [45,46].

9.3.4 Natural language processing technology applications in healthcare

NLP-based technologies have become more and more numerous in the field of healthcare, especially in hands-free communication. These technologies were incorporated in many AI-based apps, especially for those Chatbots destined for remote patient monitoring through some wearable technology. Some NLP applications in health include

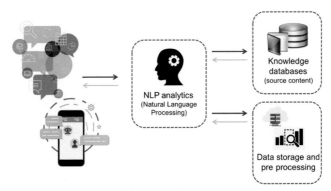

Figure 9.5 A general description of a natural language processing system.

Recording notes: As for NLP applications in education field where various mobile apps can help the student with note-taking and summarization, some NLP-based apps can also help the patient with the key point notation during their appointment with the doctor. The incorporated NLP approach basically uses sophisticated speech recognition algorithms that allow summarizing and extracting pertinent information. Doctors and nurses can also use NLP-based mobile apps for recording verbal updates, for example, during surgical interventions, the surgeon can verbally record findings and easily communicate with the other team members [47].

Remote device controlling: NLP technology offers a hands-free communication for individuals with certain disabilities. The NLP technology incorporated in the mobile app could mimic the human speech and interact with the users via artificially intelligent Chatbots.

Natural language processing applications in the Chatbots industry: m-health applications.

Chatbots consist of smart conversational apps that use sophisticated AI algorithms to interpret and react to what the users say by mimicking a human narrative. Nowadays and in the near future, these Chatbots will mimic medical professionals that could provide immediate medical help to patients. Many Chatbot solutions in healthcare are showing promise [48].

Babylon health: This medical advice app incorporates many advanced AI algorithms that allow patients with the possibility of remotely consulting doctors and healthcare professionals. Consultations are assigned to patients based on their medical history in and other aspects about their health using adequate AI algorithms. This app also offers the possibility for users to have live video chats with doctors depending on their health profile. NLP is integrated in this app within a symptom checker Chatbot that could provide relevant information to patients including recommending subscriptions and solutions for their health problems. This app is currently providing a limited free service in collaboration with the UK's National Health Service (NHS) to accommodate the health needs of many patients.

Buoy Health: Focusing on diagnosis, Buoy Health app assists patients/users in identifying the cause of their illness. Using developed AI algorithms that are trained on large medical datasets, this app offers a conversational Chatbot that can assist patients in real time. The Chatbot provides the patients with a highly accurate understanding of their condition using the entered symptoms and also provides them with suitable medical advice for their health condition.

9.3.5 Clinical decision support

In the past few years, clinical decision support systems (CDSSs) have been developed and designed for home tele monitoring of patients with chronic diseases such as chronic obstructive pulmonary disease [49]. The main objective of the CDSS is the management of the health condition of the patient through daily monitoring, therefore any recurrence or aggravation of the disease can be identified, hence prevented at an early stage [50]. CDSS relies on computable biomedical knowledge, patient-specific data, and incorporated AI models that make use of these knowledge and data to generate relevant information for clinicians and help in their decision-making [51]. The most common and developed CDSS is those that address medication safety adherence. These CDSSs are basic systems that mainly incorporate electronic prescribing systems in a clinical decision support. In general, they do not require complex AI algorithms to provide the support; only the available data is enough to support clinicians' decisions (Fig. 9.6). However, as these CDSS archive enormous amounts of patient data [52], AI tools are developing in parallel with an increasing availability in rich data sources. For instance, ML algorithms and neural networks are designed to handle huge datasets using developed specific features selection methods [53]. The use of AI to process large-scale data allows CDSS to operate on different levels. For instance, the AI algorithms incorporated in a CDSS can make use of the hospital's electronic health records to predict emergency admissions on many levels. This would allow the hospital to adopt better decision-making strategies for emergency room management, improve the emergency care service

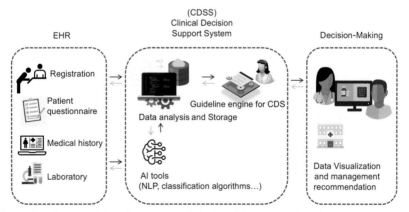

Figure 9.6 A general description of a clinical decision support system.

for patients, and also reduce the related costs. Moreover, these systems can also improve the patients' quality of life by allowing the app or device to track their health condition through early detection and of future exacerbations [50].

However, the presence of a CDSS does not always mean that the clinical staff will use it as designed. In recent studies, nurses have reported that they tend to rely on CDSS when they are new to their occupation, but once they gain experience, they just use the system to "double check" and they base their decisions mainly on their own experience [54]. Some nurses report that these system's recommendations only stifle their professional judgment and when poorly designed, they are only in conflict regarding their experienced decisions [55]. This implies that when designing and evaluating any CDSS, some strict guidelines should be followed. As stated in the literature, the five rights rule should be used: (1) right information (clinical knowledge, clinical practice guidelines, and adequate AI algorithms); (2) right people (practitioners, interdisciplinary team, and patient require information to make decisions); (3) right format (alerts, prompts, order sets, templates, and information buttons to present decision data); (4) right channels (incorporating the electronic health record, portals, and mobile technology); and (5) right time (identifying the time structure of the data and include it in the decision-making process in an adequate manner [56]). When these guidelines are met and the clinician receives relevant information that does not interfere with their professional opinion nor demand additional time and effort, this will result in a satisfied medical staff.

9.3.6 Ambient assisted living

Ambient assisted living (AAL) refers to adopting new technologies in a person's daily living to handle the increasing needs for healthcare services and assistance, hence improving their quality of life. In particular, AI-based AAL systems provide a significant help to the elderly, disabled people, and individuals with chronic conditions. They offer substantial solutions in healthcare such as patient remote monitoring, self-sufficiency development, emergency assistance, and home care services [57]. For instance, smart homes can be very useful in remote monitoring, using AI-based sensors, patients can be monitored in their daily life and older people can be continuously assisted by caregivers. One key element of AAL systems is human activity recognition (HAR) [58]. It is mainly based on the IoT and AI.

Using AI approaches including big data and ML, AAL systems are able to handle massive volumes of data, analyze it with the adequate models, identify the accurate pattern, and make predictions that can be used for decision-making. Different machine algorithms are used for HAR such as Random Forest, Naïve Bayes, K-means, and IBK classifiers to detect different mobility patterns. For instance, using IoT, a variety of sensors, and AI algorithms, an AAL system can monitor in real time different activities for the patient such as sitting, laying down, standing, walking on a horizontal surface, walking downstairs, or walking upstairs [59].

Many AAL systems use the smartphone's sensors (accelerometer, gyroscope) for activity recognition. Studies by Davis-Owusu et al. [58] used a smartphone and its incorporated sensors to collect data related to old aged participants. Using three ML algorithms: SVM (support vector machine), ANNs (artificial neural networks), and a hybrid method combining HMM (Hidden Markov Models) and SVM (SVM-HMM), they were able to detect six different activities (sitting, walking, laying, standing, descending, and ascending the stairs). DL-based methods were also used for HAR in AAL systems [60]. Moreover, relevant work on using convolution neural networks (CNNs) has been done, where daily activities such as jogging, going upstairs, going downstairs, walking, sitting, and standing were recognized [61]. A variety of sensors can be used to build an AAL system. Smartphones and wearable sensors are commonly used, especially when they are not intrusive and are low-cost making very efficient AAL systems such as e-textile, smart homes, and assistance robots [62].

Syed et al. proposed a smart healthcare framework for AAL to monitor the physical activities of old people using AI and IoT [63]. To develop this framework, they had to place several sensors on different parts of the participant's body. Data was collected from sensors on the person's left ankle, right arm, and chest and transmitted via IoT equipment. For the data analysis, advanced AI algorithms such as MapReduce and multinomial classifiers were used for HAR tasks. The developed framework was able to remotely predict 12 physical activities with an accuracy of 97.1%. In this work, the user interface consists of a mobile health application that allows remote monitoring of the patients by visualizing data and generating reports. A remote expert app allows experts to remotely monitor their patients and produce reports based on the classified daily activities as shown in Fig. 9.7.

Based on the physical activities classification results, experts could take the right decisions and generate the necessary reports. The second

Figure 9.7 Example of expert interfaces for remote monitoring using mHealth app. (A) Start screen, (B) login screen, (C) patients list, (D) patient details, (E) classification results, and (F) results visualization. *Reprinted with permission from Syed L, Jabeen S, Manimala S, Alsaeedi A. Smart healthcare framework for ambient assisted living using IoMT and big data analytics techniques. Future Gener Comput Syst 2019;101:136–51.*

interface shown in Fig. 9.8 is intended for caregivers, who receive notification alerts concerning treatment recommendations to be able to measure eventual risks and provide the necessary interventions in the case of emergencies such as heart attacks or fall detection.

Figure 9.8 Example of caregiver interfaces for remote monitoring using mHealth app. *Reprinted with permission from Syed L, Jabeen S, Manimala S, Alsaeedi A. Smart healthcare framework for ambient assisted living using IoMT and big data analytics techniques. Future Gener Comput Syst 2019;101:136—51.*

Such a framework can be considered as an optimal solution for AAL for the elderly and people with disabilities since they could recognize different physical activities, thus remotely monitor the health condition of the patient.

9.4 Diabetes prediction and monitoring using artificial intelligence

Diabetes constitutes a highly prevalent condition with more than 463 million adults living with diabetes and incurs a heavy economic burden on the healthcare system [64]. Although a variety of treatment options exist, there is an unmet need in improving the quality of life for people with diabetes and providing disease management to prevent diabetes-related complications. Diabetes management is more than just tracking markers such as glycated hemoglobin (HbA_{1C}) and includes feeling better, and understanding or following the condition [65]. Here, the developments within device technology as well as digital technology present an opportunity to rethink the way diabetes is treated and managed. The large number of publications on AI applications for diabetes care from the last decade indicates the immense interest in the area and demonstrates many different approaches of using AI to improve the quality of life for people with diabetes. For instance, such systems can be used for screening, diagnostics, treatment, and management of people with diabetes [66]. In this section, we will have a closer look at one of these remote patient monitoring solutions for diabetes management, DiaBits.

9.4.1 DiaBits

There are multiple marketed diabetes management solutions for remote patient monitoring and one of these is the BCT app, DiaBits, a mobile app the allows people with diabetes to monitor their blood glucose levels in real time and predict future fluctuations in glucose levels up to 1 hour ahead in time. DiaBits uses data from CGM as well as smartphone sensor data and manual user inputs to predict future blood glucose fluctuations applying various ML techniques. This feature provides patients the opportunity to take corrective measures ahead of time to maintain blood glucose levels with the preferred range. Furthermore, this is combined with tools for analysis of historical data to keep track of and manage the condition and gain better control of blood glucose [67].

The DiaBits app is used both for people with type 1 and type 2 diabetes and provides personalized diabetes management where the app builds a model of your blood sugar. It further helps the user in better understanding the impact of different stimuli such as foods on the blood sugar trend to mitigate hypo- and hyperglycemia. The app is used together with the Dexcom continuous glucose monitors, or Nightscout, a third-party software, to access continuous glucose monitor data. It can also be connected with Fitbit or Apple Watch for further data integration including heart rate, meals, and step count. Such additional inputs help increase predictive accuracy. The ML algorithm of DiaBits takes 3−7 days to familiarize with the users' physiology and gradually becomes more accurate at predicting blood glucose as it accumulates data from the user. Here, both SVM and unsupervised DL algorithms are used depending on the amount of data and whether it is data from individuals or populations. Autoregressors are useful but only for predicting blood glucose values up to 30 minutes ahead in time [68].

BCT has successfully implemented its technology at British Columbia Children's hospital in Canada and has demonstrated a prediction accuracy of up to 95% [69]. The study involved testing the DiaBits app together with a CGM and Fitbit tracker in eight pediatric patients over a period of 60 days collecting data on physical activity, heart rate, and blood glucose. Here, the learning algorithm was trained on the effects of physical activity and heart rate on blood glucose during the first 30 days and was subsequently used to predict the user's near-future values of blood glucose. The study demonstrated that future blood glucose values can be predicted up to 60 minutes ahead in time based on data recorded from CGM and Fitbit. Currently, there is ongoing work on integrating the algorithms of DiaBits into the patient healthcare app of Premier Health, a medical care provider. Premier Health consists of 290 clinics with almost three million registered patients in their ecosystem. Premier Health will further integrate clinical decision support tools into their Electronic Medical Records (EMR) software to enhance health-related decisions improve overall patient care. In the pilot studies, they are using patients who already use the Premier Health app and monitor heart rate, blood pressure in addition to their age, sex, and weight. It will be interesting to see how the app performs on a larger clinical population. Currently, a majority of diabetes management apps have only been tested in a small to medium size patient population to demonstrate whether glucose control can be improved. Long-lasting studies involving a large number of patients are necessary to establish proper evidence for clinical outcome and safety [70].

9.4.2 Other diabetes remote monitoring apps and devices

There are various diabetes monitoring products under development using AI-based approaches for applications including patient self-management, decision support tools for clinicians, personalized risk assessment and detection of diabetic retinopathy as well as other late-stage complications.

The recent developments in new devices including glucose sensors, insulin pumps, and health tracking consumer electronics have resulted in a substantial number of clinical trials using real-time CGM in combination with AI algorithms for assessing and improving glycemic control and reducing hypoglycemic events. In several of these studies, an artificial pancreas device was used consisting of a closed-loop setup (Fig. 9.3) combined with AI-powered monitoring and assessment [71−73].

9.5 Cardiac monitoring using artificial intelligence

AI has several promising applications in cardiovascular medicine mainly in diagnosis, treatment, and patient remote monitoring. These applications can be divided into two main lines of research: physical and virtual. The virtual area of studies include the application of ML, DL, and NLP for patient monitoring and automated CDSSs. AI applications in the physical branch encompass mainly robotics interventional operations.

9.5.1 Virtual application of artificial intelligence in cardiology

9.5.1.1 Imaging interpretation

The application of AI in the field of imaging concerns three main aspects: image analysis, reconstruction, and interpretation (Fig. 9.9). The use of sophisticated AI programs on huge amounts of imaging data has simplified image analysis through developing advanced structural layouts for image interpretations [74]. Many AI-based solutions were proposed to automate certain image analysis processes. In a study by Ciusdel et al. [75], deep neural networks were trained on 17,800 coronary angiograms of 3900 patients and validated using 27,900 angiograms of 6250 patients, to fully automate cardiac phase and end–diastolic frame. Other AI frameworks are based on ML approaches, for instance, researchers from Emory University have developed an ML method for intravascular ultrasound image segmentations. This method enables online automation of the process of calculating lumen area and plaque burden. This method provided promising results for image segmentation and was positively supported when presented to the field experts [76].

Researchers hypothesize that in the near future, integrated DL and ML methods in smart AI solutions will enable an automated diagnosis and

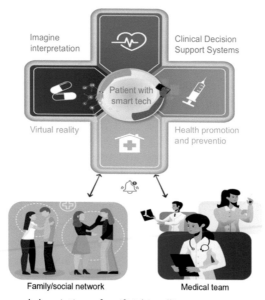

Figure 9.9 A general description of artificial intelligence remote monitoring systems in cardiology.

monitoring of several imaging-based pathologies in a way that largely imitates the work of an expert [77].

9.5.1.2 Clinical decision support systems

In cardiovascular medicine, CDSSs are also evolving toward self-learning systems using ML, DL, and NLP to simulate a decision-making process similar to the reasoning of a human expert. For instance, IBM's Watson for health cognitive technology uses large amounts of data from medical reports (EMR, lab, and imaging reports) and online data sources [78]. Mainly based on ML and DL, these new AI solutions are designed to mimic the human brain process of decision-making. In cardiovascular medicine, IBM is developing a smart assistant for cardiologists and radiologists to help their decision-making. The Medical Sieve project developed by IBM includes several functions for cardiac monitoring, especially in imaging such as the automated detection of coronary stenoses in angiography [77]

9.5.1.3 Artificial intelligence in virtual reality, augmented reality, and voice powered virtual assistants

One of the major applications of AI approaches in cardiology is their combination with virtual reality (VR) and augmented reality (AR) technologies. Main VR frameworks are designed for periprocedural and pre-procedure modalities such as heart interventions and stress management for patients prior to an intervention [79]. Similarly, AR platforms are also used to assist the interventionist during the heart procedures and provide them with necessary real-time information via several monitoring screens [80]. On the other hand, voice powered virtual assistants, for example, Google Assistant, Apple's Siri, and Amazon's Alexa, use highly advanced AI programs for speech recognition. Using these voice-based virtual assistants, physicians and operators can search data in the EMR system or obtain online information in a very fast, effective, and convenient way by using only their voice signature passwords [81].

9.5.1.4 "BIG DATA" for predictive analysis

In the presence of larger amounts of heterogeneous and valuable data such as for sequencing data, social media, and cardiovascular images, data-sets are becoming more complex and very difficult to analyze using traditional methods of statistics. AI approaches using BIG Data methods for automatic hypotheses generation have shown promising results in their analytic and predictive abilities [82]. For instance, deep neural networks can explore very complex datasets of some cardiovascular diseases, extract the relevant risk factors, and provide an accurate prediction analysis.

For patient monitoring, ML-based techniques can be used to predict stroke risk and mortality risk associated with some heart interventions. In cardiology, ML was used to predict mortality in patients with heart failure within a 1-year window and 5-year mortality for patients with coronary artery disease [83].

9.6 Neural applications of artificial intelligence and remote patient monitoring

9.6.1 Dementia

Recently, the demographic pattern of most western countries has witnessed a significant increase in life expectancy and a considerable drop in the birth rate, which means that societies are aging. At the same time, challenges are becoming more difficult to manage, since the medical staff are facing many shortages, this is especially true for doctors, nurses, and caregivers for the elderly [84]. This difficulty in ensuring a good quality nursing care impacts people with age-related diseases, such as those with dementia. Today, 50 million people worldwide are suffering from dementia, and and every year there are approximately 10 million new cases [85]. Therefore a serious effort is needed to improve the nursing care services and provide dementia patients with sufficient care and security so that they can independently lead a normal life. Most dementia patients are treated in home care, where their individual needs for assistance change in different ways depending on the stage of the disease. These needs can sometimes be temporary, but they can also be time-consuming when the required resources and efforts are beyond that of the caregivers [86]. Therefore dementia management requires a multidisciplinary and multi-sectoral coordination. However, in reality, a delay or a lack in information sharing between actors in care professions make the coordination less sufficient [87]. This motivates the establishment of common support systems that integrates primary care, remote care, and informal care provided by family members. Using a similar support system, anomalies in the usual behavior of dementia patients or urgent situations can be automatically recognized, thus the necessary interventions are easily coordinated. An example of similar support systems is the German research project "SAKI" that aims to tackle the challenges with dementia and to address the needs of the patients at their homes [3].

9.6.1.1 Artificial intelligence in dementia monitoring

Current AI-based support systems for dementia monitoring aim to track the disease progression over time [88] and provide the necessary support for patients to maintain their usual activities. An example of a home-monitoring system for dementia disease is "the Cognitive Orthosis for Assisting Activities" (COACH) system [89]. This system was built to assist dementia patients through the hand washing procedures as a usual activity in their daily lives. This system is mainly based on AI techniques and computer vision which allows the monitoring of the actual stage of activity and then decides on delivering what verbal or visual prompts are necessary when dementia patients are washing their hands. This work has promoted the concept of "zero effort technologies," that is, technologies that require a minimal effort to no effort at all to operate, which motivated many systems that followed [90]. The common concept of these technologies is to carry out data collection, analysis, and the application of all necessary ICT (information and communication technologies) including computer vision, AI approaches, sensor fusion, and the IoT without being intrusive in people's lives. Recently, more advanced work is focusing on integrating AI models for individual emotion and identity analysis in those support systems to make them easily fit in the lives of users [91].

9.6.1.2 Smart homes to support dementia patients

The integration of AI and IoT-based support systems at people's homes created the concept of smart homes (see Fig. 9.10). These recent developments in AI and IoT made smart homes more advanced, by introducing novel concepts of support systems based on activities like cooking and dressing [92]. An early model of smart homes for people with dementia is the Gloucester Smart Home [93]. It consists of bath and cooker monitors, an automated nightlight system, objects locator, and a digital message board to deliver visual and verbal instructions. This model also used voice prompts to alert patients if some dangerous situations occur such as when cookers being left on or in the case of night-time wandering. This autonomous remote monitoring system was evaluated within a smart home installation in a care home to assess its impact on the life of someone with severe dementia. By analyzing data from sensors and questionnaires, this technology has shown promising results in improving the patient's independence. This smart home system helped the client regain urinary continence, improved his sleep by gaining 2 more hours of sleep per night, and also it reduced the number of night-time wanderings by 50% [94].

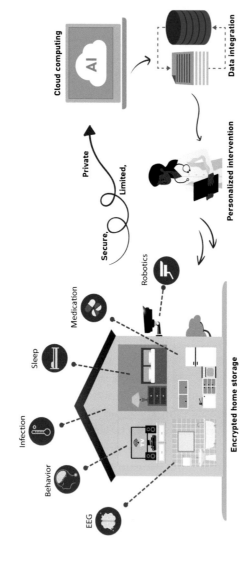

Figure 9.10 A general scheme showing the concept of smart homes for remote patient monitoring.

Nowadays, the advancements in AI algorithms have made these smart homes more intelligent by allowing more interactivity with the users. These complex AI algorithms allow the smart home components to learn the user's habits and update their programs to automatically adjust to people's needs. Human-like (in behavior and not in looks) robots are now the major focus of the research community, in fact in the future, it is expected that most smart homes will be equipped with artificially intelligent robots that can help people in many aspects of their daily life. In healthcare, Japan has already launched loveable and friendly nurse robots to take care of patients [95]. These smart robots are not only designed to help people in physical tasks, but they can also take care of the emotional well-being of the patient, an important dimension in life of people in need such as dementia patients, for example, the PARO robot [96] which is an interactive and lovely seal toy that provides its users with the benefits of animal therapy by comforting them and reducing stress.

9.6.2 Migraine

Migraine is a chronic neurological disorder characterized by persistent headache attacks accompanied by symptoms like nausea, vomiting, and high sensitivity to light and sound [97]. It is globally ranked as the third most prevalent disease and the seventh most disabling condition among neurological disorders. However, migraine is a multifactorial complex disease that can often be misdiagnosed due to its overlapping symptoms with other diseases such as epilepsy and tension headaches. In the last few years, several studies have attempted to address the challenge of migraine diagnosis identification where flash simulation technique has shown promising results. The concept of this method is to analyze the patient's neural responses to flash stimulation at different frequencies using a multichannel electroencephalography (EEG) where it was found that flash frequency at 4 Hz provides the most accurate results [98]. Moreover, support systems were also proposed for clinical diagnosis of tension headache and eventual migraine. They provided slightly accurate results based on performance metrics such as recall rate, precession rate, F-score, and accuracy [99].

There are several AI-based efforts for managing migraine via a mobile app or smart devices. AI-based approaches of Big Data made use of spontaneous data of migraine patients [100]. Wavelet-based features were used to identify migraine signals. Then, neural networks were applied to classify the EEG patterns into two categories: headache-free and during attack

[100]. Another study has proposed a decision support system that integrates ML models when classifying headache types [101]. This system provided accurate results in the classification of headaches into several types including tension-type headache and migraine headache. Although many challenges are faced in the field of migraine identification—mainly due to its symptomatic characteristic, it has been shown that new technologies incorporating AI and IoT can provide promising results in the presence of large amounts of data relative to migraine patients [102].

Recently, researchers in the migraine field are more focused on creating predictive mobile health applications that could alert patients of episodes of migraine. These AI-based mobile apps use different Big Data algorithms to build adequate predictive models. These models are more accurate when large datasets with various types of data are collected. For instance, an ongoing research study conducted by collaborators from Mayo Clinic, University of Southern California and Second Opinion Health (a digital health company) has initially resulted in a mobile health app called *MigraineAlert*. This app uses multivariate models from ML to develop personalized predictive models that can determine migraine attacks for an individual on a daily basis. The AI models in this app were trained on a large dataset where data about migraine symptoms, daily exposures, and triggers in addition to other physiological data (from Fitbit) were collected from the clinical trial conducted by the other collaborators [103,104].

There are many other mobile health mobile apps that offer quality solutions to help reduce migraine based on Big Data and ML algorithms, for example, Migraine Buddy and Manage My Pain Pro [105]:

9.6.2.1 Migraine Buddy
Using a diary style questionnaire, this tracking app was designed by experts in neurology and data science fields to help individuals track their potential everyday triggers and lifestyle factors that could lead to the migraine attacks. Through a user friendly interface, this app has shown to be an efficient remote monitoring tool for the doctors who can suggest various relief solutions based on the assessment of their patients condition (https://migrainebuddy.com/).

9.6.2.2 Manage My Pain Pro
This application uses data about migraine pain gathered from the users to help them track their symptoms by providing a smart personalized visualization of graph, charts, and calendar views for their migraine condition.

Based on the same concept, there are mobile health applications for migraine monitoring including Migraine Insight and Migraine Monitor.

There are other apps that are not mainly designed for migraine tracking but are more focused on pain relieving methods. Examples include Blue Light Filter and Night Mode app, which aim to minimize exposure to blue light that can be considered as a migraine trigger. Other examples provide mindful solutions through hypnosis and guided meditations using relaxing melodies such as *Relax Melodies: Sleep Sounds* or Waking Up by Sam Harris.

9.7 Conclusions

More and more people are affected by chronic conditions that require lifestyle and medication management and there is an increasing interest in remote monitoring solutions that can facilitate disease management, both for the patients and for the healthcare practitioners. Such monitoring tools allow users to collect health data, at their home or "on the go," and store or transmit this data to the healthcare professionals for clinical support. Here, we discussed a few notable applications of remote patient monitoring including glucose monitoring for diabetes, movement monitoring for Parkinson's, and ECG monitoring for heart failure, but there are many more applications that are already implemented or under development. Despite its enormous potential, limited evidence is available demonstrating improved health outcomes and cost benefit with remote patient monitoring interventions. It is believed that large-scale implementation is required to really document the benefits of such solutions across different sectors and populations in the healthcare ecosystem.

Advances in data science have led to the application of ML in nearly all areas of remote monitoring in the past few years and the results are encouraging. ML tools can be used to process the large datasets often collected from remote monitoring solutions including medical devices, wearables, and apps and can provide enhanced prediction, classifications, and decision support. Increased health outcomes are already observed for many remote monitoring applications including areas such as heart failure, migraine, and diabetes management, using AI-based tool. However, there are still numerous challenges to assess including adaption barriers, clinician training, data reliability, privacy and security, and integration into routine healthcare services. We believe that development of wearable technology and introduction of affordable software-based solutions will have a great impact on the future patient monitoring solutions and medical treatment strategies.

References

[1] Wickramasinghe N, John B, George J, Vogel D. Achieving value-based care in chronic disease management: intervention study. JMIR Diabetes 2019;4(2):e10368.

[2] Centre for Diseases Control. Available from: <https://www.cdc.gov/chronicdisease/about/index.htm>.

[3] Wickramasinghe N, Troshani I, Tan J, editors. Contemporary consumer health informatics. Springer; 2016.

[4] Wickramasinghe N. Essential considerations for successful consumer health informatics solutions. Yearb Med Inform 2019;28(01):158−64.

[5] Vegesna A, Tran M, Angelaccio M, Arcona S. Remote patient monitoring via non-invasive digital technologies: a systematic review. Telemed e-Health 2017;23 (1):3−17.

[6] Malasinghe LP, Ramzan N, Dahal K. Remote patient monitoring: a comprehensive study. J Ambient Intell Humanized Comput 2019;10(1):57−76.

[7] Albahri OS, Zaidan AA, Zaidan BB, Hashim M, Albahri AS, Alsalem MA. Real-time remote health-monitoring systems in a medical centre: a review of the provision of healthcare services-based body sensor information, open challenges and methodological aspects. J Med Syst 2018;42(9):164.

[8] Klersy C, De Silvestri A, Gabutti G, Regoli F, Auricchio A. A meta-analysis of remote monitoring of heart failure patients. J Am Coll Cardiol 2009;54 (18):1683−94.

[9] Gopalsami N, Osorio I, Kulikov S, Buyko S, Martynov A, Raptis AC. SAW micro-sensor brain implant for prediction and monitoring of seizures. IEEE Sens J 2007;7 (7):977−82.

[10] Yoo EH, Lee SY. Glucose biosensors: an overview of use in clinical practice. Sensors. 2010;10(5):4558−76.

[11] Rodbard D. Continuous glucose monitoring: a review of successes, challenges, and opportunities. Diabetes Technol Ther 2016;18(S2):S2−3.

[12] Chen C, Zhao XL, Li ZH, Zhu ZG, Qian SH, Flewitt AJ. Current and emerging technology for continuous glucose monitoring. Sensors 2017;17(1):182.

[13] Savarese G, Lund LH. Global public health burden of heart failure. Card Fail Rev 2017;3(1):7.

[14] Bashi N, Karunanithi M, Fatehi F, Ding H, Walters D. Remote monitoring of patients with heart failure: an overview of systematic reviews. J Med Internet Res 2017;19(1):e18.

[15] Ong MK, Romano PS, Edgington S, Aronow HU, Auerbach AD, Black JT, et al. Effectiveness of remote patient monitoring after discharge of hospitalized patients with heart failure: the better effectiveness after transition−heart failure (BEAT-HF) randomized clinical trial. JAMA Intern Med 2016;176(3):310−18.

[16] Block VA, Pitsch E, Tahir P, Cree BA, Allen DD, Gelfand JM. Remote physical activity monitoring in neurological disease: a systematic review. PLoS One 2016;11 (4):e0154335.

[17] Lonini L, Dai A, Shawen N, Simuni T, Poon C, Shimanovich L, et al. Wearable sensors for Parkinson's disease: which data are worth collecting for training symptom detection models. NPJ Digit Med 2018;1(6).

[18] DiMatteo MR. Variations in patients' adherence to medical recommendations: a quantitative review of 50 years of research. Med Care 2004;200−9.

[19] Hamine S, Gerth-Guyette E, Faulx D, Green BB, Ginsburg AS. Impact of mHealth chronic disease management on treatment adherence and patient outcomes: a systematic review. J Med Internet Res 2015;17(2):e52.

[20] Baig MM, GholamHosseini H, Moqeem AA, Mirza F, Lindén M. A systematic review of wearable patient monitoring systems—current challenges and opportunities for clinical adoption. J Med Syst 2017;41(7):115.

[21] Milenković A, Otto C, Jovanov E. Wireless sensor networks for personal health monitoring: issues and an implementation. Comput Commun 2006;29(13):2521−33.

[22] Deshmukh SD, Shilaskar SN. Wearable sensors and patient monitoring system: a review. In: International Conference on Pervasive Computing (ICPC). IEEE; 2015.

[23] Available from: <https://www.pewresearch.org/internet/fact-sheet/mobile/>.

[24] Vangeepuram N, Mayer V, Fei K, Hanlen-Rosado E, Andrade C, Wright S, et al. Smartphone ownership and perspectives on health apps among a vulnerable population in East Harlem, New York. mHealth 2018;4:31.

[25] Gibbons MC, Shaikh Y. Introduction to consumer health informatics and digital inclusion. Consumer Informatics and Digital Health. Cham: Springer; 2019. p. 25−41.

[26] Dall T, West T, Chakrabarti R, Iacobucci W. The complexities of physician supply and demand: projections from 2013 to 2025. Washington, DC: Association of American Medical Colleges; 2015.

[27] Available from: <https://www.aamc.org/>, <https://www.aamc.org/system/files/c/2/481964-newdocument.pdf>.

[28] Buerhaus PI. Current and future state of the US nursing workforce. JAMA 2008;300 (20):2422−4.

[29] Apple. Apple Watch Series 4. Available from: <https://www.apple.com/apple-watch-series-4/>.

[30] Goel M., Saba E., Stiber M., et al. Spiro-Call: measuring lung function over a phone.

[31] Coppetti T, Brauchlin A, Müggler S, et al. Accuracy of smartphone apps for heart rate measurement. Eur J Prev Cardiol 2017;24:1287−93.

[32] Saleheen N, Ali AA, Hossain SM, et al. PuffMarker: a multi-sensor approach for pinpointing the timing of first lapse in smoking cessation. Proc ACM Int Conf Ubiquitous Comput 2015;2015:999−1010.

[33] mpatica. Embrace2 seizure detection. Available from: <https://www.empatica.com/>.

[34] Statista. Wearable user penetration rate in the United States, in 2017, by age. Available from: <https://www.statista.com/statistics/739398/us-wearable-penetration-by-age/>.

[35] AliveCor. KardiaMobile. Available from: <https://www.alivecor.com/>.

[36] MC10. BioStamp nPoint: wearable healthcare technology & devices. Available from: <https://www.mc10inc.com>.

[37] Gao J, Ertin E, Kumar S, al'Absi M. Contactless sensing of physiological sig- nals using wideband RF probes. In: Forty-Seventh Asilomar Conference on Signals, Systems & Computers, Pacific Grove, CA, November 3−6, 2013. Piscataway, NJ: Institute of Electrical and Electronics Engineers; 2013. p. 86−90.

[38] McLaren R, Joseph F, Baguley C, Taylor D. A review of e-textiles in neurological rehabilitation: how close are we? J Neuroeng Rehabil 2016;13:59.

[39] Hafezi H, Robertson TL, Moon GD, Au-Yeung KY, Zdeblick MJ, Savage GM. An ingestible sensor for measuring medication adherence. IEEE Trans Biomed Eng 2015;62:99−109.

[40] Brooks GC, Vittinghoff E, Iyer S, et al. Accuracy and usability of a self-administered 6-minute walk test smartphone application. Circ Heart Fail 2015;8:905−13.

[41] Singh S, Xu W. Robust detection of Parkinson's disease using harvested smartphone voice data: a telemedicine approach. Telemed e-Health 2020;26::327−34.

[42] Sim I. Mobile devices and health. N Engl J Med 2019;381(10):956−68.

[43] Miller E, Polson D. Apps, avatars, and robots: the future of mental healthcare. Issues Ment Health Nurs 2019;40(3):208−14.

[44] Nadkarni PM, Ohno-Machado L, Chapman WW. Natural language processing: an introduction. J Am Med Inform Assoc 2011;18(5):544−51.

[45] Senders JT, Karhade AV, Cote DJ, Mehrtash A, Lamba N, DiRisio A, et al. Natural language processing for automated quantification of brain metastases reported in free-text radiology reports. JCO Clin Cancer Inform 2019;3:1−9.

[46] Peyrou B, Vignaux JJ, André A. Artificial intelligence and health care. Digital Medicine. Cham: Springer; 2019. p. 29−40.

[47] Dimitrov DV. Medical internet of things and big data in healthcare. Healthc Inform Res 2016;22(3):156−63.

[48] Cameron G, Cameron D, Megaw G, Bond R, Mulvenna M, O'Neill S, et al. Towards a chatbot for digital counselling. Proceedings of the 31st British Computer Society Human Computer Interaction Conference. BCS Learning & Development Ltd; 2017. p. 24.

[49] Sanchez-Morillo D, Fernandez-Granero MA, Leon-Jimenez A. Use of predictive algorithms in-home monitoring of chronic obstructive pul- monary disease and asthma: a systematic review. Chron Respir Dis 2016;13(3):264−83. Available from: https://doi.org/10.1177/1479972316642365.

[50] Iadanza E, Mudura V, Melillo P, Gherardelli M. An automatic system supporting clinical decision for chronic obstructive pulmonary disease. Health Technol 2019;1−12.

[51] Mills S. Electronic health records and use of clinical decision support. critical care nursing. Clinics 2019;31(2):125−31.

[52] Lilly CM, Zubrow MT, Kempner KM, et al. Critical care telemedicine: evolution and state of the art. Crit Care Med 2014;42(11):2429−36.

[53] Kindle RD, Badawi O, Celi LA, Sturland S. Intensive care unit telemedicine in the era of big data, artificial intelligence, and computer clinical decision support systems. Crit Care Clin 2019;35(3):483−95.

[54] Dowding D, Randell R, Mitchell N, et al. Experience and nurses use of computerised decision support systems. Stud Health Technol Inf 2009;146:506−10.

[55] Ernesäter A, Holmström I, Engström M. Telenurses' experiences of working with computerized decision support: supporting, inhibiting and quality improving. J Adv Nurs 2009;65(5):1074−83.

[56] Borum C. Barriers for hospital-based nurse practitioners utilizing clinical decision support systems: a systematic review. Comput Inf Nurs 2018;36(4):177−82.

[57] Darwish M, Senn E, Lohr C, Kermarrec Y. A comparison between ambient assisted living systems. In: International Conference on Smart Homes and Health Telematics. Cham: Springer; 2014. p. 231−7. June 25.

[58] Davis-Owusu K, Owusu E, Bastani V, Marcenaro L, Hu J, Regazzoni C, et al., Activity recognition based on inertial sensors for ambient assisted living; 2016.

[59] Chetty G, White M, Akther F. Smart phone based data mining for human activity recognition. Procedia Comput Sci 2015;46:1181−7.

[60] Ronao CA, Cho S-B. Human activity recognition with smartphone sensors using deep learning neural networks. Expert Syst Appl 2016;59:235−44.

[61] la Hoz-Franco ED, Colpas PA, Quero JM, Espinilla M. Sensor-based datasets for human activity recognition—a systematic review of literature. IEEE Access 2018;6:59192−210.

[62] Majumder S, Mondal T, Deen MJ. Wearable sensors for remote health monitoring. Sensors 2017;17(1).

[63] Syed L, Jabeen S, Manimala S, Alsaeedi A. Smart healthcare framework for ambient assisted living using IoMT and big data analytics techniques. Future Gener Comput Syst 2019;101:136−51.

[64] Available from: <https://www.idf.org/aboutdiabetes/what-is-diabetes/facts-figures.html>.

[65] Fagherazzi G, Ravaud P. Digital diabetes: perspectives for diabetes prevention, management and research. Diabetes Metab 2019;45(4):322—9.

[66] Dankwa-Mullan I, Rivo M, Sepulveda M, Park Y, Snowdon J, Rhee K. Transforming diabetes care through artificial intelligence: the future is here. Popul Health Manag 2019;22(3):229—42.

[67] HAYERI A. 922-P: Diabits—an AI-powered smartphone application for blood glucose monitoring and predictions.

[68] Interview of Diabits COO, Stephen Dodge. 27/01/2020.

[69] Available from: <https://www.globenewswire.com/news-release/2019/03/13/1752302/0/en/Premier-Health-Partnering-with-Bio-Conscious-Technologies-to-bring-AI-Powered-Remote-Patient-Monitoring-to-its-App.html>.

[70] Cvetković B, Janko V, Romero AE, Kafalı Ö, Stathis K, Luštrek M. Activity recognition for diabetic patients using a smartphone. J Med Syst 2016;40:256.

[71] DeJournett L, DeJournett J. In silico testing of an artificial-intelligence-based artificial pancreas designed for use in the intensive care unit setting. J Diabetes Sci Technol 2016;10:1360—71.

[72] ClinicalTrials.gov. Adult accuracy study of the Elite 3 Glucose Censor (E3). Available from: <https://www.clinicaltrials.gov/ct2/show/study/NCT02246582?term = NCT02246582&rank = 1&sect = X0123456>.

[73] Thabit H, Tauschmann M, Allen JM, et al. Home use of an artificial beta cell in type 1 diabetes. N Engl J Med 2015;373:2129—40.

[74] Henglin M, Stein G, Hushcha PV, Snoek J, Wiltschko AB, Cheng S. Machine learning approaches in cardiovascular imaging. Circ Cardiovasc Imaging 2017;10:e005614.

[75] Ciusdel C, Turcea A, Puiu A, et al. An artificial intelligence based solution for fully automated cardiac phase and end-diastolic frame detection on coronary angiographies. Paper presented at: Transcatheter Cardiovascular Therapeutics Symposium (TCT) 18; September 21—25, 2018; San Diego, CA.

[76] Molony D, Hosseini H, Samady H. Deep IVUS: a machine learning framework for fully automatic IVUS segmentation. Paper presented at: Transcatheter Cardiovascular Therapeutics Symposium (TCT) 18; September 21—25, 2018; San Diego, CA.

[77] Sardar P, Abbott JD, Kundu A, Aronow HD, Granada JF, Giri J. Impact of artificial intelligence on interventional cardiology: from decision-making aid to advanced interventional procedure assistance. JACC Cardiovasc Interv 2019;12(14):1293—303.

[78] Syeda-Mahmood T. Role of big data and machine learning in diagnostic decision support in radiology. J Am Coll Radiol 2018;15(Pt B):569—76.

[79] Project Brave Heart: studying the impact of virtual reality preparation and relaxation therapy. Lucile Packard Children's Hospital Stanford website, <http://www.stanford-childrens.org/en/innovation/virtual-reality/anxiety-research> [accessed 10.10.18].

[80] Improving visualization and interaction during transcatheter ablation using a mixed reality system: first-in-human experience. SentiAR Technology website, <https://sentiar.com/#!info-graph> [accessed 10.10.18].

[81] Steinhubl SR, Topol EJ. Now we're talking: bringing a voice to digital medicine. Lancet 2018;392:627.

[82] Jiang F, Jiang Y, Zhi H, et al. Artificial intelligence in healthcare: past, present and future. Stroke Vasc Neurol 2017;2:230—43.

[83] Johnson KW, Torres Soto J, Glicksberg BS, et al. Artificial intelligence in cardiology. J Am Coll Cardiol 2018;71:2668—79.

[84] He W, Goodkind D, Kowal PR. An aging world; 2015.

[85] Bohlken J, Schulz M, Rapp MA, Baetzing-Feigenbaum J. Pharmacotherapy of dementia in Germany: Results from a nationwide claims database. Eur Neuropsychopharmacol 2015;25(12):2333—8.

[86] Prince M, Comas-Herrera A, Knapp M, Guerchet M, Karagiannidou M. World Alzheimer report 2016: improving healthcare for people living with dementia: coverage, quality and costs now and in the future.

[87] Hamper A, Eigner I. Dementia monitoring with artificial intelligence. Contemporary Consumer Health Informatics. Cham: Springer; 2016. p. 53–71.

[88] Kaye JA, Maxwell SA, Mattek N, Hayes TL, Dodge H, Pavel M, et al. Intelligent systems for assessing aging changes: home-based, unobtrusive, and continuous assessment of aging. J Gerontol Ser B: Psychol Sci Soc Sci 2011;66(Suppl_1): i180–90.

[89] Mihailidis A, Boger JN, Craig T, Hoey J. The COACH prompting system to assist older adults with dementia through handwashing: an efficacy study. BMC Geriatr 2008;8(1):28.

[90] Mihailidis A, Boger J, Hoey J, Jiancaro T. Zero effort technologies: considerations, challenges, and use in health, wellness, and rehabilitation. Synthesis Lectures on assistive, rehabilitative, health-preserving technologies, vol. 1. Morgan & Claypool; 2011. p. 1–94.

[91] Robillard JM, Hoey J. Emotion and motivation in cognitive assistive technologies for dementia. Computer 2018;51(3):24–34.

[92] Burleson W, Lozano C, Ravishankar V, Lee J, Mahoney D. An assistive technology system that provides personalized dressing support for people living with dementia: capability study. JMIR Med Inform 2018;6(2):e21.

[93] Orpwood R, Gibbs C, Adlam T, Faulkner R, Meegahawatte D. The Gloucester smart house for people with dementia—user-interface aspects. Designing a more inclusive world. London: Springer; 2004. p. 237–45.

[94] Orpwood R, Adlam T, Evans N, Chadd J, Self D. Evaluation of an assisted-living smart home for someone with dementia. J Enabling Technol 2008;2(2):13.

[95] Available from: <https://nurse.org/articles/nurse-robots-friend-or-foe/>.

[96] Available from: <http://www.parorobots.com/>.

[97] Olesen J. The international classification of headache disorders. Headache J Head Face Pain 2008;48(5):691–3.

[98] Akben SB, Subasi A, Tuncel D. Analysis of repetitive flash stimulation frequencies and record periods to detect migraine using artificial neural network. J Med Syst 2012;36(2):925–31.

[99] Yin Z, Dong Z, Lu X, Yu S, Chen X, Duan H. A clinical decision support system for the diagnosis of probable migraine and probable tension-type headache based on case-based reasoning. J Headache Pain 2015;16(1):29.

[100] Akben SB, Tuncel D, Alkan A. Classification of multi-channel EEG signals for migraine detection. Biomed Res 2016;27(3):743–8.

[101] Krawczyk B, Simic D, Simic S, Wozniak M. Automatic diagnosis of primary headaches by machine learning methods. Open Med 2013;8(2):157–65.

[102] Subasi A, Ahmed A, Aličković E, Hassan AR. Effect of photic stimulation for migraine detection using random forest and discrete wavelet transform. Biomed Signal Process Control 2019;49:231–9.

[103] Available from: <https://clinicaltrials.gov/ct2/show/NCT02910921>.

[104] Available from: <https://www.somobilehealth.com/about.html>.

[105] Available from: <https://www.healthline.com/health/migraine/top-iphone-android-apps#blue-light-filter>.

CHAPTER 10

Security, privacy, and information-sharing aspects of healthcare artificial intelligence

Jakub P. Hlávka
Health Policy and Management Department of the Price School of Public Policy and Schaeffer Center for
Health Policy & Economics, University of Southern California, Los Angeles, CA, United States

This chapter addresses key topics in security, privacy, and information sharing related to the emergence of artificial intelligence (AI)-based solutions in healthcare. It addresses security and privacy concerns, including the risks and opportunities associated with emerging AI-based technologies in healthcare and potential challenges to its adoption due to regulatory interventions. It also discusses potential benefits and challenges associated with information sharing in an era of AI in healthcare, describing its value, technological, and other constraints, and shows how it can improve patient experience throughout the disease cycle. The chapter concludes with a discussion of desirable policy responses to security, privacy, and information-sharing challenges in healthcare AI applications around the world, as countries continue to introduce AI-based technologies [1] to their healthcare systems.

10.1 Introduction to digital security and privacy

Healthcare is a field awash with data, and digital security and privacy concerns are paramount given the sensitivity and personal nature of these data. Software aiming to integrate and analyze information from multiple sources and for different audiences is certain to face multiple privacy and security challenges, ranging from practical (making sure that data can be extracted, decrypted, and analyzed) to regulatory (ensuring that privacy of patients and healthcare providers is protected, and data are used for legal activities). New AI-based solutions add to the complexity of such challenges and as this chapter concludes, are simultaneously going to exacerbate some of the risks associated with electronic information exchange while increasing the clinical utility of patient-level information and the productivity of healthcare professionals.

Artificial Intelligence in Healthcare
DOI: https://doi.org/10.1016/B978-0-12-818438-7.00010-1

While discussions about data privacy, security, and data sharing are commonplace in public and private institutions, policy-makers tend to be a leap behind technological advances and have historically focused on privacy concerns in general. As a result, regulation is not very specific to novel technologies and solutions, posing both challenges for enforcement and technical standard development. In the United States, for example, the Health Insurance Portability and Accountability Act (HIPAA) Privacy Rule requires the protection of individuals' medical records and other "personal health information" recorded in any medium. [2] Entities covered by the Privacy Rule are "health plans, healthcare clearinghouses, and to any healthcare provider who transmits health information in electronic form" [3] and the protections defined in it are not technology-specific. Similarly, in the European Union, Regulation (EU) 2016/679 (also known as the "General Data Protection Regulation," or GDPR) covers "all data pertaining to the health status of a data subject which reveal information relating to the past, current or future physical or mental health status of the data subject" and requires explicit and unambiguous consent for the use and sharing of such data, among others. [4] It also provides patients with the right to access one's personal data and the "right to be forgotten."

In both the United States and Europe (and elsewhere in the world), enforcement of existing privacy rules is an uphill battle: identifying and documenting privacy breaches is as elusive as ever, with many successful breaches undetected by their targets for long periods of time, and attributing data privacy breaches to specific bad actors remains difficult. More specific to healthcare—existing regulation does not generally provide sufficient guidance on best practices for collection and transmission of personal information in healthcare, creating challenges for providers and innovators in an era of rapid innovation as they develop appropriate privacy control regimes that result in a sustainable balance between privacy, flexibility, and clinical utility.

The increasing diversity among healthcare providers adds to the complexity of protecting data security and privacy in health, no less because of the disruptive impact new technologies, including AI, have on the industrial organization. It is not uncommon for emerging health technology companies, ranging from e-Health and m-Health (electronic health and mobile health) services to genetic test providers, to emphasize data privacy and security in their public statements, only to be found later to use sensitive data in ways their consumers did not authorize or anticipate, such as by sharing information with law enforcement. [5] Fitness trackers, for example, are commonly not subject to health-specific privacy rules, despite collecting sensitive health information such as one's heart rate data, unless they transmit data to a covered entity. Finally, data privacy and security challenges continue to affect traditional healthcare providers who increasingly depend on electronic tools for data collection, storage, and analysis and are required to rapidly share information with other entities, ranging from medical notes to medical imaging data and laboratory results.

10.2 Security and privacy concerns in healthcare artificial intelligence

With rapid advances in information technologies, healthcare creates a fertile ground for future innovation due to the inefficiencies resulting from gaps in data sharing. While electronic medical records have been broadly adopted in most developed countries, it is not uncommon for hospitals to not be able to release records electronically when requested by a patient, not to include nursing notes in electronic form, [6] and for different healthcare providers to use incompatible systems that cannot reliably share documents with each other. This is driven not just by legal and technological challenges, but also the competitive nature of the market which discourages information sharing [7] In 2018, frequent fax usage was documented between hospitals in the United States, possibly due to technological hurdles and to achieve compliance with HIPAA's Privacy Rule, which considers fax an acceptable technology for personal data transfer. [8] The Centers for Medicare and Medicaid (CMS) Administrator has challenged doctor's offices to be fax-free by 2020. [9]

Digital tools are already catalyzing greater and more rapid information sharing within and among key stakeholders in healthcare, thus producing new opportunities as well as risks to privacy and data security. In addition, the use of AI-based tools (and thus greater reliance on electronic data capture, storage, transfer, and disclosure) may create new opportunities for bad actors to reidentify sensitive personal data or intercept sensitive data in transit using hacking techniques. Solutions to such challenges require a coordinated approach by policy-makers, healthcare providers, and innovators, starting with an agreement on core tenets of data privacy and security in healthcare and AI-specific regulation that ensures data privacy and security are protected while incentives for innovation and information sharing in healthcare are protected.

10.2.1 Defining privacy

Defining concepts such as privacy and security is far from straightforward. Privacy, for example, has been associated with the "control over and safeguarding of personal information" and thought of as "an aspect of dignity, autonomy, and ultimately human freedom". [10] Here, the focus on *personal* information, rather than *any* data, is vital: privacy concerns people, confidentiality concerns data. [11] In an economic sense, privacy is primarily considered in its informational dimension, as the cumulative "trade-offs arising from protecting or sharing of personal data". It is important to note that privacy is not opposite of data sharing, but in our context describes the level of control over what information is shared. Data sharing, thus, is the direct byproduct of a data privacy regime in which individuals and other entities are allowed to exchange information with predetermined constraints.

The pace of technological advances in healthcare often produces new solutions that are yet to be considered by policy-makers who develop and introduce privacy regulations, whether within a single institution or on a regional, national, or international level. Although countries in Europe and North America have traditionally had the strongest voice in data privacy regulation more generally, [12] more recently, countries like Brazil and China introduced new regulations addressing data privacy issues (both in 2018). While the Brazilian General Data Privacy Law mirrors Europe's GDPR, the Chinese National Standard on Personal Information Protection has been seen as even more onerous than GDPR itself. [13] The Chinese law requires data to be "de-identified before sharing or else have prior notice and consent from individuals" and imposes strict limits on secondary data use. In countries like China, however, enforcement of privacy laws is subject to greater uncertainty, given the emphasis on national security aspects of data protection and an opaque legal system. The United States lacks a comprehensive federal data privacy policy as new rules are being adopted in other parts of the world, putting the United States at a potential disadvantage in building data privacy-compliant innovation.

10.2.2 Privacy and data sharing

The nature of healthcare services requires personal information exchange between multiple stakeholders—in some cases public and private—and is fraught with challenges. Data sharing requires a certain level of "interoperability"—defined by the Healthcare Information and Management Systems Society (HIMSS) as

The ability of different information systems, devices or applications to connect, in a coordinated manner, within and across organizational boundaries to access, exchange and cooperatively use data amongst stakeholders, with the goal of optimizing the health of individuals and populations. [14]

In the United States, the position of the National Coordinator for Health Information Technology was formed in 2004 with an initial task of "developing a nationwide interoperable health information technology infrastructure". [15] The National Coordinator published the first *Nationwide Interoperability Roadmap* in 2015 which generates high-level objectives for a "learning health system"—"an ecosystem where all stakeholders can securely, effectively and efficiently contribute, share and analyze data". [16] With respect to privacy, the US-focused *Roadmap* emphasizes the need to "protect privacy and security in all aspects of interoperability and respect individual preferences" and demands that "federal and state privacy and security requirements that enable interoperability" be clarified and aligned.

In the European Union, the European eHealth Interoperability Roadmap was published in December 2010 but contains few specific recommendations for data privacy and security. [17] In April 2019, European Union's High-Level Expert Group on Artificial Intelligence published its report, *Ethics Guidelines for Trustworthy AI*, which generally recommend that "AI systems must guarantee privacy and data protection throughout a system's entire lifecycle," that "the integrity of the data must be ensured," and that "data protocols governing data access should be put in place". [18] However, little detail is provided regarding the actual implementation, division of responsibilities, and oversight at the EU level, leaving a lot of room to define privacy- and information sharing-related regulations to member states.

10.2.2.1 Toward data interoperability in healthcare

Interoperability has a critical function in healthcare, enabling a timely and accurate exchange of data and services between key stakeholders. Given the unique characteristics of healthcare systems, however, varying degrees of healthcare interoperability have been achieved around the world. In Estonia, for example, 99% of health data and prescriptions have been digitized, and e-Health records are accessible to authorized individuals [Estonia uses an electronic ID card and a Keyless Signatures Infrastructure (KSI) blockchain technology to ensure data integrity and mitigate threats] and are used in a secure way to collect national-level statistics. [19] In turn, the US healthcare system suffers from significant fragmentation, with limited data exchange between stakeholders despite the fact that the vast majority of physicians have adopted electronic health records (EHR). [20] It is unlikely that a new solution, blockchain included, will provide an easy fix in a healthcare ecosystem with a great number of stakeholders, technological standards, and complex privacy and reporting requirements.

Monetary and other costs are associated with both the sharing of inappropriate information (such as nonessential data that could be used for discrimination) and the sharing of information with the wrong party, such as the general public. Conversely, the lack of information sharing may lead to information asymmetry and suboptimal decision-making, including in life-saving situations. As a result, *no* privacy and *perfect* privacy are two extremes that are rarely desirable when more than one party is involved. In addition, there are both positive and negative externalities that stem from data sharing: aggregated and deidentified data may be used to detect important health and other trends, for example. On the other hand, costs of protecting one's privacy increase when data sharing is required to receive specific services and negative privacy externalities are more prevalent when increasing volumes of data are collected and shared.

Data interoperability has many forms and varying implications for data privacy, as noted in Table 10.1.

There are important implications for future interoperability efforts in healthcare in light of emerging AI solutions. First, AI will require greater interconnectivity between systems and datasets (foundational interoperability), increasing opportunities for data intercept and theft, and will naturally require standards for data exchange, possibly making data and network integrity attacks easier to conduct. As a result, solutions that utilize AI and result in better foundational and structural interoperability must be accompanied by safeguards against intercept, theft, and alteration in the process. Furthermore, greater semantic and organizational interoperability will motivate attacks using system and operator weaknesses, making investments in network upgrades, rigorous risk assessment, and proper training of a broad range of personnel even more urgent than today.

In December 2018, the US Department of Health and Human Services (HHS) released its "Health Industry Cybersecurity Practices (HICP): Managing Threats and Protecting Patients" publication. [21] Drawing on the work of a specialized Task Group, it recommends 10 practices to mitigate cybersecurity threats in healthcare, ranging from e-mail and endpoint protection systems to

Table 10.1 Interoperability implications for data privacy.

Type of interoperability	Key focus	Example data privacy and integrity threats
Foundational interoperability	Establishing interconnectivity requirements for two systems to exchange data	Data intercept and theft
Structural interoperability	Defining the structure or format of data exchange to preserve data purpose and meaning	Data and network integrity attacks
Semantic interoperability	Developing ability to exchange and interpret information using data codification and standardization	Decryption and reidentification attacks
Organizational interoperability	Developing technical, policy, and organizational means to achieve timely and efficient data exchange	Network integrity attacks, unauthorized data use

[a]Health Information Systems Management Society. HIMSS redefines interoperability. Retrieved March 5, 2020, from: https://www.himss.org/resource-news/himss-redefines-interoperability; 2019. *Source*: HIMSS,[a] as adapted by author.

better medical device security. [22] A 2019 *Roundtable on Sharing and Utilizing Health Data for AI Applications*, cohosted by the Center for Open Data Enterprise (CODE) and the Office of the Chief Technology Officer (CTO) at the HHS, has identified challenges associated with AI applications in healthcare, including inconsistent restrictions on data use and overly restrictive interpretations of HIPAA, and recommended several relevant solutions, such as providing guidance on deidentification methods to protect data privacy and security, developing credentialing systems for controlled access to sensitive health data, and developing specific guidelines for entities not covered by HIPAA. [23] AI is highly likely to increase the exposure to different data privacy and integrity risks that various healthcare stakeholders will face.

In March 2020, the *21st Century Cures Act: Interoperability, Information Blocking, and the ONC Health IT Certification Program Final Rule* was published by the Office of the National Coordinator for Health Information Technology, aiming to standardize the format for data exchange and reduce data blocking. [24] Within 36 months after publication, it requires compliance of health IT developers with specific provisions related to data interoperability and aims to prevent information blocking. It is expected that the new interoperability standards will increase the utility of smart-phone and other digital tools given the ability to draw on accessible and standardized application programming interfaces (APIs) and to link claims, demographic, and other relevant data together. Yet, challenges to data security and privacy in healthcare will remain acute.

10.2.3 Safety and security in an era of emerging technologies in healthcare

Data safety breaches have been reported by most industrial sectors in recent years, ranging from banking to entertainment and from manufacturing to healthcare. A 2019 analysis has indicated the average cost per stolen record amounted to $150 and the mean time to identify a data breach was 206 days, with a mean time to contain it of additional 73 days—both experiencing an increase from the previous years. [25] In healthcare, the mean time to identify a data breach was 236 days and 93 days to contain (both highest in the study). Moreover, healthcare was reported to have the largest per-capita cost of a breach of any sector, amounting to $429 (the second highest, financial sector, amounted to $210 per capita)., [26]

Another study found that in the United States, a slight majority of data breaches (52%) was attributed to malicious actors, with 24% attributed to a system glitch and another 24% to human error (in all countries studied, malicious actors represented the highest share of data breaches, ranging from 40% in Italy to 59% in the Middle East). [27] An even higher share of data breaches (81%) was attributed to hackers in a 2016 study. [28] This shows the vulnerability of the sector to serious and consequential threats associated with data privacy and security. Aside from economic harm (e.g., denial of insurance coverage, job loss), it has been argued that reidentification of sensitive data may be associated with psychosocial harm (such as social stigma and difficulty in obtaining employment). [29]

Another report published in 2018, the "Verizon Data Breach Investigations Report (DBIR)" presents an analysis of 53,000 incidents and 2,200 data breaches, with 536 breaches (24%) reported in healthcare alone. [30] Healthcare has reported the highest number of incidents due to error, malware, and hacking, while data breaches were most commonly caused by error, misuse, and hacking, and is the only industry in the analysis that reports more internal (56%) than external (43%) actors behind data breaches (the remainder is caused by partners or multiple parties). Misdelivery—sending information to an unintended recipient—accounts for most of the errors (62%) reported. Most frequently, data compromised in healthcare are medical (79%), personal (37%), and payment-related (4%). The primary motive is financial (75%), followed by "fun" (13%), convenience (5%), and espionage (5%). The report itself highlights the unique challenges healthcare data are subject to: they are highly sensitive and can easily be abused, yet timely access to accurate and up-to-date data is vital, especially in emergency situations. The report recommends full-disk encryption as an effective, low-cost strategy to reduce data breaches in healthcare, in addition to developing policies for the monitoring of internal protected health information access and minimizing the potential impact of ransomware on any given network.

10.2.4 Measures to protect sensitive data in healthcare

The threats discussed above are not unique to AI technologies, although it is generally agreed that a greater use of online tools and databases, including reliance on solutions in the cloud—some of which may fall under AI—and ever-increasing and cheaper computing power, will increase frequency and complexity of threats and data breaches in healthcare and other sectors. To counter these threats, multiple efforts are underway to protect sensitive data, using both traditional techniques and, more recently, AI-based solutions.

10.2.4.1 Traditional techniques

Typically, three types of tools are deployed to ensure digital data privacy and security: (1) masking measures, (2) disclosure risk measures, [31] and (3) encryption.

Masking measures have different characteristics and may or may not lower the quality of data in question. These measures can be (1) perturbative, (2) nonperturbative, or rely on (3) synthetic data generators. *Perturbative* methods are based on slight alterations to data, controlling information loss by parameters, and may include record swapping and adding random noise. [32] Such technique has been used by companies like Apple and their *local differential privacy* solutions to generate insights based analysis of sensitive personal data—these are altered by adding statistical noise to sensitive data before they leave a user's device (also called data perturbation)—hence, preventing future attempts to reproduce the original data by masking them. [33] Data perturbation is used by the National Health Interview Survey in the United States, for example. [34]

In contrast, *nonperturbative* methods such as cell suppression and global recording do not change the data, but rather collapse certain variables—leading to information loss. These have several disadvantages, including potentially making time-series analysis worthless. Synthetic data generators are useful in recreating datasets that are statistically similar but do not contain real-world observations and can use machine learning (ML) to produce synthetic equivalents of underlying datasets. [35] These tools are currently under development given their promise to ensure a high degree of data protection while increasing utility of future data analyses, and applications drawing on Generative Adversarial Networks that create synthetic data are being developed for health data generation (e.g., using Medical Generative Adversarial Network, or medGAN [36], image-to-image translation, medical imaging, anomaly detection in drug and diagnostic research, and other applications. [37] A general rule, however, holds: the more complex the original dataset, the more difficult it will be to produce statically comparable data, often requiring substantive computing power and in many cases, accepting a decrease in quality.

Disclosure measures address two types of hazard: attribution disclosure and identity disclosure. [38] Attribution disclosure pertains to a specific attribute of an individual (such as their infectious disease status) while identity disclosure to their identification in a data file. Masking is typically preferable to encryption: the latter keeps data inaccessible outside of a safe environment, limiting medical or healthcare research applications. [39] Hence, encryption is typically used for data in transfer while masking is used when data are shared with external stakeholders.

10.2.4.2 Emerging techniques

Active research is underway with respect to the role of AI, including deep learning, as regards data security and privacy. [40] Secure and private AI (SPAI) as a concept was first presented in late-2018. Its two individual components are presented in Table 10.2.

Techniques used today to ensure privacy of data in AI applications are of particular interest. Both Fully Homomorphic Encryption (HFE) and Differential Privacy approaches have been tested in attacks targeting private data. While FHE allows any computable function to be performed on the encrypted data directly, Differential Privacy adds noise to query responses, making deductions with high confidence unlikely. While FHE is seen as particularly strong, its computational demands are still a large burden in most settings. A new approach termed "anonymized generative adversarial networks" (AnomiGAN) is being tested, which could allow

Table 10.2 Secure and private AI.

Component	Focus	Threat types	Attack techniques used	Defense techniques used
Secure AI	Attacks and defense within AI systems (models)	• Subversion of a learning process • Inductions of false predictions (e.g., adversarial sample injections)	*Subversion*, such as gradient-based techniques (white-box attacks) *False predictions* (black-box attacks)	• Adversarial training • Gradient masking • GAN • Statistical approaches
Private AI	Preservation of data privacy, including in transfer	• Data theft • Data misuse	*Model inversion* *Membership test* *Data integrity attacks* *Privacy violations* in information silos *Attacks on clouds* storing private data	Privacy-preserving techniques, such as • Fully homomorphic encryption • Differential privacy • Secure multiparty computation (SMC)

[a]Bae H, Jang J, Jung D, Jang H, Ha H, Yoon S. Security and privacy issues in deep learning. arXiv Preprint arXiv 2018;1807.11655.
Source: Adapted from Bae et al. (2018).[a]

users to control the "level" of anonymization and could allow for the anonymization of genomic data, for instance.

10.3 Artificial intelligence's risks and opportunities for data privacy

While there are many benefits of using AI for the protection of data privacy and integrity, new AI-based tools also produce negative externalities in the form of susceptibility to data privacy and integrity attacks and a more limited level of control over data ownership. We discuss both AI's risks and opportunities for data privacy, as shown in Table 10.3.

Table 10.3 Artificial intelligence-related risks and opportunities for healthcare data privacy.

AI's risks for data privacy	AI's opportunities for data privacy
• Susceptibility to data privacy and integrity attacks	• Improving data utility
• Reidentification of patients using AI	• Better attribution of privacy violations
• Data integrity and bias	• Improving physician productivity and data sharing
• Inadvertent disclosure	
• More limited level of control over data ownership and handling	
• Less intuitive understanding of one's data privacy	
• Use of sensitive health data outside of healthcare	
• Privacy externalities	

10.3.1 Artificial intelligence's risks for data privacy

10.3.1.1 Susceptibility to data privacy and integrity attacks

The introduction of AI technology to healthcare increases the number of opportunities for data privacy-related attacks. The dominant risks include the reidentification of patients using AI, data integrity and bias attacks, and inadvertent disclosure.

10.3.1.1.1 Reidentification of patients using artificial intelligence

AI is likely to increase the likelihood of successful reidentification (deductive disclosure) of patients in anonymized datasets. This was shown, for example, by Liangyuan et al. (2018) who successfully reidentified physical activity data of more than 90% adult individuals using support vector machines and a random forest method (one of the most successful and accurate ML techniques). [41] It has even been argued by some that AI advances have made standards like HIPAA obsolete, given AI's capacity in overwhelming deidentification requirements and HIPAA's focus on healthcare providers. This, however, does not imply that no regulation of data privacy is sufficient: in contrast, a more comprehensive, data-focused privacy regulatory framework should be developed.

Consider a dataset consisting of entries from multiple sources: a hospital, an insurer, and a government agency. While today's anonymization using traditional methods may be considered adequate, ML could be potentially used to infer—with a high degree of confidence—that individual patients are by combining health and other data (such as social media site information). Combined with traditional network threats, including network intrusions, new algorithms pose a potent threat to data privacy and integrity in healthcare and beyond.

Aside from new regulations, which should address the handling of sensitive information by nontraditional healthcare providers, among others, advanced techniques to address data privacy threats can make the use of AI-based analytical and research tools safer and ultimately expedite their adoption. Such techniques include masking measures, disclosure risk measures, and encryption, as described earlier.

10.3.1.1.2 Data integrity and bias

While AI-based tools are more adept at overcoming the weaknesses of existing causal inference methods, they may be susceptible to inductive (learning) bias. For example, a lack of demographic diversity in learning datasets may affect the accuracy and utility of models using ML, and must be accounted for. [42] In addition, using disparate data sources may produce skewed results where data quality is affected in other ways, such as due to a sampling or training data bias. [43] It is incumbent upon policymakers around the world to develop ethical guidelines and rules for data privacy and equity alike, similar to the nonbinding guidelines published by the European Commission in April 2019.

ML may also sidestep the protections of antidiscrimination laws, such as the Americans with Disabilities Act, by using inferences about predispositions to disability rather than observed disability to socially sort individuals. [44] New regulations should take the risk of such inferences into account, and preventing ML and other AI-based methods from putting any social group into disadvantage.

10.3.1.1.3 Inadvertent disclosure

Significant gaps exist in the existence of disclosure protocols in the United States and other countries. [45] Both routine and nonroutine (augmented or happening outside of regular operations) disclosures pose data privacy and integrity breach risks. It has been suggested that more rigorous protocols must be developed by relevant stakeholders to verify the following:

- Appropriateness of the disclosure (whether authorized by law or specific policy);

- Integrity of the information/data being disclosed (whether it still accurately reflects the underlying information once disclosed);
- Identity of the person receiving the disclosed information (whether they are authorized to receive the information);
- Security of the mode of information transmission (whether data transfer itself is secure).

Healthcare and other valuable data are at an increased risk of being maliciously targeted by bad actors who often use a variety of methods to deceive data owners or managers. A number of these, such as ransomware, malware, phishing, and denial-of-service attacks—have been well-documented. [46] The use of AI-based solutions in healthcare will create additional vulnerabilities and increasingly advanced defensive methods will be needed.

10.3.1.2 More limited level of control over data ownership and handling

With respect to both publicly and privately held healthcare data, it is clear that greater automation and the use of AI-based solutions will improve the utility and value of existing datasets, but potentially also lead to reduced control over data ownership and data handling (including storage, access, sharing, archival, and disposal). Concerns surrounding the use of healthcare data by private entities, such as IT and social media companies, have been raised around the world. [47]

Cases of data integration in the real world have already led to breaches of privacy protections. In the United Kingdom, DeepMind—an Alphabet company—obtained data of more than 1.6 million patients in violation of the local Data Protection Act, according to the Information Commissioner's Office (ICO), a government privacy watchdog. [48] Another privacy breach involving unconsented voice recognition of more than 7 million records was identified by ICO in May 2019, involving the government's tax and customs authority, Her Majesty's Revenue and Customs (HMRC). HMRC allegedly collected millions of voiceprints without consent. [49] These examples indicate that both public and private stakeholders may find themselves at odds with privacy protections, especially when novel digital technologies are used, and better regulation and enforcement will be required to protect data privacy and security.

While solutions are being developed to access data remotely without the need for it to leave its secure storage (under the umbrella of "data virtualization"), these do not completely eliminate risks to data privacy and

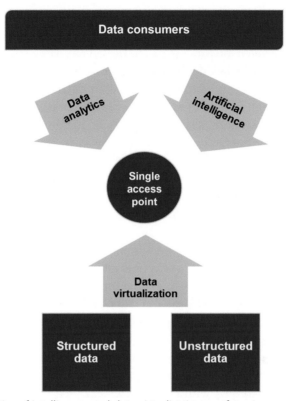

Figure 10.1 Use of intelligence and data virtualization use for privacy.

integrity. Nonetheless, it has been argued that data virtualization is beneficial for the handling of structured and unstructured data [50] and may improve both conventional data analytics and emerging AI-based applications (schematically shown in Fig. 10.1). [51]

10.3.1.3 Less intuitive understanding of one's data privacy

ML takes advantage of both commonly collected and self-disclosed data about individual patients and will make it more difficult to understand "what others can know about them based on what they have explicitly and implicitly shared" A more complex data environment presents unique challenges as patients may be left in the dark about how their private information is used beyond the immediate clinical setting. Technical and regulatory controls should be implemented that require providers and data aggregators to disclose and obtain explicit consent for any nonessential use of sensitive personal information and allow patients to selectively approve data use for different purposes without compromising their access to receive timely and comprehensive care. Achieving the ultimate objective—empowerment of patients to choose which data can be shared for what purposes—will also depend on patients' ability to understand complex privacy regimes and distinctions between different levels of data disclosure and use (in both clinical and nonclinical settings). To achieve this, patient-friendly solutions will be required, in addition to a greater need for public and private stakeholders to engage in increasing public awareness of emerging tools and their consequences for privacy.

10.3.1.4 Use of sensitive health data outside of healthcare

Public perceptions on the dissemination of sensitive healthcare data, such as genetic information, vary widely by survey and application. A 2018 survey studying the use of direct-to-consumer genetic testing data in police investigations found that 91% of respondents approved of using genealogical websites that match DNA to relatives to identify perpetrators of violent crimes (this survey was fielded after such technique was used to identify a murder suspect at large in California) [52] while a 2009 survey of US military veterans found that only 13% approve of sharing genetic data with law enforcement. [53] Another survey of genetic testing customers found that 96% were against sharing of genetic data with insurers and employers, and 89% with law enforcement. [54] It is likely that as AI is introduced, better data linkages are introduced, making the use of healthcare data more desirable by commercial entities as well as law enforcement. Without protections in place that ensure informed consent or a judicial order is obtained prior to using healthcare data for nonhealth purposes, data collection and integration would be hindered, further restricting the benefits of AI healthcare applications.

10.3.1.5 Privacy externalities

Similarly to other fields where data is collected on a massive scale, healthcare is subject to "privacy externalities"—potential inferences about individual patients by studying increasingly rich data about other similar patients. [55] While seemingly innocuous, ML will accelerate the ability of providers, insurers, and other entities to make ever-more precise predictions about individual patients, and thus undermine their privacy in ways that did not use to be possible just in the recent past. It has been argued by some that a privacy model based on *notice and consent* may not be sufficient, and rules should be in place to protect individuals against privacy-violating membership inferences (e.g., whether an individual is a member of a sensitive dataset), [56]

10.3.2 Artificial intelligence's opportunities for data privacy

Although AI-based tools are expected to increase challenges to data privacy protection, they may also create new opportunities that ultimately foster greater privacy in healthcare. Such benefits are discussed next, ranging from increased utility of data with expanded privacy protections to increased access to healthcare, especially among marginalized populations.

10.3.2.1 Improving data utility

While it has been suggested that a dichotomy exists between data privacy and data utility, the two are not necessarily incompatible. [57] New methods such as homomorphic encryption and federated learning can access and analyze data in disparate locations and without the need for centralized access. Federated learning reduces the risk of data integrity and safety breaches by only exchanging and updating data models based on raw data, which never leave their secure repository where models are trained and retrained. It will require, however, data standardization and sufficient computing infrastructure at individual sites, such as hospitals, to realize full benefits of such techniques.

AI tools offer advantages to standard causal inference methods, which are difficult to apply when data are both qualitative and quantitative, and often unstructured. [58] It is expected that this may have several benefits: (1) radically improving diagnosis by drawing on complex EHR databases, rather than select variables, (2) expediting image recognition and reducing workload for radiologists and pathologists, and (3) improving diagnostic accuracy by eliminating medical errors and recommending most suitable testing.

Similarly, AI may more efficiently deal with missing data (relative to standard estimations) and is thus highly advantageous in healthcare applications where incomplete data are common at the patient level. Several reasons for this exist: AI-based tools may have access to a broader set of confounding variables by integrating multiple data sources in a secure way, thus improving their ability to identify important causal patterns (this, however, poses privacy concerns). Second, ML algorithms may use various available data to create proxies where missing data would prevent conventional regression-based analysis. Third, AI tools may be more flexible in rapidly applying different methods to existing data in search for the most robust model, thus increasing the value of its outputs. In addition, ML techniques are being developed to identify and adjust for measurement error often contained in healthcare data (e.g., metabolic data accuracy issues).

For example, new research on "confidential machine learning" suggests future data analysis may be conducted on encrypted data, possibly even in the cloud using neural networks, [59] while still producing highly useful results. [60] This will likely require significant computing power and result in complexities related to data ownership and access, all of which are yet to be resolved.

In 2018, the National Academy of Medicine's "Strategic Initiative on Health Data as a Core Utility for the Common Good" was launched in the United States, expected to produce a special publication that describes "the promise, development, and deployment of AI/ML models to improve health." One of its intended chapters focuses on regulatory and policy considerations for the use of AI in healthcare. [61]

10.3.2.2 Better attribution of privacy violations
Across industries, the use of digital technologies and more recently, ML-based tools, produces additional complexities in attributing data and privacy breaches to individual actors. Similarly, a distinction in different levels of culpability—from intentional to inadvertent—may be required in new regulations to reflect the complicated nature of using powerful, AI-based tools which may weaken patients' privacy with varying degrees of human oversight and based on the level at which it is applied. At the same time, greater benefits due to more readily available data using AI-based tools may outweigh some of their costs, including limited privacy loss, and may be permitted to some extent under future regulatory systems.

10.3.2.3 Improving physician productivity and data sharing
Among the largest potential benefits of applying AI tools to medicine is the opportunity for physicians to spend more time with the patient rather than in data entry. [62] This could be done by autopopulating some data fields and transcribing voice to text, thus giving physicians more face-to-face time with a patient. This, potentially, could have significant benefits on lowering physician burnout but also increase physician productivity and lead to better data sharing among different providers. It is difficult to quantify the potential impact of such advances, but estimates suggesting

that physicians spend over half of their time on data entry into EHR during and after clinic hours suggest productivity gains and data sharing implications of utilizing AI tools could be large. [63]

New tools will not be immune to data breaches experienced by current medical transcription platforms, such as a 2017 data breach involving 45,000 patient records in Massachusetts, [64] and advanced solutions to protect the confidentiality of data will be required to realize the full benefits of AI tools on physicians' productivity without violating patient privacy. [65] In addition, there are additional challenges related to the understanding that physicians have about data and results originating in AI-based tools which may require additional training and pose ethical dilemmas. [66]

10.4 Addressing threats to health systems and data in the artificial intelligence age

It has been argued that the health sector is less prepared than other critical infrastructure domains such as finance to face digital threats, and efforts have been underway to catch up with emerging threats all around the world. [67] The cost implications of data breaches range from lost business to HIPAA fines, lawsuits, and forensics. [68] The requirement to adequately invest in prevention will only increase as more data becomes available and databases connected using AI-based tools.

Around the world, countless incidents of cyberattacks on healthcare systems are reported every year, with major cases including a 2018 WannaCry ransomware attack on multiple institutions, including English and Scottish National Health Service hospitals, paralyzing 200,000 computers and producing losses of up to £100 million in the United Kingdom alone (of which £19 million was due to lost output). [69] The Department of Health and Social Care has pledged to invest £150 million in safeguarding its systems between 2016 and 2019.

Addressing health system's vulnerability to data breaches, theft and misuse pose a significant challenge, given the number of stakeholders involved, the complexity of the digital health ecosystem, and the lack of awareness of data security best-practices and threats. It has been suggested that better data protection could prevent hundreds of billions of dollars in losses [70] and that outdated technology software is a common vulnerability of healthcare institutions globally, including in the United Kingdom during the 2018 attack. [71]

10.4.1 Addressing cyberattacks and data breaches in healthcare

Increasingly, cyberattacks are targeting electronic databases that hold sensitive health-related information. In the United States, this was first generally acknowledged by the Presidential Decision Directive (PDD) 63, resulting in the formation of Information Sharing and Analysis Centers (ISACs). One of these—the National Health ISAC, later renamed as Health Information Sharing and Analysis Center (H-ISAC) to reflect its international membership, is a global platform for health sector stakeholders to share information on cyber and physical threats to sensitive health data. [72] Despite efforts to share best practices globally, however, significant vulnerabilities to data privacy and security exist in all healthcare systems around the world. AI tools make defense against threats in the cyber space more complex given its reliance on complex data sources and predictive methods. In turn, techniques such as deep learning are used for adversarial sample detection, malware detection, and network intrusion detection and help defend systems against attacks that humans may not detect themselves. [73] A short discussion about deep learning and its implications for data privacy is included in the Box.

Deep learning and data privacy

Deep learning is one of several common ML methods, relying on artificial neural networks to train its own capabilities, which are then used for inference from raw data. Given the requirement for deep learning to comb through large amounts of data, encryption of sensitive information may create delays in algorithm processing, limiting its deployment in areas like healthcare. Deep learning algorithms are susceptible to common attack types, such as adversarial perturbation (caused by their linear nature), and may be more difficult to interpret than conventional risk calculations, for example. Conversely, deep learning is also used to improve system security, such as by detecting network intrusions. AI still needs refinement in this area, however: it suffers from a high false-positive incidence, a lack of generalizability, and has significant infrastructure demands due to growing network speeds and sizes. [74]

10.4.2 Government regulation of security and privacy in healthcare

AI's benefits do not apply to resource-rich environments only. Starting in 2017, the International Telecommunications Union, a United Nations agency, has convened the *AI for Good Global Summits*, addressing the applications of AI in multiple social domains, including healthcare and the challenges related to data privacy and data protection. [75]

Governments may choose to pursue various levels of control and over-sight over the collection, use and exchange of sensitive healthcare data in the age of AI. [76] In the United States, for example, it has been suggested that it was more effective to regulate the use rather than collection and analysis of sensitive data and that new policies should focus on defining their own purpose, rather than prescribing specific technical mechanism. [77] In addition, education and awareness building are highly describe among the general public and healthcare workers alike, given how many social sectors data privacy and security issues permeate.

Moreover, expanding the protections granted by privacy regulations (such as HIPAA) has been proposed given the growing diversity of service providers using sensitive personal data. [78] A failure to do so could result in adverse effects, including AI-enabled discrimination using one's disability status, for example. We discuss desirable responses to AI's proliferation with respect to privacy protection by stakeholder below.

10.5 Defining optimal responses to security, privacy, and information-sharing challenges in healthcare artificial intelligence

As new applications of AI are developed in healthcare, new opportunities and risks for data security, privacy, and information sharing emerge. Different stakeholder will be faced with a need to adopt strategies that minimize AI's risks and maximize its opportunities, possibly relying on decision-makers who have a limited understanding of AI's technical foundations. In this section, we discuss key recommendations based on existing scholarship and analysis presented in this chapter.

There is no unanimous agreement on an optimal response to AI applications in healthcare, although several high level for a have been convened. For instance, the OECD adopted the Recommendation of its Council on Artificial Intelligence in May 2019, outlining key principles "for responsible stewardship of trustworthy AI" [79]:

- Inclusive growth, sustainable development and well-being,
- Human-centered values and fairness,
- Transparency and explainability,
- Robustness, security and safety, and
- Accountability.

Additional considerations should be given to Elinor Ostrom's research on governing the commons, [80] which may apply to the emerging "medical information commons" [81]:

1. Clearly defined boundaries: Individuals or households who have rights to withdraw resource units from the common-pool resources (CPRs) must be clearly defined, as must the boundaries of the CPR itself.

2. Congruence between appropriation and provision rules and local conditions: Appropriation rules restricting rime, place, technology, and/or quantity of resource units are related to local conditions and to provision rules requiring labor, material, and/or money.

3. Collective-choice arrangements: Most individuals affected by the operational rules can participate in modifying the operational rules.

4. Monitoring: Monitors, who actively audit CPR conditions and appropriator behavior, are accountable to the appropriators or are the appropriators.

5. Graduated sanctions: Appropriators who violate operational rules are likely to be assessed graduated sanctions (depending on the seriousness and context of the offense) by other appropriators, by officials accountable to these appropriators, or by both.

6. Conflict-resolution mechanisms: Appropriators and their officials have rapid access to low-cost local arenas to resolve conflicts among appropriators or between appropriators and officials.

7. Minimal recognition of rights to organize: The rights of appropriators to devise their own institutions are not challenged by external governmental authorities.

8. Nested enterprises (when resources are parts of larger systems): Appropriation, provision, monitoring, enforcement, conflict resolution, and governance activities are organized in multiple layers of nested enterprises.

While such principles present an important starting point, it is clear that domain-specific principles for AI must be developed (especially in highly specific areas like healthcare), and specific attention given to security, privacy, and information-sharing challenges.

In Fig. 10.2, key stakeholders and their select functions with respect to data privacy, security, and sharing are highlighted under a proposed "Privacy-Centered Information Sharing Model for Healthcare" (PCISM).

In this model, the four key stakeholder groups play both shared and distinct privacy and security roles:

• *Governments* (and other policy-makers) are the ultimate standard setters, and in some cases (e.g., the European Union), have far-reaching impact beyond their immediate jurisdiction. They can develop and

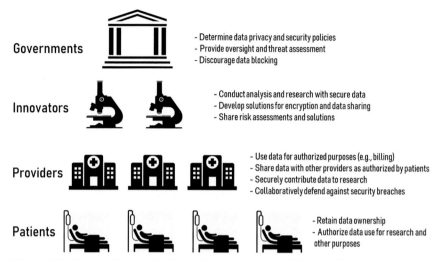

Governments
- Determine data privacy and security policies
- Provide oversight and threat assessment
- Discourage data blocking

Innovators
- Conduct analysis and research with secure data
- Develop solutions for encryption and data sharing
- Share risk assessments and solutions

Providers
- Use data for authorized purposes (e.g., billing)
- Share data with other providers as authorized by patients
- Securely contribute data to research
- Collaboratively defend against security breaches

Patients
- Retain data ownership
- Authorize data use for research and other purposes

Figure 10.2 Privacy-Centered Information Sharing Model for Healthcare.

enforce data privacy and security policies and provide other incentives (e.g., tax- or permit-based) to service providers to protect patients' private information.

- *Innovators*, including drug manufacturers, payers, and software developers, both develop and deploy new technological tools, and have a significant role in the design of future technologies and processes that ensure patient privacy as well as in the sharing of threat assessments with regulators and providers and building resilience to data security breaches.

- *Providers*, including hospitals, physicians, and e-health service providers, have the most direct access to sensitive patient information and may, on both individual and aggregate levels, cause the most expensive leaks or be targets of bad actors. They must balance the protection of privacy with maximizing clinical utility of data available to them, while managing differential privacy regimes based on patients' consent and the public good stemming from using sensitive information in research, such as in clinical trials.

- *Patients* are the ultimate owners of sensitive medical data and provide authorization for the data to be used for research and other purposes. They must be sufficiently informed to make educated decisions about the sharing of their personal information and be protected when they are unable to make such decisions but some privacy concessions are necessary (e.g., when access to multiple data sources with sensitive information may result in a life-saving intervention).

We discuss desirable responses by individual stakeholders in more detail below.

10.5.1 Governments

10.5.1.1 Regulation

Regulation of new technologies is commonly fraught with challenges, including the unintentional impacts it may have on innovation, but it is clear that AI is radically changing healthcare privacy expectations and protection around the world. While it has been recommended that new regulations are context- rather than AI-specific, an understanding of the positive and negative implications AI has on data privacy, security, and information sharing is vital in developing future regulations. Although existing regulations (such as GDPR in the European Union and HIPAA in the United States) do not directly affect businesses not deemed covered entities, it is likely that as AI-based tools are developed further, they will continue to blur the line between covered and uncovered entities. The requirement for deidentification in current regulations is not up to par with emerging tools. Given the interdisciplinary nature of modern companies, such as those that are in the business of digital marketing, a broader view toward regulating the privacy of health data may be required.

In the United States, healthcare privacy regulations assume stable and knowable data characteristics and uses, which in the era of ML and other tools does no longer hold. As such, absent any privacy reform, AI will likely result in new leaks of sensitive personal information given its ability to infer patients' health status or identity using linked data. In most jurisdictions, an agreement on acceptable solutions to ensure data privacy, security, and integrity must be reached, and governments should enact and enforce laws that ensure patients' sensitive information is protected, especially when business motives may erode data privacy protections.

10.5.1.2 Oversight and enforcement

Given the nature of modern AI algorithms, segmentation of patients into ever-smaller groups may present significant clinical benefits, but also undermine the privacy and safety of already vulnerable patient populations, including those with multiple diseases. [82] New mechanisms for postmarketing oversight and surveillance according to values agreed upon by policy-makers will be required as the long-term implications of AI-based technologies are not going to be understood when they first become available. An agile and technically proficient enforcement mechanism must be in place in individual jurisdiction as it is less likely that existing privacy watchdogs will have the capacity and erudition to deal with the ever-more technical aspects of data privacy protection in healthcare.

10.5.1.3 National strategy development

Ultimately, national strategies for regulating AI in healthcare and other industries will be required. To achieve success, they should include considerations for data security, privacy, and information sharing, among others. They should also contribute to the development of ethical principles, accountability, and oversight mechanisms to prevent innovators and other stakeholders from exploiting the greater availability of data, and to reconcile disagreements and adjudicate privacy breaches and violations in a way that puts patient needs first. Specific incentives to prevent data blocking should also be considered.

10.5.2 Innovators

10.5.2.1 Build sustainable solutions

Facing new regulations while advancing technology that can revolutionize many industries presents innovators with dilemmas. Many may opt to exploit the lack of AI-specific regulation (with respect to privacy and other domains, such as liability) and build products that will achieve fast adoption. Others may choose a more complex path, requiring confidence building with end users (providers and patients alike) as well as regulators, and ensuring that privacy and safety of patients are prioritized. While the latter may require greater investment and will inevitably take longer, recent trends suggest that "privacy-compliant innovators" will achieve more sustainable success, especially if they carefully calibrate the balance of privacy protections and more permissive, albeit secure data-sharing using AI-based and other digital tools.

10.5.2.2 Adopting privacy principles

Laws may be adopted to ensure that AI-based tools do not violate important ethical principles, including fairness and equal opportunity, and that the privacy of vulnerable patient populations is protected, but innovators will play a vital role in their implementation.

10.5.3 Developing privacy-centered solutions

While AI-based techniques are proliferating in clinical settings, such as in diagnostic tools, significant gaps exist in foundational technologies, including data encryption and virtualization. An emphasis on developing privacy standards and decision protocols compliant with modern privacy protections should be at the core of new tool development and could be further incentivized by hospitals and payers.

10.5.4 Providers

10.5.4.1 Internal review

While new regulations may affect provider practices when implementing AI tools in delivery of care, additional steps must be taken by hospitals, clinics, and physicians to ensure patient privacy. For instance, some business have set up internal review boards to evaluate the ethical impact their work may have on their clients, and healthcare providers may do the same when it comes to working with sensitive personal data and using AI to glean new insights. Additional considerations should be given to the potential for discrimination and unequal access to privacy protections by vulnerable social groups, including ethnic minorities.

10.5.4.2 Governance

Another tool available to providers is the designation of independent data privacy officers (DPOs). In the United States, there is no such legal requirement (although this may change in the future) but such practice has become standard in the European Union following the introduction of the GDPR in 2018. The GDPR requires an independent DPO to be in place if any of the following conditions hold for the organization: "1) it is a public authority, 2) it engages in systematic monitoring of people, or 3) it processes sensitive personal data on a large scale". [83] In the United States, the term Chief Privacy Officer (CPO) is more commonly used although multinational companies may opt to have both a CPO and a DPO, with targeted duties with respect to privacy protection and different levels of independence. It is expected that the GDPR will result in 75,000 new DPO positions opening worldwide in the coming years. [84]

10.5.4.3 Education and training

In addition, new AI-based tools will require additional education of medical professionals across different domains: from engineers developing new AI-based tools to scientists conducting basic and applied research using these tools (e.g., drug developers using adaptive trial approaches) and data consumers (e.g., clinicians) using information based on new methods and techniques to improve the management and delivery of care. Aside from adequate technical proficiency, aspects of protecting data privacy and integrity will need to be included in their respective curricula, and new coursework will be required for physicians to "understand AI methods and systems sufficiently to be able to trust an algorithm's predictions". [85]

10.5.4.4 Communication

Providers must ensure that they explain privacy implications of relevant technologies to patients prior to obtaining their consent, where possible, and be trained to understand the policies regulating data security and permissible information sharing. While this will require upfront investment, it is likely that the benefits of new AI-based tools will far exceed the costs of retraining and compliance with privacy regulations.

10.5.5 Patients

10.5.5.1 Data ownership

In most cases, personal healthcare data collected by providers and payers about patients do not have clear ownership. Given differential privacy and intellectual property regimes, ownership of personal data may by region. In the United States, for example, it is understood that only one state—New Hampshire—gives patients explicit ownership of their data. In the future, patients should be granted a more transparent and explicit custody of their health records and be allowed to grant differential access rights to specific entities. Control and access to one's health records may increase both patients' satisfaction [86] as well as their autonomy in an increasingly complicated data ecosystem. It can also incentivize providers to better share information with each other, rather than blocking data access due to commercial reasons. [87] Negative externalities, however, should be considered (e.g., by allowing patients to opt in or opt out of specific data uses, the validity of national-level health datasets may be affected), and provisions for emergency situations when patients are not capable of making informed decisions developed.

10.5.5.2 Consent granting

Better definition of patient control and ownership of sensitive healthcare data is closely linked to consent granting: in the future, patients should be able to limit data disclosure to select entities or for select purposes. Such right will also come with greater responsibility and individual accountability, which may produce disparities in healthcare access. As a result, patient advocacy and interest protection groups should consider best practices and guidelines that define optimal "default" options for data sharing and control. Thus, only patients who are in unique circumstances or have specific data protection objectives may decide to update their default consent options, without the need to educate each patient to the same degree on data privacy protections.

Patient autonomy will likely increase the demands on providers and innovators to provide transparency and accountability in the handling of sensitive personal information using healthcare AI: patients are generally reluctant to trust technology as much as their healthcare provider is. A 2017 survey in the United Kingdom found that 63% of adults were not comfortable with delegating some tasks performed by medical professionals to AI. [88] Expectations are muted in other settings—healthcare professionals in the United States believe AI will not meet current expectations and will produce errors, though will improve the delivery of care. [89]

Case study
European Union's data privacy leadership

For many years, the European Union has been a leader in data privacy and protection, with the GDPR, replacing the 1995 Data Protection Directive 95/46/EC (DPD), adopted in 2016 and implemented in 2018. At the time of the adoption of the Data Protection Directive, no other region or country, including the United States, had a comparable privacy regulation in place, although the health-care specific HIPAA was signed into law in August 1996 in the United States. Europe's GDPR has at least two key distinctions from the DPD: it applies to EU citizens worldwide and governs data transfer with outside entities as well as internally, thus effectively setting new global standards for data privacy, and it harmonizes data privacy laws across the EU. [90] Noncompliant actors face multimillion euro fines for violating the GDPR. GDPR governs the protection of "any information concerning an identified or identifiable natural person"—a concept that AI-based tools are making broader and less specific. The enforcement of GDPR in deidentified healthcare datasets will pose new challenges as better AI-based tools will make reidentification increasingly easier. However, the regulation explicitly excludes "anonymous information, including for statistical or research purposes". EU member states regulate the implementation of EU law, increasing the complexity of regulatory analysis on the continent.

Building on its experience in developing new privacy rules, the European Union is currently debating new ethical principles related to AI and has suggested that it must increase public and private funding of research and development (R&D) in AI, prepare for socioeconomic changes brought about by AI, and develop an appropriate ethical and legal framework in light of the disruption AI is expected to cause. [91] In 2018, the European Group on Ethics in Science and New Technologies published a statement including a discussion of data protection and privacy issues, highlighting the need for informed consent and the right to private life, including the "right to be free from surveillance". [92] It further addresses two potential other rights for consideration by policymakers: "the right to meaningful human contact" and "the right to not be profiled, measured, analyzed, coached, or nudged". In April 2019, the European Commission launched the Communication on Building Trust in Human-Centric Artificial Intelligence, [93] which listed seven key requirements for trustworthy AI in the communication:

- Human agency and oversight
- Technical robustness and safety
- Privacy and data governance
- Transparency
- Diversity, nondiscrimination, and fairness
- Societal and environmental well-being
- Accountability

As part of its discussion of privacy and data governance, the Communication highlights the need to guarantee privacy and data protection at "all stages of the AI system's life cycle," provide individuals with full control over their own data, and guarantee that personal data will not be used to harm or discriminate against them. Additional emphasis is on avoiding bias, inaccuracies, and errors due to data quality, integrity, and access challenges.

It is expected that future discussions about the regulation of AI in Europe will include aspects of data privacy, security, and information transfer in healthcare and other sectors in more detail, potential expanding the reach of European regulation on data privacy even further.

10.6 Conclusions

There are many unknowns in the application of AI tools to healthcare, but several key conclusions can be made. First, protecting data integrity and privacy will continue to be a challenge in electronic exchanges, whether AI-based or not, and AI software is likely to expedite the complexity and challenges with data privacy and security in healthcare. On the other hand, AI-based technologies have the potential to provide solutions to many of the data concerns discussed in this chapter, given their capability to work with deidentified or synthetic data with comparable statistical properties as the original data. As such, AI is most likely to make existing data privacy and security-related challenges more acute and will accelerate the emergence of new solutions for them, but is unlikely to result in a fundamental shift resulting in much better (or worse) data privacy and security on its own. Policy-makers, innovators, providers, and patient all have a role to play in defining how new technologies are used, and the role of patients in determining how their personal data are used is likely to grow.

Our PCISM outlines general principles that should be considered as new regulatory rules for AI in healthcare are developed in an era of increasing complexity of interactions between regulators, healthcare innovators, providers, and patients. With the increase number of entities involved in one's healthcare and the collection of sensitive data from healthcare and other sources, patients' needs and preferences should be reflected in the design of future decision-frameworks affecting privacy, security, and information sharing in healthcare. The empowerment of patients through regulation as well as institutional policies ought to foster the creation of a more sustainable privacy regime: one that will ensure patient needs are prioritized and that actionable and practical improvement to healthcare delivery can be realized. The most significant role remains with policy-makers who need to develop regulations and guidelines that foster privacy and integrity of sensitive healthcare data as new tools are developed and provide independent oversight. The significant technological complexity of AI as well as the fragmentation of some healthcare systems (especially in the United States) will require investment in skills training and the information infrastructure. An adoption of privacy-centered technical and regulatory standards in healthcare, however, is expected to produce significant efficiency gains if they result in better data protection and more agile information sharing.

The issues of data privacy, security, and information sharing in healthcare will continue to evolve as more systems and datasets are linked and researchers find innovative applications in different contexts, from drug discovery to reimbursement and quality control. All of these will require a careful analysis that considers the benefits and costs of new tools and applications and ensures that benefits to patients outweigh their risks. We should expect AI applications to accentuate the need for a more explicit, agile, and sustainable privacy regime in healthcare in the future.

Acknowledgements

The author wishes to thank Jason Jones (Health Catalyst), Paul Tarini (Robert Wood Johnson Foundation), and Sharon F. Terry (Genetic Alliance) for their comments and input. Any outstanding errors and omissions are the author's sole responsibility.

References

[1] In this chapter, we use the terms artificial intelligence (AI) and machine learning (ML) to describe related emerging technologies based on computer-based learning systems. AI is a higher-order term and may include technologies not based on predictive statistical methods that are at the core of ML (e.g., expert systems). It holds that existing ML methods are also AI methods. In addition, deep learning is typically considered a subset of machine learning.

[2] Retrieved March 5, 2020, from Government info: https://www.govinfo.gov/content/pkg/CRPT-104hrpt736/html/CRPT-104hrpt736.htm; 1996.

[3] Office for Civil Rights. Summary of the HIPAA privacy rule. Retrieved March 5, 2020, from Health Information Privacy: https://www.hhs.gov/hipaa/for-professionals/privacy/laws-regulations/index.html; 2013.

[4] European Union. Regulation (EU) 2016/679 of the European Parliament and of the Council of 27 April 2016 on the protection of natural persons with regard to the processing of personal data and on the free movement of such data, and repealing Directive 95/46/EC. Retrieved March 5, 2020, from European Union Law: https://eur-lex.europa.eu/legal-content/EN/TXT/?uri = celex%3A32016R0679; 2016.

[5] Haag M. DNA testing company admits to secretly sharing people's genetic data with FBI. Retrieved March 5, 2020, from Independent: https://www.independent.co.uk/news/world/americas/dna-testing-fbi-data-breach-privacy-family-tree-bennett-greenspan-a8763521.html; 2019.

[6] Lye CT, Forman HP, Gao R, et al. Assessment of US Hospital Compliance with regulations for patients' requests for medical records. JAMA Netw Open 2018;1(6):e183014. Available from: https://doi.org/10.1001/jamanetworkopen.2018.3014.

[7] Wu H, LaRue E. Linking the health data system in the U.S.: challenges to the benefits. Int J Nurs Sci 2017;4(4):410−17.

[8] Withers R. Why in the world do doctor's offices still use fax machines? Retrieved March 5, 2020, from Slate: https://slate.com/technology/2018/06/why-doctors-offices-still-use-fax-machines.html; 2018.

[9] CMS. Remarks by Administrator Seema Verma at the ONC Interoperability Forum in Washington, DC. Retrieved March 5, 2020 from: https://www.cms.gov/newsroom/press-releases/speech-remarks-administrator-seema-verma-onc-interoperability-forum-washington-dc; 2018.

[10] Acquisti A, Taylor C, Wagman L. The economics of privacy. J Econ Lit 2016;54(2):442−92.

[11] Elliot M, Domingo-Ferrer J. The future of statistical disclosure control. Washington: Government Statistical Services. Retrieved March 5, 2020, from https://arxiv.org/ftp/arxiv/papers/1812/1812.09204.pdf; 2018.

[12] Graf M, Hlávka J, Triezenberg B. A change is in the air: emerging challenges for the Cloud Computing Industry. RAND Working Paper; 2016.

[13] Sacks S. New China data privacy standard looks more far-reaching than GDPR. Retrieved March 5, 2020, from The Center for Strategic and International Studies: https://www.csis.org/analysis/new-china-data-privacy-standard-looks-more-far-reaching-gdpr; 2018.

[14] Health Information Systems Management Society. HIMSS redefines interoperability. Retrieved March 5, 2020, from: https://www.himss.org/resource-news/himss-redefines-interoperability; 2019.

[15] Bush GW. Executive order 13335: incentives for the use of health information technology and establishing the position of the national health information technology coordinator. Washington, DC. Retrieved March 5, 2020, from: https://www.govinfo.gov/content/pkg/WCPD-2004-05-03/pdf/WCPD-2004-05-03-Pg702.pdf; 2004.

[16] Office of the National Coordinator for Health IT. Connecting health and care for the nation: a shared nationwide interoperability roadmap. Retrieved March 5, 2020, from: https://www.healthit.gov/sites/default/files/hie-interoperability/nationwide-interopera-bility-roadmap-final-version-1.0.pdf; 2015.

[17] Calliope Network. EU eHealth Interoperability Roadmap. Europe: CALL for Interoperability. Retrieved March 5, 2020, from http://www.ehgi.eu/Download/European%20eHealth%20Interoperability%20Roadmap%20[CALLIOPE%20-%20published%20by%20DG%20INFSO].pdf; 2010.

[18] European Commission. Ethics guidelines for trustworthy AI. Retrieved March 5, 2020, from: https://ec.europa.eu/digital-single-market/en/news/ethics-guidelines-trustworthy-ai; 2019.

[19] e-Estonia Briefing Centre. e-Health records. Retrieved March 5, 2020, from: https://e-estonia.com/solutions/healthcare/; 2018.

[20] CDC. Electronic medical records/electronic health records (EMRs/EHRs). Retrieved March 5, 2020, from: https://www.cdc.gov/nchs/fastats/electronic-medi-cal-records.htm; 2017.

[21] US Department of Health & Human Services (HHS). HHS, in partnership with industry, releases voluntary cybersecurity practices for the health industry. Retrieved March 5, 2020, from: https://www.hhs.gov/about/news/2018/12/28/hhs-in-part-nership-with-industry-releases-voluntary-cybersecurity-practices-for-the-health-industry.html; 2018.

[22] US Department of Health and Human Services. Health industry cybersecurity prac-tices: managing threats and protecting patients. Retrieved March 5, 2020, from: https://www.phe.gov/Preparedness/planning/405d/Documents/HICP-Main-508.pdf; 2018.

[23] Center for Open Data Enterprise. Sharing and Utilizing Health Data for AI Applications. Retrieved March 5, 2020, from: https://www.hhs.gov/sites/default/files/sharing-and-utilizing-health-data-for-ai-applications.pdf; 2019.

[24] Department of Health and Human Services. HHS finalizes historic rules to provide patients more control of their health data. News (March 9, 2020). Retrieved March 10, 2020, from: https://www.hhs.gov/about/news/2020/03/09/hhs-finalizes-his-toric-rules-to-provide-patients-more-control-of-their-health-data.html; 2020.

[25] Ponemon and IBM Security. Cost of a data breach report. Retrieved March 5, 2020, from: https://www.ibm.com/downloads/cas/ZBZLY7KL; 2019.

[26] The cost included in the analysis includes detection and escalation costs (such as forensic and investigative activities), post-data breach response (such as legal expenses), notification costs (such as communication with regulators and affected individuals), and lost business cost (including diminished goodwill). This assessment identifies only a fraction of the total problem—it draws on a sample of 477 compa-nies in 15 countries and regions (in Europe, for instance, only Germany, Italy, France, the United Kingdom, and Turkey were included).

[27] Ponemon and IBM Security. 2017 Cost of a data breach study. Retrieved March 5, 2020, from: https://www.ibm.com/downloads/cas/ZYKLN2E3; 2017.

[28] CynergisTek. 2016 breach report. Retrieved March 5, 2020, from: https://cynergis-tek.com/cynergistek-resources/breach-report-2016/; 2016.

[29] Lane C, Schur CJ. Balancing access to health data and privacy: a review of the issues and approaches for the future. Health Serv Res 2010;45(5p2):1456−67 CynergisTek. 2016 Cost of data breach study; 2016.

[30] Verizon. 2018 data breach investigations report. Retrieved March 5, 2020, from: https://enterprise.verizon.com/resources/reports/DBIR_2018_Report.pdf; 2018.

[31] Torra V, Navarro-Arribas G. Big data privacy and anonymization. In: Lehmann A, Whitehouse D, Fischer-Hübner S, Fritsch L, Raab C, editors. Privacy and identity

management. Facing up to next steps. Privacy and identity 2016. IFIP advances in Information and Communication Technology, vol. 498. Cham: Springer; 2016.

[32] Antal L, Enderle T, Giessing S. Harmonised protection of census data in the ESS. Specific Grant Agreement (SGA); 2017.

[33] 2020, from Apple: https://www.apple.com/privacy/docs/Differential_Privacy_Overview. pdf; 2020.

[34] Mooney SJ, Pejaver V. Big data in public health: terminology, machine learning, and privacy. Annu Rev Public Health 2018;39:95−112.

[35] Koperniak S. Artificial data give the same results as real data—without compromising privacy. Retrieved March 5, 2020, from MIT News: http://news.mit.edu/2017/artificial-data-give-same-results-as-real-data-0303; 2017.

[36] Baowaly MK, Lin CC, Liu CL, Chen KT. Synthesizing electronic health records using improved generative adversarial networks. J Am Med Inform Assoc 2019;26 (3):228−41. Available from: https://doi.org/10.1093/jamia/ocy142.

[37] Kolbasin V. Usage of generative adversarial networks in healthcare. Retrieved March 5, 2020, from: https://aiukraine.com/wp-content/uploads/2017/10/2_7-Kolbasin. pdf; 2017.

[38] Torra V. Privacy models and disclosure risk: integral privacy. Retrieved March 5, 2020, from: http://www.ppdm.cat/dp/slides/slides.torra.dpm.2017.pdf.

[39] Motiwalla L, Li X-B. Developing privacy solutions for sharing and analyzing healthcare data. Int J Bus Inf Syst 2013;13:2. Available from: https://doi.org/10.1504/IJBIS.2013.054335.

[40] Bae H, Jang J, Jung D, Jang H, Ha H, Yoon S. Security and privacy issues in deep learning. arXiv Preprint arXiv 2018;1807.11655.

[41] Na L, Yang C, Lo C, Zhao F, Fukuoka Y, Aswani A. Feasibility of reidentifying individuals in large national physical activity data sets from which protected health information has been removed with use of machine learning. JAMA Netw Open 2018;1(8): e186040. Available from: https://doi.org/10.1001/jamanetworkopen.2018.6040.

[42] Dutchen S. The importance of nuance: artificial intelligence may seem objective, but it's subject to human biases. Retrieved March 5, 2020, from Harvard Medicine: https://hms.harvard.edu/magazine/artificial-intelligence/importance-nuance; 2018.

[43] Chung Y, Haas PJ, Upfal E, Kraska T. Unknown examples & machine learning model generalization. arXiv Preprint arXiv 2018;1808.08294.

[44] Horvitz E, Mulligan D. Data, privacy, and the greater good. Science 2015;349 (6245):253−5.

[45] Myers J, Frieden TR, Bherwani KM, Henning KJ. Ethics in public health research: privacy and public health at risk: public health confidentiality in the digital age. Am J Public Health 2008;98(5):793−801.

[46] Groot JD. The history of data breaches. Retrieved March 5, 2020, from Digital Gurdian: https://digitalguardian.com/blog/history-data-breaches; 2019.

[47] Jones K. Doctor or patient? Who owns medical records? Retrieved March 5, 2020, from AAFP: https://www.aafp.org/news/blogs/freshperspectives/entry/doctor_or_-patient_who_owns.html; 2016.

[48] ICO. Royal Free—Google DeepMind trial failed to comply with data protection law. Retrieved March 5, 2020, from: https://ico.org.uk/about-the-ico/news-and-events/news-and-blogs/2017/07/royal-free-google-deepmind-trial-failed-to-comply-with-data-protection-law/; 2017.

[49] Lomas N. UK tax office ordered to delete millions of unlawful biometric voiceprints. Retrieved March 5, 2020, from Tech Chrunch: https://techcrunch.com/2019/05/10/uk-tax-office-ordered-to-delete-millions-of-unlawful-biometric-voiceprints/; 2019.

[50] Patrizio A. What is data virtualization? Retrieved March 5, 2020, from Datamation: https://www.datamation.com/big-data/what-is-data-virtualization.html; 2019.

[51] Alexander R. Data virtualization: unlocking data for AI and machine learning. Retrieved March 5, 2020, from Microsoft: data virtualization: unlocking data for AI and machine learning; 2017.

[52] Guerrini CJ, Robinson JO, Petersen D, McGuire AL. Should police have access to genetic genealogy databases? Capturing the Golden State Killer and other criminals using a controversial new forensic technique. PLoS Biol 2018;16:10.

[53] Kaufman D, Murphy J, Erby L, Hudson K, Scott J. Veterans' attitudes regarding a database for genomic research. Genet Med 2009;11(5):329−37.

[54] Bollinger JM, Green RC, Kaufman D. Attitudes about regulation among direct-to-consumer genetic testing customers. Genet Test Mol Biomarkers 2013;17(5):424−8.

[55] MacCarthy M. How to address new privacy issues raised by artificial intelligence and machine learning, Brookings Institution. www.brookings.edu/blog/techtank/2019/04/01/how-to-address-new-privacy-issues-raised-by-artificial-intelligence-and-machine-learning/; 2019 [accessed 04.03.20].

[56] Shokri R, Stronati M, Song C, Shmatikov V. Membership inference attacks against machine learning models. In: 2017 IEEE Symposium on Security and Privacy (SP). IEEE; 2017. p. 3−18.

[57] Hao K. A little-known AI method can train on your health data without threatening your privacy. Retrieved March 5, 2020, from Technology Review: https://www.technologyreview.com/s/613098/a-little-known-ai-method-can-train-on-your-health-data-without-threatening-your-privacy/; 2019.

[58] Obermeyer Z, Emanuel EJ. Predicting the future—big data, machine learning, and clinical medicine. N Engl J Med 2016;375(13):1216−19. Available from: https://doi.org/10.1056/NEJMp1606181.

[59] Gilad-Bachrach R, Dowlin N, Laine K, Lauter K, Naehrig M, Wernsing J. Cryptonets: applying neural networks to encrypted data with high throughput and accuracy. In: International conference on machine learning; 2016. p. 201−10.

[60] Graepel T, Lauter K and Naehrig M. ML confidential: machine learning on encrypted data. In: International Conference on Information Security and Cryptology. Berlin: Springer;2012. p. 1−21.

[61] National Academy of Medicine. Digital Learning Collaborative (DLC) Meeting, January 2019. Retrieved March 5, 2020, from: https://nam.edu/wp-content/uploads/2019/01/DLC-_01162019_Meetingsummary_FINAL.pdf; 2019.

[62] Nundy S, Hodgkins ML. The application of AI to augment physicians and reduce burnout. Health Aff Blog. Retrieved March 5, 2020, from: https://www.healthaffairs.org/do/10.1377/hblog20180914.711688/full/; 2018.

[63] Arndt BG, Beasley JW, Watkinson MD, Temte JL, Tuan WJ, Sinsky CA, et al. Tethered to the EHR: primary care physician workload assessment using EHR event log data and time-motion observations. Ann Family Med 2017;15(5):419−26.

[64] Donovan F. 45,000 patient records exposed in nuance healthcare data breach. Retrieved March 5, 2020, from Health IT Security: https://healthitsecurity.com/news/45000-patient-records-exposed-in-nuance-healthcare-data-breach; 2018.

[65] Char DS, Shah NH, Magnus D. Implementing machine learning in health care—addressing ethical challenges. N Engl J Med 2018;378(11):981.

[66] Ngiam KY, Khor W. Big data and machine learning algorithms for health-care delivery. Lancet Oncol 2019;20(5):e262−73.

[67] Office of the National Coordinator for Health IT. Connecting health and care for the nation: a shared nationwide interoperability roadmap. Retrieved March 5, 2020, from: https://www.healthit.gov/sites/default/files/hie-interoperability/Interoperibility-Road-Map-Supplemental.pdf; 2015.

[68] Medicine S. The democratization of health care: Stanford medicine 2018 health trends report. Retrieved March 5, 2020, from: https://med.stanford.edu/content/

dam/sm/school/documents/Health-Trends-Report/Stanford-Medicine-Health-Trends-Report-2018.pdf; 2018.

[69] UK Government. Securing cyber resilience in health and care. Department of Health and Social Care. Retrieved March 5, 2020, from: https://assets.publishing.service.gov.uk/government/uploads/system/uploads/attachment_data/file/747464/securing-cyber-resilience-in-health-and-care-september-2018-update.pdf; 2018.

[70] ForgeRock. U.S. consumer data breach report 2019. Boston: ForgeRock. Retrieved March 5, 2020, from: https://www.forgerock.com/resources/view/92170441/industry-brief/us-consumer-data-breach-report.pdf; 2019.

[71] Carol A. NHS 'could have prevented' WannaCry ransomware attack. Retrieved March 5, 2020, from BBC: https://www.bbc.com/news/technoogy-41753022; 2017.

[72] H-ISAC. Website. Retrieved March 5, 2020, from: https://h-isac.org/; 2019.

[73] KPMG. Fighting Cyber With Cyber: Deep Learning Threats Demand Deep Learning Solutions. Retrieved March 5, 2020, from: https://advisory.kpmg.us/content/dam/advisory/en/pdfs/fighting-cyber-with-cyber.pdf; 2018.

[74] Vinayakumar R, Alazab M, Soman KP, Poornachandran P, Al-Nemrat A, Venkatraman S. Deep learning approach for intelligent intrusion detection system. IEEE Access 2019;7:41525−50.

[75] ITU. AI for Good Global Summit 2018. Retrieved March 5, 2020, from: https://www.itu.int/en/ITU-T/AI/2018/Pages/default.aspx; 2018.

[76] Crigger E, Khoury C. Making policy on augmented intelligence in health care. AMA J Ethics 2019;21(2):188−91.

[77] Abouelmehdi K, Beni-Hessane A, Khaloufi H. Big healthcare data: preserving security and privacy. J Big Data 2018;5(1):1.

[78] University of California-Berkeley. Artificial intelligence advances threaten privacy of health data. Retrieved March 5, 2020, from Science Daily: https://www.sciencedaily.com/releases/2019/01/190103152906.htm; 2019.

[79] OECD. Recommendation of the Council on Artificial Intelligence. Retrieved March 5, 2020, from: https://legalinstruments.oecd.org/en/instruments/OECD-LEGAL-0449; 2019.

[80] Ostrom E. Governing the commons: the evolution of institutions for collective action. Cambridge: University Press; 1990.

[81] Bollinger JM, Zuk PD, Majumder MA, Versalovic E, Villanueva AG, Hsu RL, et al. What is a medical information commons? J Law Med Ethics 2019;47(1):41−50. Available from: https://doi.org/10.1177/107311051984048.

[82] Vayena E, Blasimme A, Cohen IG. Machine learning in medicine: addressing ethical challenges. PLoS Med 2018;15:11.

[83] PWC. 10 considerations to help position the GDPR data protection officer for success. Retrieved March 5, 2020, from: https://www.pwc.com/us/en/services/consulting/cybersecurity/general-data-protection-regulation/data-protection-officer-10-considerations.html; 2018.

[84] PricewaterhouseCoopers. 10 considerations to help position the GDPR data protection officer for success. Retrieved March 5, 2020, from PWC: https://www.pwc.com/us/en/services/consulting/cybersecurity/general-data-protection-regulation/data-protection-officer-10-considerations.html; 2018.

[85] American Medical Association. Augmented intelligence in health care: report 41 of the AMA Board of Trustees. Retrieved June 22, 2019, from: https://static1.squarespace.com/static/58d0113a3e00bef537b02b70/t/5b6aed0a758d4610026a719c/1533734156501/AI_2018_Report_AMA.pdf.

[86] Delbanco T, Walker J, Bell SK, Darer JD, Elmore JG, Farag N, et al. Inviting patients to read their doctors' notes: a quasi-experimental study and a look ahead. Ann Intern Med 2012;157(7):461−70.

[87] Pecora A. Importance of data sharing in healthcare. Retrieved March 5, 2020, from AJMC: https://www.ajmc.com/peer-exchange/oncology-stakeholders-summit-fall-2016/importance-of-data-sharing-in-healthcare; 2017.

[88] Fenech M, Strukelj N, Buston O. Ethical, social, and political challenges of artificial intelligence in health. Future Advocacy. Retrieved March 5, 2020, from: http://futureadvocacy.com/wp-content/uploads/2018/04/1804_26_FA_ETHICS_08-DIGITAL.pdf; 2018.

[89] Intel. Overcoming barriers in AI adoption in healthcare. Retrieved March 5, 2020, from: https://newsroom.intel.com/wp-content/uploads/sites/11/2018/07/healthcare-iot-infographic.pdf; 2018.

[90] Schildhaus A. EU's General Data Protection Regulation (GDPR): key provisions and best practices. Chicago: American Bar Association. Retrieved March 5, 2020, from: https://www.americanbar.org/groups/international_law/publications/international_law_news/2018/winter/eu-general-data-protection-regulation-gdpr/; 2018.

[91] European Commission. Artificial intelligence: Commission outlines a European approach to boost investment and set ethical guidelines. Retrieved March 6, 2020, from: https://ec.europa.eu/commission/presscorner/detail/en/IP_18_3362; 2018.

[92] European Group on Ethics in Science and New Technologies. Statement on artificial intelligence, robotics and 'autonomous' systems. Retrieved March 6, 2020, from European Commission: http://ec.europa.eu/research/ege/pdf/ege_ai_statement_2018.pdf; 2018.

[93] European Commission. Communication: building trust in human centric artificial intelligence. Retrieved March 5, 2020, from Europa: https://ec.europa.eu/digital-single-market/en/news/communication-building-trust-human-centric-artificial-intelligence; 2018.

CHAPTER 11

The impact of artificial intelligence on healthcare insurances

Rajeev Dutt
AI Dynamics Inc., Redmond, WA, United States

11.1 Overview of the global health insurance industry

In 2018, the three largest health insurance companies in the world, all US based, measured by market capitalization had a total capitalization of over $155B [1]. In total direct premiums, they collected about $239B in 2016 [1]. These figures do not include state sponsored healthcare and health insurance. In the United States alone, healthcare spending reached $3.5T in 2017 or about 17.9% of gross domestic product (GDP) [2]. Healthcare spending globally is anticipated to grow from $7.7T in 2017 to about $10T by 2022 [3]. In the United States, it is likely that from 2018 to 2027, healthcare expenditures will increase [4] by an average of 5.5% per year.

In 2017 in the United States, personal healthcare spending reached a total of $2.9T of which out-of-pocket expenses accounted for a mere $365B. Of the amount paid by health insurance, about 44% came from private health insurance companies [5]. From 2010 to 2017, the compound annual growth rate in private health insurance expenditure was 4.6%.

In 2016, the average margin for the health insurance industry globally was about 6.1% [6]. Administrative costs accounted for about 12% of gross premiums (smaller players such as those found in Estonia accounted for about 19% of premiums). This figure is not too far from the United States where insurers are expected to follow the "80/20" rule. 80% of collected premiums are expected to be spent on healthcare costs and quality improvement activities while the rest can be spent on administrative/back-office [7].

In terms of both the size and growth rate of healthcare expenditures, there is a pressing need for the entire health insurance industry to identify

Artificial Intelligence in Healthcare
DOI: https://doi.org/10.1016/B978-0-12-818438-7.00011-3
271

and implement ways to achieve higher efficiencies to control costs while at the same time offering their customers premiums that are still affordable. Based on the current state and direction of the healthcare insurance industry, there are two areas where significant improvements can be made—identifying ways to reduce claims while maintaining high-quality patient outcomes and reducing back-office costs through increased efficiencies.

Because of the size of the US health insurance market, we will be devoting much of this chapter on the use of artificial intelligence (AI) in the US healthcare industry while still covering some of the innovations in other countries.

11.2 Key challenges facing the health insurance industry

Some of the key challenges facing the health insurance industry ranging from risk assessment, premium assessments, fraud, preventative medicine, rising cost of healthcare, competition, and customer care.

There are many challenges facing the healthcare insurance industry. According to Allianz and Fineos [8,9], the top three issues are

1. Cost of healthcare;
2. Increased expectations of consumers regarding choices, outcomes, and quality of services that they would like to see from providers and insurers;
3. Regulatory overhead is growing in most countries (especially in the United States with the myriad rules and uncertainty regarding the current and future state of the Affordable Care Act).

11.2.1 Cost of healthcare

The cost of healthcare globally has been growing. For OECD countries, on average, healthcare expenditures currently account for about 8.8% of GDP. In 2010, by comparison, the average was 8.7% of GDP. The United States has seen a much sharper growth from 17.4% of GDP in 2010 to 17.9% of GDP today (Fig. 11.1).

In real terms, the average healthcare expenditures for OECD countries have grown by nearly 18.7% (adjusted for inflation) from 2010 to 2018 [10]. Some of the driving factors for the growth in healthcare costs include (but not limited to)

- Aging population—in 2014 about 17.9% of the population of the G-7 countries was over 65 [10].

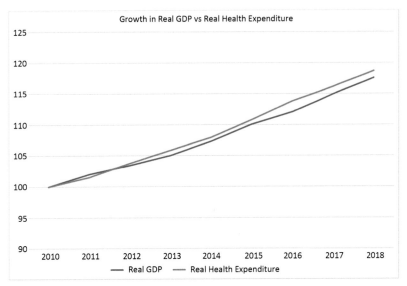

Figure 11.1 Growth in real GDP compared to real health expenditure.

- Increasing drug prices—from 2008 to 2016, the cost of injectable drugs grew by 15.1% annually and oral brand name drugs increased by 9.2% annually [11].
- Increase in noncommunicable diseases (NCDs)—of the top 10 causes of death globally, according to the WHO, six of them are preventable cardiovascular diseases or diabetes [12,13].
- Fraud—according to the Economist [14], in 2011 between $82B and $272B was lost due to fraud. The sheer size of the losses means that it is difficult to staff a department of fraud investigators.

11.2.2 Increased expectations of customers

One of the interesting changes in the United States in the last few years is the decline of the Health Maintenance Organization (HMO) and the growth of "consumer directed health plans" known as Preferred Provider Organizations (PPOs). The HMO was designed to restrict choices and force patients into low-cost primary care, which was often of lower quality. PPOs, on the other hand, offer high deductibles that are often linked to a Health Savings Account. For patients, PPOs offer more choice, for example, choosing providers outside of the plan or choosing nongeneric drugs. For employers, PPOs offer employers the chance to transfer the risks and cost of healthcare onto the employees subscribing to PPOs [15].

According to an Ernst and Young report [16], there are several trends related to increasing customer expectations:
- Greater focus on outcomes and value of services;
- "M-Health" or mobile health, which relies increasingly on mobile technologies, IoT, and social media to increase interaction between patients and their providers;
- The proliferation of choice through increased transparency and availability of information.

11.2.3 Regulatory overhead

Healthcare is often seen as a public good and so there are many controls on health insurance such as who can sell insurance, who can be covered, what can be covered, how prices are set, and how providers are paid. To support these controls, governments use legislation and licensing, active monitoring and auditing, and finally intelligence and analytics [17].

The regulatory overhead is growing globally as more people are becoming insured, the supply of providers is not keeping pace with demand, and the tax base for many countries is shrinking due to aging populations. To ensure broader coverage, governments impose even more controls on the health industry, which in turn impacts health insurance companies.

11.3 The application of artificial intelligence in the health insurance industry

11.3.1 Overview and history of artificial intelligence

The concept of AI, arguably, has been around for hundreds of years: In Ovid's play, the mythological Pygmalion (who the play was named after) created a statue of a beautiful woman who Aphrodite brought to life; Mary Shelly imagined the creature of Dr. Frankenstein in her classic novel; according to legend, the rabbi Judah Loew ben Bezalel created a golem out of clay and brought it to life using Hebrew incantations; in his play R.U.R, written in 1920, Karel Čapek wrote about artificial automata; and, of course, the 1927 Fritz Lang classic film *Metropolis* featured the robot Maschinenmensch. There are, of course, many other examples.

It is debatable as to the origins of modern AI. In 1943, McCulloch and Pitts [18] modeled a simple neural network with electrical circuits but most argue that the work to create true AI began at the Dartmouth

College workshop of 1956 [19], attended by Marvin Minsky, Arthur Samuelson, Allen Newell, and others. Several innovations were introduced in the 1950s including the perceptron of 1956 by Frank Rosenblatt [20]. The doldrums of AI followed between 1960 and 1980 with the exception of the first true commercial use of AI using expert systems in the late 1970s and early 1980s. In the 1980s in a series of important papers including Hopfield [21,22], the modern era of neural networks began. A second winter followed for most of the 1990s until the foundations for deep learning were laid by three key papers on long short-term memory [23], gradient learning [24], and convolutional neural networks [25]. In 2015, LeCun published a paper in Nature that described deep learning which encapsulated where we are going with AI in the next 10 years at least [26].

Globally, the AI industry is growing at a furious pace (270% over 4 years) [27] but these are still early days. Currently, only 37% of companies are using AI in *some* way. For most organizations, this means utilizing web services offered by Google, Amazon AWS, IBM, and Microsoft. However, there is an acute talent shortage that is impeding further reach of AI within organizations and often organizations require highly customized solutions that generic services that are offered by the likes of the big four simply do not offer.

11.3.2 Artificial intelligence in health insurance

There are no silver bullets; however, AI is increasingly being used by health insurance providers. Insurers are relying on big data [16] to identify trends, control pricing, identify fraud, encourage healthy behavior, optimize disbursements, and identify risks.

According to a Gartner report [28], the top three uses of AI in the Insurance industry are

1. Fraud analysis;
2. Chatbots;
3. Process optimization.

Other uses of AI are rapidly coming to the fore as the benefits are increasingly being realized.

It is estimated that insurers can save up to $7B over 18 months using AI [29] by making administrative tasks more efficient and reducing waste. About 72% of insurance executives claimed that they would be looking at AI within a year (Fig. 11.2).

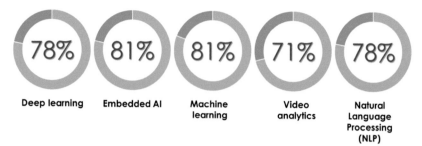

| 78% | 81% | 81% | 71% | 78% |
| Deep learning | Embedded AI | Machine learning | Video analytics | Natural Language Processing (NLP) |

Figure 11.2 Percentage of insurers planning to investment in these areas over the next 3 years [30].

There are several ways that AI has and can be used to reduce cost [31,32]. We can broadly classify the AI applications into five categories based on the benefits they provide

1. Improving customer experience;
2. Reducing fraud;
3. Improving back-office efficiency and reducing cost;
4. Reducing risk and optimizing premiums;
5. Creating new business opportunities.

While AI is already starting to make inroads into other industries, the health insurance industry, because of its size, the impact on society and the economy, and the rapid changes in healthcare, stands to benefit significantly.

11.3.2.1 Improving customer experience

AI can improve customer experience in a number of ways including improving communications by encouraging more frequent interaction between the patient and the insurer; anticipate and understand customer needs; anticipating changing life needs.

Improving communications can be achieved by the use of bots, which allows the customer to check on the status of claims, apply for insurance, create a new claim, ask questions regarding billing, and so on. An insurance company called ZhongAn Tech, the largest insurer in China, uses Chatbots to field close to 97% of calls while routing only 3% to a human being. Currently, about 68% of insurance companies are already using Chatbots of one type or another. It is estimated that by 2030 there will be 70%–90% fewer humans interacting with customers [33]. While this certainly benefits the insurer from a cost perspective, it also helps the customer by being able to access services more easily without the customary wait times associated with call-centers.

AI can also be used to adjust to customer's needs. A software company called Clover Health [34] uses data analytics to customize health insurance for seniors by identifying gaps in their care. The patient can then select doctors to fill these gaps. Another software company called Accolade uses their platform to help patients choose the best health coverage and helps them with lifestyle choices to promote better health.

11.3.2.2 Reducing fraud

Fraud, Waste, and Abuse (FWA), as mentioned above, is a serious concern for health insurers (indeed for all insurers). Several companies have built solutions that claim to address some of these issues [32] such as Kirontech, which was founded in 2014 [35], that claims to be able to identify FWA issues using its CaseHound software. Kirontech leverages public health data published on healthdata.gov, a Federal Government website that gives access to many datasets that can be used for training AI models. Another example of a company claiming to help reduce FWA is Azati [36], founded in 2001, that has trained their platform on previous claims to identify unusual claims.

11.3.2.3 Improving back-office efficiency and reducing cost

Close to 70% of claims require a human adjuster to intervene [37] because they have been flagged as potentially incorrect. Up to 10% of claims have been filed incorrectly. The challenge is that to recognize whether a claim is genuinely incorrect or fraudulent relies heavily on human effort. As an example, a mid-size German insurer with about 1.5 million members receives about 700,000 claims each year, which means that the effort expended to identify incorrect claims or fraudulent claims is monumental.

An AI-based claims system can be trained to identify incorrect claims and weed them out successfully; moreover, the system can be improved over time as misidentified claims can be added to the training set. The impact is clearly significant.

In terms of implementation of such a system, the following steps would need to be taken:

1. Digitize the claims process;
2. Initially use humans to classify claims as either correct or not;
3. Construct a training set based on the collected incorrect claims;
4. Train a claims model.

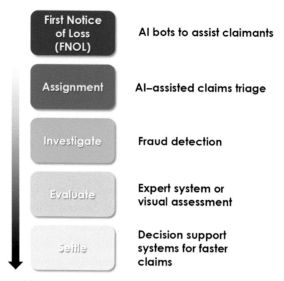

Figure 11.3 Use of AI in the claims process.

AI can be used throughout the claim process and can result in significant cost reductions starting with AI bots to assist claimants to decision support systems for faster claims [38] (Fig. 11.3).

11.3.2.4 Reducing risk and optimizing premiums

Health insurers can use AI to identify clients with higher risks. Swiss Re and Max Bupa Health entered into a partnership with an Indian startup called GOQii Health [39]. GOQii health claims to offer health coaching and health management tools such as health assessments. Swiss Re is providing underwriting assistance to Max Bupa to create risk assessment models. The benefit to Max Bupa is that they can reduce claims cost, increase customer engagement, promote healthy living, and access a broader market of technology enthusiasts.

If the customer is volunteering to release a continuous stream of health data to the insurer, then the insurer can customize premiums by assessing individual risk based on customer health and lifestyle. However, this does pose some ethical issues (see below).

The process for optimizing claims is shown in Fig. 11.4.

11.3.2.5 Creating new business opportunities

Traditional market segmentation uses Segmentation, Targeting, and Positioning (STP), which has traditionally broken down the users of a

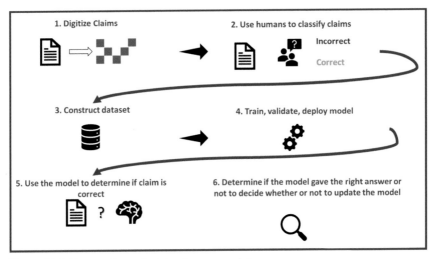

Figure 11.4 Process for optimizing insurance claims.

service or product into personas. An example could be a health insurer breaking down its patients into segments based on age, sex, health condition, and so on. Increasingly, however, this way of segmenting customers is being replaced by behavioral methods [40]. The reason for this is that there is an increasing realization that although two customers belong to different market segments, they may have more in common with each other than with customers in the same segment. Furthermore, partitioning customers into personae or distinct segments causes loss of details that might indicate a market opportunity. Finally, there is a realization that customer segments are not discrete but are part of a continuum. There are simply too many ways to correctly segment customers.

A behavioral approach is only possible with the use of data science and machine learning. Analysis of patterns of behavior such as claims, frequency of claims, type of claims, movement, choices of sport, and so on may yield patterns that were not obvious. These patterns may yield unique markets or offer unique ways to market to certain customers. Some examples of creating new market opportunities could be to expand the customer base to include customers who were previously identified as very high risk, or to offer additional types of packages for parents, expanded healthcare packages to frequent travelers, or targeted healthcare products for tech savvy customers who are more likely to use digital health offerings such as wearables.

11.4 Case studies
11.4.1 Successes

In 2010, Aetna introduced a bot named Ann, developed by a firm Next IT based out of Spokane, Washington State, who helps users of their website to perform simple tasks such as assisting with forgotten passwords, registration, coverage info, and status of claims [41]. After it was rolled out, there were over 2500 chat sessions per day between Ann and the users. Since Aetna, many insurance companies have followed suit such as Premera (the largest health insurer in the Pacific Northwest in the United States) and ZhongAn, and there has been a substantial investment in that area [31].

Another example of a success story is Collective Health [32], which claims to have integrated population data and claims to offer personalized recommendations to both employers and employees such as providers and health insurance policies that best match their needs. They claim to have reached over 70,000 employees and dependents.

11.4.2 Failures

In October 2013, IBM announced that MD Anderson, a cancer treatment center belonging to the University of Texas, was using IBM Watson to work toward the eradication of cancer. In a subsequent audit report, it was revealed that after $62M in spending, the program had been a failure [42].

While this is not a failure in the health insurance industry, it is a case study of a failure of AI in healthcare and is instructive of how AI may not always work as expected. One of the challenges in the case of MD Anderson was that for many types of cancer there was a lack of data [43]. According to David Howard, a faculty member in the Department of Health Policy, although there are many published articles, there are only a handful of high-quality, randomized trials.

11.5 Moral, ethical, and regulatory concerns regarding the use of artificial intelligence
11.5.1 Regulatory challenges

In the United States, the Health Insurance Portability and Accountability Act and its implementing regulations (HIPAA) have been in place since 1996 and they place severe restrictions on the use of patient data, which is a challenge when building AI models as vast quantities of data are needed to train AI models [44].

In Europe, the General Data Protection Regulation (GDPR) also has an impact on how AI models are designed and implemented. For example, GDPR has placed requirements on firms who use AI to make decisions regarding customers (such as determining the premiums they have to pay) including having to explain the decision. This is a major shortcoming of machine learning. Machine learning is often a black box but new advances are seeking to "open the box"; nevertheless, for many problems explainability remains an unsolved problem.

11.5.2 Ethical challenges

Aside from regulatory questions, there are ethical dilemmas that insurance companies have to manage. One such dilemma is using metadata gleaned from public sources to make conclusions about prospective customers. Suppose, for example, a customer claims that she does not smoke or drink but we have trained an AI model to analyze images for smoking or drinking and then apply that model to the customer's public social media site or even request access to her social media pages.

Determining health insurance premiums or whether to offer a policy to a prospective customer is fundamentally a decision about risk. One of the applications of AI is to predict risk in potential insures based on available data. Some of the predictions are straightforward and have been known for many years such as smoking increasing the risk of cancer or obesity leading to cardiovascular issues. However, suppose it is possible to determine other types of risks using publicly available metadata such as analyzing social media images to determine that an insurance policy holder has a predilection for dangerous sports or that he is sexually promiscuous based on available information. The question becomes how intrusive should the insurer be?

Recently, some insurance companies such as John Hancock have started to offer customers lower premiums in exchange for the customers wearing wearables [45]. Ostensibly, the insurer can argue that the customer consents to having the data collected by the wearable to be sent to the insurer. However, unlike other types of medical data where the purpose is well defined or well understood, the data collected by the wearable may not directly have a clear purpose but through data analytics can reveal, by indirect methods, whether a patient may have a medical condition. The same applies to the use of medical apps to monitor patient mental health (so-called neurotechnologies), physical health, or prescription adherence. In and of themselves, their purpose can be said to be salubrious; however, the data harvested may be used to construct analytic models that harm the patients. Furthermore, as AI models are inherently probabilistic, if the AI model does make the prediction that a given patient has a heart condition with a probability of 80%, how actionable is the result?

If the outcome is to encourage the insuree to go through a more rigorous checkup for his benefit then the insurer may find this acceptable, whereas if a conclusion is drawn from the AI model and the insurer decides to raise the premiums of the insuree then this is surely questionable.

11.6 The limitations of artificial intelligence

11.6.1 Lack of talent

One of the biggest roadblocks to implementing AI is lack of talent. Annual salaries for AI engineers routinely exceed $300,000 [46] and it is estimated that there are only about 300,000 AI engineers globally while there are millions of roles to be filled. Furthermore, anecdotally, most engineers prefer to work for the larger AI companies in the industry such as Microsoft, Google, Facebook, Apple, and Amazon, while avoiding non-AI players who are simply users of AI.

There are few options open to companies wishing to use AI for their businesses. These options are AutoML- and AugmentedML-based technologies [47], retraining existing staff, recruiting machine learning talent, and outsourcing.

AutoML and AugmentedML is a branch of machine learning that uses a variety of techniques to choose the appropriate machine learning algorithm to solve a problem, identify the correct parameters (called hyperparameters) that the algorithms use, and train the resulting model. There are a number of platforms in the industry that use some type of AutoML-based technique and the large players—especially Google have invested heavily in this field. The advantage of AutoML is that a machine creates the AI so that humans do not have to. The challenge with AutoML is that even though the mechanics of creating the AI are managed, preparing the training data or evaluating the resulting models is not easy and often requires the same set of skills or some statistical experience.

Although there are a plenty of resources available to teach AI to people, the fact remains that many of the practitioners in machine learning hold advanced degrees because, unlike conventional software engineering, there is a fundamental disparity in skillsets that are needed. Machine learning is mathematics heavy, depending heavily on linear algebra and calculus, which is not a skill that is easily taught.

Hiring skilled machine learning engineers is difficult for reasons mentioned above and is likely to remain so because the skillsets needed often require advanced degrees and strong mathematical skills, which is unlikely to change due to the decline in the number of students, especially in the West, studying STEM subjects [48,49]. As a result, for many organizations, the remaining option is outsourcing, which is expensive because the professional services or system integration organizations suffer from the same talent issues as the company who is outsourcing. Furthermore, many of the outsourcing companies are not motivated to find simple solutions because they earn by the hour and will often take substantially longer.

11.6.2 Lack of data or poor data quality

To understand the role of data in machine learning, one must first understand the process of creating machine learning models.

The elephant in the room for AI is data [50]. To train AI models using algorithms such as neural networks, vast quantities of data are necessary. Although new methods are being developed to reduce the amount of data needed to train AI models, large amounts of data are still necessary. In fact, charting the growth of training datasets over time, we see that the growth has been exponential [51]. There are a few technical and nontechnical approaches [52]. Nontechnical approaches include leveraging existing data for training or leveraging humans for generating data.

One example is a corporate customer of ours who wanted to train an AI model to differentiate between inappropriate and appropriate comments on their forums. Fortunately, they had a team of moderators who had been classifying the comments and so they had amassed a sizeable amount of comments and associated flags. This data could be used for training.

There are also a number of services available to crowd source data collection such as Snovio [53] and Amazon Mechanical Turk [54]. These services act as a broker between someone who needs data and a distributed workforce who are paid for the amount of data they generate or who are paid for the specific task.

There are tremendous amounts of data generated in the world [50]; it is estimated that there will be about 180 ZB of data by 2025 but about 80% of that data will be unstructured. Only 1% of data has been analyzed. The opportunity is enormous but the challenge is that structuring or preparing the data for training is time consuming and error prone. High-quality data has the following five characteristics:

- *Accuracy*—that the data represents reality;
- *Completeness*—that data does not contain any missing elements. An example would be if we are collecting the age, sex, and weight for a group of people and for half of the people, we forget to collect the weight;
- *Consistency*—that if the data contains constraints, these constraints are respected. For example, if someone has an Id in one table then he should have the same Id in a different table;
- *Uniqueness*—if different tables refer to the same entity then they should be consistent with each other;
- *Timeliness*—the ground truth may change with time. For example, if one is building an AI model to predict asset values then the model will eventually become outdated as economic conditions change.

Collecting data in the health insurance industry is also fraught with other issues such as patient privacy, organizational boundaries between different providers, inconsistent data formats, security, and correctness of data. These factors act as barriers to collect data for building AI models, which in turn slows down major AI projects.

Healthcare privacy laws are covered in more detail elsewhere [55], but this chapter will focus on some of the key points that are discussed in more detail below:

- Health record privacy—every region has its own set of privacy laws. This is discussed in more detail in Section 11.5.
- Collection of data—rules regarding what data can or cannot be collected from patients.
- Maintenance of data—security controls that protect the data from malicious third parties. There may also be rules regarding how long the data can be kept.
- Use of data in analytics—anonymization of data that is used to train models.
- Access to data—there may be intra- or interorganizational boundaries that prevent or prohibit access to certain types of data on security or regulatory grounds.

The most significant barrier to entry for organizations wanting to build AI solutions is to collect, curate, and manage data that can be used for training AI models. Health insurance companies are further hampered by regulatory constraints that impose severe penalties on organizations that violate the rules. Every OECD country has some measure of privacy protection for its patients and this can further impose barriers for implementing solutions.

11.6.3 Artificial intelligence accuracy, bias, and relevance

Even if a model has been correctly trained and validated in the lab, it may not necessarily reflect reality. To understand this statement, we need to do a little math. Before starting, we will lay out the following scenario.

Suppose we have a test that has the following characteristics. If a person develops pancreatic cancer then it correctly predicts this 95% of the time. On the other hand, if a person does not develop pancreatic cancer, it falsely predicts that the person will develop it 8% of the time. The lifetime risk of someone developing pancreatic cancer is 1 out of 64 (1.6%) that means that 98.4% of the people will not develop pancreatic cancer. The question is, how good is this model?

We can use something called Bayes Theorem to determine the actual effectiveness of the test. Bayes Theorem uses something called conditional probability denoted $P(A|B)$, which can be read as "the probability that A has occurred given that B has occurred." It is a description of how A is correlated with B. However, we may also want to know how B is correlated with A, that is, calculate $P(B|A)$. To calculate this, we need to know the likelihood that A is correlated with B and relate this to all the possible ways for A to happen (whether B has happened or not). If B is a rare event then it is possible that relative to other events that might be correlated with A, our observations might be a fluke.

Bayes Theorem is shown in Fig. 11.5.

For our example of pancreatic cancer above, we can express the Bayes Theorem as Fig. 11.6.

From the expression in Fig. 11.6, we have

$P(\text{cancer}) = .016;$

$$P(B|A) = \frac{P(A|B)P(B)}{P(A|B)P(B) + P(A|\sim B)P(\sim B)}$$

Figure 11.5 Bayes Theorem.

$$P(\text{cancer} \,|+) = \frac{P(+|\text{cancer})P(\text{cancer})}{P(+|\text{cancer})P(\text{cancer}) + P(+|\text{no cancer})P(\text{no cancer})}$$

Figure 11.6 Probability that a positive test result means that the patient will develop pancreatic cancer.

$P(+ \,|\, \text{cancer}) = .95;$
$P(+ \,|\, \text{no cancer}) = .08;$
$P(\text{no cancer}) = .984.$

If we plug these in Fig. 11.6, we can determine that the probability that someone will actually develop cancer given a positive test result is 16.2%! Clearly, this is an ineffective test even though on paper, the statistics look promising. The reason why the test is so inaccurate is because cancer is rare and the false positives were too high. In order for the test to be meaningful (e.g., having an accuracy over 90%) given these statistics, the false positive rate would have to be below 0.17%.

If a health insurance company was to deny a patient access to insurance as a result of this test then, aside from the moral implications, it would be denying itself significant revenue opportunities.

Aside from the harsh realities imposed by Bayes Theorem, there are several other issues with the relevance of AI models. One such example is bias.

Suppose we have a medical test that was trained on a dataset where 90% of the participants were male. This means that women are severely underrepresented in the dataset and it is likely that the test will be inaccurate for women. The sad reality is that while training the model, the machine learning engineer would, most likely, have been happy with himself because of the way many machine learning models are validated today (Fig. 11.7).

In addition to the validation data, we will also need to test the model "in the wild." What this means is that we should test our model in a superset of data to correct for sampling biases to see if the assumptions we made in the training data were accurate. Even then, we may find that our model is still not effective.

Suppose that we are diligent and we build a model with 50% male and 50% female, the model may still have significant issues. For example, it may not be applicable to certain ethnic groups. As a result, the model may be applicable in some scenarios and not in others.

Finally, AI models cannot be generalized. Suppose a researcher develops an AI model to predict the effects of a given drug at certain dosages.

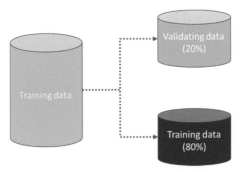

Figure 11.7 When preparing data for machine learning, it is typically broken down as data to train the model ("Training Data") and data to validate the model ("Validating Data").

The problem is that the model is only applicable to the ranges that the model has been trained on and is completely ineffective outside this range. Within the range, the model may be quite accurate.

The conclusion we can draw from this is that building an AI model is difficult and the biggest challenge of all is to select a training set that is truly representative of the population to which the model is applicable. Furthermore, it is not enough to simply release a model and call it a day; the model accuracy has to be monitored and maintained. For example, in the above case, if the demographics of the population changes, the accuracy of the model may drop.

11.6.4 Artificial intelligence as part of the decision process

Another challenge, following from the previous section, is the reliability of the AI models. This is by far one of the biggest issues. Using AIs to make decisions about people is fraught with risk because this involves several aspects—the quality of the training data, applicability of the model to a certain demographic, and drawing false conclusions.

Using AI models in decision processes entails multiple layers of trust:

- *Source of the training data*: AI models need data to train—how trustworthy is the data that was used to train the data? Does the data contain biases? Where was the data collected? How was it collected?, and who collected it? There may also be questions related to the confidence that the model was truly trained on the data that the engineer claims were used to train the data.
- *Feature engineering*: In order to use data for training, it must often go through substantial "preprocessing" such as removal of extraneous

results, combining results when data is sparse, statistical methods such as normalization, and sometimes just manual selection based on the intuition or knowledge of the one training the model. How one does feature engineering can have profound implications on the quality and predictive capabilities of the resulting AI model.

• *Pretrained models*: Often a technique called "transfer learning" is used to expand the capabilities of an existing AI model or to use an existing model to create highly accurate models in a related domain. For example, suppose a model has been trained to detect cats and dogs, it is possible to expand that model by training it using hamster images to detect hamsters too (Fig. 11.8). Doing so, however, requires trust that the pretrained model used must have been trustworthy.

• *Statistical analysis of the model*: Assuming that one has confidence in the data, there are many ways to train a model and each of these methods may result in the AI model displaying different characteristics such as its false positive or false negative rate. There are many statistical measures used to assess AI models such as area under the curve and receiver operating characteristic curve. Understanding these statistical methods is vital to understanding the integrity of the model but is beyond the scope of this chapter.

To use AI effectively and correctly, decision-makers will need to understand the process of training machine learning models, which will entail an understanding of the chain of trust. Consequently, they must demand more transparency from vendors and suppliers of AI to determine the extent to which they can be confident in the results of inferences made by AI models.

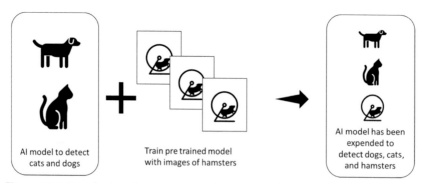

Figure 11.8 Transfer learning can be used to expand the capabilities of an existing model.

11.6.5 Process control

These are still early days for AI and so many of the issues that all technologies face as they mature simply have not yet materialized; nevertheless, in a few years, the landscape will be different and operational control of AI systems will become critical. There are some core issues:

- *Deprecation of AI models*: Over time, the ground truth will shift and AI models will lose their accuracy from initial baselines and will have to be retrained or replaced. Deprecating models seamlessly becomes essential.
- *Security*: Very little thought has gone into the security of AI systems. By modifying the training data, a model can be made to look like it is accurate even though it contains hidden flaws that may enable fraudulent behavior. Conversely, AI models can be fooled if its behavior can be understood—so-called adversarial attacks. Furthermore, there are decisions needed regarding who has the rights to train the data and separation of concerns between those who own the data and those who create the models.
- *Operational control*: The creators of AI systems are AI specialists but the engineers supporting the AI systems may not be. As a result, diagnostic systems need to be put into place to alert support engineers when something is amiss.
- *Cost control*: Understanding the ROI of machine learning models is still not fully grasped. Aside from the pure cost of training, deploying and supporting the AI model, there may be costs associated with collecting, storing, and maintaining the training data. There are also the issues of "not built here." Machine learning talent is scarce and expensive and often the motives of machine learning engineers, who are after higher salaries or more interesting problems, may not coincide with organization objectives.
- *Version control of the data*: AI models depend on the training data and so for operational control to be maintained, the data needs to be placed under some sort of control to reproduce the AI model, diagnose issues with the model, or to track differences between versions of the model.

11.7 The future of artificial intelligence in the health insurance industry

Currently, AI adoption in the health insurance industry is nascent but is expected to grow dramatically in the next 18 months. If there is a central theme, it is to provide more personalized health insurance options such customizing premiums and providing more choices to the customer.

Insurance companies will be increasingly motivated to lower costs using AI because of an aging population, the growth of preventable diseases

In addition, AI can enable insurers to automate their interaction with customers using bots, which have the added benefit to customers of lowering wait times. Insurers can also encourage insuree to lead healthier lifestyles and preemptively avoid healthcare emergencies leading to lower premiums. Insurers can also reduce claim times while reducing fraud. Data science also offers insurers the opportunity to identify new markets and identify emerging trends faster.

Finally, with the growth of wearables and apps, insurers can receive real-time data about their customers which can be used to predict emergency room visits, potential health issues, and encourage healthier lifestyles. Increasingly, genetics and the use of biomarkers are also used by insurance companies to adjust premiums, which, though fraught with ethical issues, can lead to lower costs for a segment of the population.

While many of these opportunities can prove to be revolutionary, risks abound. Most critically are questions about privacy, ethics, and usability. AI is, for all the hype surrounding it, in its infancy and there are serious concerns regarding model accuracy, availability of data, bias, explainability, and process control.

Because of the myriad of issues involved in the use of AI in the healthcare and insurance sectors ranging from the technical merits of AI, based on questions such as accuracy and bias, to the ethical boundaries of gaining insights about the patient population, it is also likely that we will see increasing pressure on regulators to clamp down on how AI can be used. Already, the Food and Drug Administration in the United States is proposing a regulatory framework for the use of AI in certain medical settings [56] and GDPR has implications for the use of AI in the healthcare and health insurance industry by letting subjects have access to data concerning them, by having to explain outcomes such as calculation of health premiums, by restricting data access from third parties, and by providing traceability and transparency so that subjects can know whether and how third-party entities are involved in making decisions about them [57].

References

[1] Adam H. Top 10 insurance companies by the metrics. Available from Investopedia: <https://www.investopedia.com/articles/active-trading/111314/top-10-insurance-companies-metrics.asp>; 2019.

[2] Rama A. National Health Expenditures, 2017: the slowdown in spending growth. American Medical Association; 2019.

[3] Allen S. 2019 Global health care outlook. Deloitte; 2019.

[4] CMS Office of the Actuary. 2018−2027 projections of National Health Expenditures. Centers for Medicare & Medicaid Services; 2019.

[5] Center for Medicare and Medicaid Services. Historical National Health Expenditure Data. Available from cms.gov: <https://www.cms.gov/Research-Statistics-Data-and-Systems/Statistics-Trends-and-Reports/NationalHealthExpendData/NationalHealthAccountsHistorical.html>; 2018.

[6] Laas L. Costs in the insurance industry today. Available from medium.com: <https://medium.com/blackinsurance/costs-in-the-insurance-industry-today-b86e225eea0b>; 2018.

[7] healthcare.gov. Rate review & the 80/20 rule. Available from: <https://www.healthcare.gov/health-care-law-protections/rate-review/>; 2019.

[8] Allianz Care. 3 challenges and 3 opportunities for international health insurance brokers in 2018. Available from alianzcare.com: <https://www.allianzcare.com/en/employers/employer-blogs/2018/08/broker-challenges.html>; 2018.

[9] Newman G. The top 5 claims challenges for life, accident and health insurers. Available from fineos.com: <https://www.fineos.com/blog/top-5-claims-challenges-life-accident-health-insurers/>; 2016.

[10] Organisation for Economic Cooperation and Development (OECD). OECD data. Available from data.oecd.org: <https://data.oecd.org/>; 2019.

[11] Haas CE. Drug price increases: here we go again? Available from pharmacytimes.com: <https://www.pharmacytimes.com/publications/health-system-edition/2019/march2019/drug-price-increases-here-we-go-again>; 2019.

[12] Chan M. Ten years in public health 2007−2017. Geneva: World Health Organization; 2017.

[13] American Heart Association. Nearly half of all Americans have cardiovascular disease. Available from ScienceDaily: <https://www.sciencedaily.com/releases/2019/01/190131084238.htm>; 2019.

[14] The Economist. The $272 billion swindle. Available from: <https://www.economist.com/united-states/2014/05/31/the-272-billion-swindle>; 2014.

[15] Japsen B. HMOs decline, consumer plans rise as health insurance option. Available from forbes.com: <https://www.forbes.com/sites/brucejapsen/2012/09/17/hmo-declines-consumer-plans-rise-as-health-insurance-option/#59f5aa1815bc>; 2012.

[16] Mulder J, Fera B, Crawford S, Jaggi G, Costanzo J, Delaney C, et al. The future of healthcare. Ernst and Young; 2015.

[17] Motaze N, Chi C, Ongolo-Zogo P, Ndongo J, Wiysonge C. Government regulation of private health insurance. Cochrane Database Syst Rev 2015;CD011512.

[18] McCulloch W, Pitts W. A logical calculus of the ideas immanent in nervous activity. Bull Math Biophys 1943;5:115.

[19] Russell S, Norvig P. Artificial intelligence: a modern approach. 3rd ed. Pearson Education; 2011.

[20] Rosenblatt F. The perceptron: a probablistic model for information storage and organization in the brain. Psychol Rev 1958;386−408.

[21] Hopfield JJ. Neural networks and physical systems with emergent collective computational abilities. Proc Natl Acad Sci USA 1982;79:2254−558.

[22] Rumelhart D, Hinton G, Williams R. Learning representations by back-propagating errors. Nature 1986;323:533−6.

[23] Hochreiter S, Schmidhuber J. Long short-term memory. Neural Comput 1997;9.8:1735−80.

[24] LeCun Y, Bottou L, Bengio Y, Haffner P. Gradient-based learning applied to document recognition. Proc IEEE 1998;86:2278−324.

[25] LeCun Y, Bengio Y. Convolutional networks for images, speech, and time series. Handb Brain Theory Neural Netw 1995;3361(10).

[26] LeCun Y, Bengio Y, Hinton G. Deep learning. Nature 2015;426.

[27] Katie Costello. Gartner survey shows 37 percent of organizations have implemented AI in some form. Available from gartner.com: <https://www.gartner.com/en/newsroom/press-releases/2019-01-21-gartner-survey-shows-37-percent-of-organizations-have>; 2019.

[28] Lopez J, Sau M. Toolkit: strategic industry maps of AI use cases. Gartner; 2019.

[29] Accenture. The era of the intelligent insurer. Available from accenture.com: <https://www.accenture.com/us-en/insight-insurance-technology-vision-2017>; 2017.

[30] Jubraj R, Sachdev S, Tottman S. How smarter technologies are transforming the insurance industry. Accenture; 2018.

[31] Insights Team. Can AI cure what ails health insurance? Available from forbes.com: <https://www.forbes.com/sites/insights-intelai/2019/02/11/can-ai-cure-what-ails-health-insurance/#6d1b7a3c2d59>; 2019.

[32] Senaar K. Artificial intelligence in health insurance—current applications and trends. Available from emerj.com: <https://emerj.com/ai-sector-overviews/artificial-intelligence-in-health-insurance-current-applications-and-trends/>; 2019.

[33] Balasubramanian R, Libarikian A, McElhaney D. Insurance 2030—the impact of AI on the future of insurance. McKinsey & Company; 2018.

[34] Clover Health. Clover. Available from cloverhealth.com: <https://www.clover-health.com/en/>; 2019.

[35] Kirontech. Kirontech main page. Available from: <https://www.kirontech.com/>; 2019.

[36] Azati. Azati Main Page. Available from: <https://azati.ai>; 2019.

[37] Hehner S, Körs B, Martin M, Uhrmann-Klingen E, Waldron J. Artificial intelligence in health insurance: smart claims management with self-learning software. Munich: McKinsey & Company; 2017.

[38] DimensionalMechanics Inc. DimensionalMehanics Inc.: Insuretech Report. Bellevue: DimensionalMechanics Inc; 2018.

[39] Terry H. Max Bupa partners with GOQii and Swiss Re for health offerings. Available from The Digital Insurer: <https://www.the-digital-insurer.com/dia/max-bupa-partners-goqii-swiss-re-health-offerings/>; 2018.

[40] Teichmann J. AI meets marketing segmentation models. Available from Towards Data Science: <https://towardsdatascience.com/data-science-powered-segmentation-models-ae89f9bd405f>; 2019.

[41] Miliard M. At Aetna.com ask 'Ann' anything. Available from Healthcare IT News: <https://www.healthcareitnews.com/news/aetnacom-ask-ann-anything>; 2010.

[42] University of Texas Audit Office. Special review of procurement procedures related to the M.D. Anderson Cancer Center Oncology Expert Advisor Project. Austin: The University of Texas System Administration; 2017.

[43] Jaklevic MC. MD Anderson Cancer Center's IBM Watson project fails, and so did the journalism related to it. Available from Health News Review: <https://www.healthnewsreview.org/2017/02/md-anderson-cancer-centers-ibm-watson-project-fails-journalism-related/>; 2017.

[44] Williams RL. Privacy Please: HIPAA and artificial intelligence. Available from Davis Wright Tremaine LLP: <https://www.dwt.com/blogs/privacy--security-law-blog/2018/03/privacy-please-hipaa-and-artificial-intelligence>; 2018.

[45] Dans E. Insurance, wearables and the future of healthcare. Available from Forbes: <https://www.forbes.com/sites/enriquedans/2018/09/21/insurance-wearables-and-the-future-of-healthcare/#3621f3961782>; 2018.

[46] MIT Technology Review. It's recruiting season for AI's top talent, and things are getting a little zany. Available from: <https://www.technologyreview.com/f/609707/its-recruiting-season-for-ais-top-talent-and-things-are-getting-a-little-zany/>; 2017.

[47] Thomas R. An opinionated introduction to AutoML and neural architecture search. Available from topbots.com: <https://www.topbots.com/an-opinionated-introduction-to-automl-and-neural-architecture-search/>; 2018.

[48] Herman A. America's high-tech STEM crisis. Available from Forbes: <https://www.forbes.com/sites/arthurherman/2018/09/10/americas-high-tech-stem-crisis/#1e7f23e2f0a2>; 2018.

[49] Kim S. New research shows declining interest in STEM. Available from GovTech.com: <https://www.govtech.com/education/k-12/New-Research-Shows-Declining-Interest-in-STEM.html>; 2018.

[50] Dutt R. AI: the challenge of data. Available from inforworld.com: <https://www.infoworld.com/article/3246706/ai-the-challenge-of-data.html>; 2018.

[51] Goodfellow I, Bengio Y, Courville A. Deep learning (adaptive computation and machine learning series). The MIT Press; 2016.

[52] Gonfalonieri A. 5 ways to deal with the lack of data in machine learning. Available from kdnuggets.com: <https://www.kdnuggets.com/2019/06/5-ways-lack-data-machine-learning.html>; 2019.

[53] Snovio. Available from snov.io: <https://snov.io/>; 2019.

[54] Amazon. Amazon Mechanical Turk. Available from mturk.com: <https://www.mturk.com/>; 2019.

[55] Rinehart-Thompson L. Introduction to health information privacy & security. 2nd ed. AHIMA; 2018.

[56] Food and Drug Administration. Proposed regulatory framework for modifications to artificial intelligence/machine learning (AI/ML)-based software as a medical device (SaMD). Available from fda.gov: <https://www.fda.gov/medical-devices/software-medical-device-samd/artificial-intelligence-and-machine-learning-software-medical-device>; 2020.

[57] KPMG. kpmg.ie. The GDPR and key challenges faced by the insurance industry. Available from: <https://assets.kpmg/content/dam/kpmg/ie/pdf/2018/03/ie-gdpr-for-insurance-industry.pdf>; 2018.

CHAPTER 12

Ethical and legal challenges of artificial intelligence-driven healthcare

Sara Gerke[1], Timo Minssen[2] and Glenn Cohen[3]
[1]The Petrie-Flom Center for Health Law Policy, Biotechnology, and Bioethics at Harvard Law School, The Project on Precision Medicine, Artificial Intelligence, and the Law (PMAIL), Harvard University, Cambridge, MA, United States
[2]Centre for Advanced Studies in Biomedical Innovation Law (CeBIL), University of Copenhagen, Copenhagen, Denmark
[3]Harvard Law School, Cambridge, MA, United States

From clinical applications in areas such as imaging and diagnostics to workflow optimization in hospitals to the use of health apps to assess an individual's symptoms, many believe that artificial intelligence (AI) is going to revolutionize healthcare. Economic forecasters have predicted explosive growth in the AI health market in the coming years; according to one analysis, the market size will increase more than 10-fold between 2014 and 2021 [1]. With this growth comes many challenges, and it is crucial that AI is implemented in the healthcare system ethically and legally. This chapter will map the ethical and legal challenges posed by AI in healthcare and suggest directions for resolving them.

We will begin by briefly clarifying what AI is and giving an overview of the trends and strategies concerning ethics and law of AI in healthcare in the United States (US) and Europe. This will be followed by an analysis of the ethical challenges of AI in healthcare. We will discuss four primary challenges: (1) informed consent to use, (2) safety and transparency, (3) algorithmic fairness and biases, and (4) data privacy. We then shift to five *legal* challenges in the US and Europe, namely, (1) safety and effectiveness, (2) liability, (3) data protection and privacy, (4) cybersecurity, and (5) intellectual property law. To realize the tremendous potential of AI to transform healthcare for the better, stakeholders in the AI field, including AI makers, clinicians, patients, ethicists, and legislators, must be engaged in the ethical and legal debate on how AI is successfully implemented in practice (Table 12.1).

Artificial Intelligence in Healthcare
DOI: https://doi.org/10.1016/B978-0-12-818438-7.00012-5

Table 12.1 Overview of this chapter.

1 Understanding AI	
2 Trends and strategies	
3 Ethical	**4 Legal**
3.1 Informed consent to use	4.1 Safety and effectiveness
3.2 Safety and transparency	4.2 Liability
3.3 Algorithmic fairness and biases	4.3 Data protection and privacy
3.4 Data privacy	4.4 Cybersecurity
	4.5 Intellectual property law

12.1 Understanding "artificial intelligence"

The term "artificial intelligence," or in abbreviated form "AI," is widely used in society but its precise meaning is contested in both scholarly work and legal documents and we will not insist on a single definition here but instead pick out a few subtypes: Machine learning (ML), a subset of AI, has been the most popular approach of current AI healthcare applications in recent times since it allows computational systems to learn from data and improve their performance without being explicitly programmed ([2], p. 2020). Deep learning, a subset of ML, employs artificial neural networks with multiple layers to identify patterns in very large datasets ([3], p. 720; [2], p. 2020). Most notably, as we will see below, there are additional ethical and legal challenges in cases where ML algorithms are closer to "black boxes" (i.e., the results are very difficult for clinicians to interpret fully) ([3], p. 727; [2], pp. 2019-2021).

12.2 Trends and strategies

In this section, we discuss the US and Europe's strategies for AI and how they strive to compete against their biggest competitor China, thereby tailoring the discussion to the ethical and legal debate of AI in healthcare and research. We will also look at AI trends and discuss some examples of AI products that are already in clinical use in the US and Europe.

12.2.1 United States

During Barack Obama's presidency, the US Government's reports on AI emphasized, among other things, the applications of AI for the public good as well as aspects of fairness, safety, and governance ([4], pp. 13, 14, and 30—34, [5,6]). One of the reports also stressed the need to improve fairness, transparency, and accountability-by-design as well as building ethical AI ([5], pp. 26, 27).

Since Donald Trump's presidency, the US AI strategy has shifted to a more free market-oriented approach [7]. The White House, for instance, hosted the AI for American Industry Summit in May 2018. One of the key takeaways from the summit breakout discussions was that the Trump Administration aims to remove regulatory barriers to AI innovations ([8], pp. 3, 5). In July 2018, the Executive Office of the President announced that American leadership in AI is one of the top Administration R&D budget priority areas for 2020 ([9], pp. 1, 2). In February 2019, Trump signed the "Executive Order on Maintaining American Leadership in Artificial Intelligence" to address the criticism that the US has taken a hands-off approach to AI in contrast to other countries such as China [10,11]. With this executive order, Trump launched a coordinated Federal Government strategy, namely, the American AI Initiative, guided by five key areas of emphasis: (1) investing in AI R&D, (2) unleashing AI resources, (3) setting AI governance standards, (4) building the AI workforce, and (5) international engagement and protecting the advantage of the US in AI [10,12].

Only recently, in January 2020, the White House published draft guidance for the regulation of AI applications. It contains 10 principles that agencies should consider when formulating approaches to AI applications: (1) public trust in AI, (2) public participation, (3) scientific integrity and information quality, (4) risk assessment and management, (5) benefits and costs, (6) flexibility, (7) fairness and nondiscrimination, (8) disclosure and transparency, (9) safety and security, and (10) interagency coordination [13]. In February 2020, the White House also published an annual report on the American AI Initiative, summarizing the progress made since Trump signed the executive order. This report, for example, highlights that the US led historic efforts on the development of the Organization for Economic Co-operation and Development (OECD) Principles of AI that were signed by over 40 countries in May 2019 to promote innovative and trustworthy AI and respect democratic values and human rights [14,15]. In June 2019, the G20 also released AI Principles drawn from the OECD Principles of AI ([14], p. 22; [16]).

The White House has also launched a new website ("AI.gov") that focuses on AI for the American people and aims to provide a platform for those who wish to learn more about AI and its opportunities.

There are also numerous AI-related bills that have been introduced in the US Congress since Trump's inauguration on January 20, 2017, such as the *SELF DRIVE Act* (H.R.3388), the *FUTURE of Artificial Intelligence Act of 2017* (H.R.4625 and S.2217), and the *AI JOBS Act of 2019* (H.R.827). The SELF DRIVE Act is the only bill that has passed one chamber (i.e., the US House of Representatives), and none of these bills are directly related to the ethical and legal aspects of AI in healthcare. However, the two bills of the FUTURE of Artificial Intelligence Act of 2017, for example, stipulate the Secretary of Commerce to set up a Federal advisory committee that shall provide advice to the Secretary [Sec. 4(a) and (b)(1)]. This committee shall also study and assess, inter alia, how to incorporate ethical standards in the development and implementation of AI [Sec. 4(b)(2)(E)] or how the development of AI can affect cost savings in healthcare [Sec. 4(b)(2)(L)]. There are also legal developments related to AI at state and local levels [17]. For instance, the State of California unanimously adopted legislation in August 2018 (ACR-215) endorsing the 23 Asilomar AI principles [17,18].

AIs are already in clinical use in the US. In particular, AI shows great promise in the areas of diagnostics and imaging. In total, the Food and Drug Administration (FDA) has already cleared or approved around 40 AI-based medical devices [19,20]. For example, in January 2017, *Arterys* received clearance from the US FDA for its medical imaging platform as the first ML application to be used in clinical practice [21,22]. It was initially cleared for cardiac magnetic resonance image analysis, but Arterys has meanwhile also received clearance from the FDA for other substantially equivalent devices [23].

IDx-DR is the first FDA-authorized AI diagnostic system that provides an autonomous screening decision without the need for a human being to interpret the image or results additionally [24,25]. In April 2018, the FDA permitted marketing of this AI-based device to detect more than a mild level of the eye condition diabetic retinopathy in adult patients (ages 22 and older) diagnosed with diabetes [24,26]. The physician uploads the images of the patient's retinas to a cloud server, and the IDx-DR software then provides the physician with the recommendation either to rescreen in 12 months or to refer the patient to an eye specialist when more than mild diabetic retinopathy is detected [24].

In May 2018, the FDA also granted marketing authorization for Imagen's software *OsteoDetect* for helping clinicians in detecting a common type of wrist fracture, called distal radius fracture, in adult patients

[27,28]. OsteoDetect uses ML techniques to analyze two-dimensional X-ray images to identify and highlight this type of fracture [27,28].

12.2.2 Europe

The European Commission adopted its AI strategy for Europe in April 2018. In this Communication, the Commission ([29], pp. 3, 13−16) launched a European initiative on AI that aims to, inter alia, ensure an appropriate ethical and legal framework, for example, by creating a European AI Alliance and developing AI ethics guidelines. The Commission ([29], p. 6) also stresses in this Communication that the entire European Union (EU) should strive to increase the (public and private) investment in AI to at least € 20 billion by the end of 2020.

The European Commission's High-Level Expert Group on AI (AI HLEG)—which was appointed by the European Commission in June 2018 and is also the steering group for the European AI Alliance—published the Ethics Guidelines in April 2019. The Guidelines promote the slogan "Trustworthy AI" and contain seven key requirements that AI systems need to fulfill in order to be trustworthy: "(1) human agency and oversight, (2) technical robustness and safety, (3) privacy and data governance, (4) transparency, (5) diversity, nondiscrimination and fairness, (6) environmental and societal well-being, and (7) accountability" ([30], p. 2). For the purpose of its deliverables, the AI HLEG also published a document on the definition of AI [31]. Further, in June 2019, the AI HLEG published another deliverable that provides "Policy and Investment Recommendations for Trustworthy AI" [32].

The European Commission ([29], p. 17) encourages all EU Member States to develop a national AI strategy, and several states have already released one such as the United Kingdom (UK) [33−35] and Germany [36]. The European Commission [37] also agreed upon a coordinated plan on AI with EU Member States, Norway, and Switzerland in December 2018 to promote the development and use of AI in Europe. The overall goal of working together is to ensure that Europe becomes the world-leading region for the development and application of "cutting-edge, ethical and secure AI" [37].

Only recently, in February 2020, the European Commission released a White Paper on AI that contains a European approach to excellence and trust. At the same time, the Commission also published a Communication on a European strategy for data [38] and a Report on the liability implications and safety of AI, the Internet of Things (IoT), and robotics [39]. The Commission's White Paper, in particular, emphasizes that "Europe can combine its technological and industrial strengths with a high-quality digital infrastructure and a regulatory framework based on its fundamental values to become a global leader in innovation in the data economy and its applications" [40].

There are also already AI health applications in Europe, and more are in the pipeline. For example, Ada [41] is an AI health app that assesses an individual's symptoms and gives guidance (e.g., suggest to the user a visit to a doctor or to seek emergency care). Ada [41] has been CE-marked (class I) in Europe—a basic requirement to putting a medical device on the market within Europe—and complies with the EU General Data Protection Regulation 2016/679 (GDPR).

In August 2018, researchers at DeepMind and Moorfields Eye Hospital in London, UK, published in Nature Medicine the study results of an AI system that can read eye scans and make referral recommendation, comprising more than 50 common diagnoses; the system was trained on 14,884 scans and showed a success rate of 94% [42]. DeepMind's health team has meanwhile transitioned to Google Health, and Moorfields Eye Hospital is "excited to work with Google Health on the next phase to further develop this AI system so it can be used by patients all around the world" [43].

Another example is Ultromics [44]. The team at the University of Oxford "is dedicated to reducing misdiagnosis and enabling earlier prevention of cardiovascular disease" [44]. Ultromics's *EchoGo Pro*, for example, is an outcome-based AI system with CE marking in Europe that predicts coronary artery disease at an early stage [44].

Corti [45] is a software developed by a Danish company that leverages ML to help emergency dispatchers make decisions. Corti can detect out-of-hospital cardiac arrests (i.e., those that occur in the public or home) during emergency calls faster and more accurately than humans by listening in to calls and analyzing symptoms, the tone of voice, breathing patterns, and other metadata in real time [45−47].

12.3 Ethical challenges

As the prior section suggests, the use of AI in the clinical practice of healthcare has huge potential to transform it for the better, but it also raises ethical challenges we now address.

12.3.1 Informed consent to use

Health AI applications, such as imaging, diagnostics, and surgery, will transform the patient—clinician relationship. But how will the use of AI to assist with the care of patients interface with the principles of informed consent? This is a pressing question that has not received enough attention in the ethical debate, even though informed consent will be one of the most immediate challenges in integrating AI into clinical practice (there is a separate question about informed consent to train AI we will not focus on here; [48]). There is a need to examine under what circumstances (if at all) the principles of informed consent should be deployed in the clinical AI space. To what extent do clinicians have a responsibility to educate the patient around the complexities of AI, including the form(s) of ML used by the system, the kind of data inputs, and the possibility of biases or other shortcomings in the data that is being used? Under what circumstances must a clinician notify the patient that AI is being used *at all*?

These questions are especially challenging to answer in cases where the AI operates using "black-box" algorithms, which may result from noninterpretable machine-learning techniques that are very difficult for clinicians to understand fully ([49]; [3], p. 727). For instance, Corti's algorithms are "black box" because even Corti's inventor does not know how the software reaches its decisions to alert emergency dispatchers that someone has a cardiac arrest. This lack of knowledge might be worrisome for medical professionals [46]. To what extent, for example, does a clinician need to disclose that they cannot fully interpret the diagnosis/treatment recommendations by the AI? How much transparency is needed? How does this interface with the so-called "right to explanation" under the EU's GDPR (discussed further in Section 4.3.2)? What about cases where the patient may be reluctant to allow the use of certain categories of data (e.g., genetic data and family history)? How can we properly balance the privacy of patients with the safety and effectiveness of AI?

AI health apps and chatbots are also increasingly being used, ranging from diet guidance to health assessments to the help to improve medication adherence and analysis of data collected by wearable sensors ([50], pp. 3, 4). Such apps raise questions for bioethicists about user agreements and their relationship to informed consent. In contrast to the traditional informed consent process, a user agreement is a contract that an individual agrees to without a face-to-face dialog ([51], p. 40). Most people do not take the time to understand user agreements, routinely ignoring them

([51], p. 40; [52]). Moreover, frequent updates of the software make it even more difficult for individuals to follow what terms of service they have agreed to [53]. What information should be given to individuals using such apps and chatbots? Do consumers sufficiently understand that the future use of the AI health app or chatbot may be conditional on accepting changes to the terms of use? How closely should user agreements resemble informed consent documents? What would an ethically responsible user agreement look like in this context? Tackling these questions is tricky, and they become even more difficult to answer when information from patient-facing AI health apps or chatbots is fed back into clinical decision-making.

12.3.2 Safety and transparency

Safety is one of the biggest challenges for AI in healthcare. To use one well-publicized example, IBM Watson for Oncology [54] uses AI algorithms to assess information from patients' medical records and help physicians explore cancer treatment options for their patients. However, it has recently come under criticism by reportedly giving "unsafe and incorrect" recommendations for cancer treatments [55,56]. The problem seems to be in the training of Watson for Oncology: instead of using real patient data, the software was only trained with a few "synthetic" cancer cases, meaning they were devised by doctors at the Memorial Sloan Kettering (MSK) Cancer Center [56]. MSK has stated that errors only occurred as part of the system testing and thus no incorrect treatment recommendation has been given to a real patient [56].

This real-life example has put the field in a negative light. It also shows that it is of uttermost importance that AIs are safe and effective. But how do we ensure that AIs keep their promises? To realize the potential of AI, stakeholders, particularly AI developers, need to make sure two key things: (1) the reliability and validity of the datasets and (2) transparency.

First, the used datasets need to be reliable and valid. The slogan "garbage in, garbage out" applies to AI in this area. The better the training data (labeled data) is, the better the AI will perform [57]. In addition, the algorithms often need further refinement to generate accurate results. Another big issue is data sharing: In cases where the AI needs to be extremely confident (e.g., self-driving cars), vast amounts of data and thus more data sharing will be necessary [57]. However, there are also cases (e.g., a narrow sentiment AI-based off text) where less data will be required [57]. In general, it always depends on the particular AI and its tasks how much data will be required.

Second, in the service of safety and patient confidence some amount of transparency must be ensured. While in an ideal world all data and the algorithms would be open for the public to examine, there may be some legitimate issues relating to protecting investment/intellectual property and also not increasing cybersecurity risk (discussed in Sections 4.4 and 4.5). Third party or governmental auditing may represent a possible solution.

Moreover, AI developers should be sufficiently transparent, for example, about the kind of data used and any shortcomings of the software (e.g., data bias). We should learn our lessons from examples such as Watson for Oncology, where IBM kept Watson's unsafe and incorrect treatment recommendations secret for over a year. Finally, transparency creates trust among stakeholders, particularly clinicians and patients, which is the key to a successful implementation of AI in clinical practice.

The recommendations of more "black-box" systems raise particular concerns. It will be a challenge to determine how transparency can be achieved in this context. Even if one could streamline the model into a simpler mathematical relationship linking symptoms and diagnosis, that process might still have sophisticated transformations beyond the skills of clinicians (and especially patients) to understand. However, perhaps there is no need to open the "black box": It may be that at least in some cases positive results from randomized trials or other forms of testing will serve as a sufficient demonstration of the safety and effectiveness of AIs.

12.3.3 Algorithmic fairness and biases

AI has the capability to improve healthcare not only in high-income settings, but to democratize expertise, "globalize" healthcare, and bring it to even remote areas [58]. However, any ML system or human-trained algorithm will only be as trustworthy, effective, and fair as the data that it is trained with. AI also bears a risk for biases and thus discrimination. It is therefore vital that AI makers are aware of this risk and minimize potential biases at every stage in the process of product development. In particular, they should consider the risk for biases when deciding (1) which ML technologies/procedures they want to use to train the algorithms and (2) what datasets (including considering their quality and diversity) they want to use for the programming.

Several real-world examples have demonstrated that algorithms can exhibit biases that can result in injustice with regard to ethnic origins and skin color or gender [59—63]. Biases can also occur regarding other features such as age or disabilities. The explanations for such biases differ and may be multifaceted. They can, for example, result from the datasets themselves (which are not representative), from how data scientists and ML systems choose and analyze the data, from the context in which the AI is used [64], etc. In the health sector, where phenotype- and sometimes genotype-related information are involved, biased AI could, for instance, lead to false diagnoses and render treatments ineffective for some subpopulations and thus jeopardize their safety. For example, imagine an AI-based clinical decision support (CDS) software that helps clinicians to find the best treatment for patients with skin cancer. However, the algorithm was predominantly trained on Caucasian patients. Thus the AI software will likely give less accurate or even inaccurate recommendations for subpopulations for which the training data was underinclusive such as African American.

Some of these biases may be resolved due to increased data availability and attempts to better collect data from minority populations and better specify for which populations the algorithm is or is not appropriately used. However, a remaining problem is that a variety of algorithms are sophisticated and nontransparent. In addition, as we have seen in the policing context, some companies developing software will resist disclosure and claim trade secrecy in their work [63,65]. It may therefore likely be left to nongovernmental organizations to collect the data and show the biases [63].

In cases of "black-box" algorithms, many scholars have argued that explainability is necessary when an AI makes health recommendations, especially also to detect biases [66]. However, does this view really hold true? Some argue that what matters is *not* how the AI reaches its decision but that it is accurate, at least in terms of diagnosis [66]. The safety and effectiveness of health AI applications that are "black boxes" could, for example, be demonstrated—similar to the handling of drugs—by positive results of randomized clinical trials.

A related problem has to do with where AI will be deployed. AI developed for top-notch experts in resource-rich settings will not necessarily recommend treatments that are accurate, safe, and fair in low-resource settings [64] (Minssen, Gerke, Aboy, Price, and Cohen, 2020) [67]. One solution would be *not* to deploy the technology in such settings. But such a "solution" only exacerbates preexisting inequalities. More thought must be given to regulatory obligations and resource support to make sure that this technology does improve not only the lives of the people living in high-income countries but also of those people living in low- and middle-income countries.

12.3.4 Data privacy

In July 2017, the UK Information Commissioner's Office (ICO) ruled that the Royal Free NHS Foundation Trust was in breach of the UK Data Protection Act 1998 when it provided personal data of circa 1.6 million patients to Google DeepMind [68,69]. The data sharing happened for the clinical safety testing of "Streams," an app that aims to help with the diagnosis and detection for acute kidney injury [68,69]. However, patients were not properly informed about the processing of their data as part of the test [68,69]. Information Commissioner's Elizabeth Denham correctly pointed out that "the price of innovation does not need to be the erosion of fundamental privacy rights" [69].

Although the Streams app does not use AI, this real-life example has highlighted the potential for harm to privacy rights when developing technological solutions ([35], p. 90). If patients and clinicians do not trust AIs, their successful integration into clinical practice will ultimately fail. It is fundamentally important to adequately inform patients about the processing of their data and foster an open dialog to promote trust. The lawsuit Dinerstein v. Google [70] and *Project Nightingale* by Google and Ascension [71] are recent case studies showing patient privacy concerns in the context of data sharing and the use of AI.

But what about the ownership of the data? The value of health data can reach up to billions of dollars, and some evidence suggests that the public is uncomfortable with companies or the government selling patient data for profit ([35], pp. 88, 89). But there may be ways for patients to feel valued that do not involve ownership per se. For example, the Royal Free NHS Foundation Trust had made a deal with Google DeepMind to provide patient data for the testing of Streams in exchange for the Trust's free use of the app for 5 years ([35], p. 89). Reciprocity does not necessarily require ownership, but those seeking to use patient data must show that they are adding value to the health of the very same patients whose data is being used [72].

Beyond the question of what is collected, it is imperative to protect patients against uses outside the doctor–patient relationship that might deleteriously affect patients, such as impacts on health or other insurance premiums, job opportunities, or even personal relationships [53]. Some of this will require strong antidiscrimination law—similar to regimes in place for genetic privacy [73]; but some AI health apps also raise new issues, such as those that share patient data not only with the doctor but also with family members and friends [53]. In contrast to the doctor who is subject to duties of confidentiality set out by governing statutes or case law, family members or friends will probably *not* have legally enforceable obligations of such kind [53]. Does this need to be changed? Another sensitive issue is whether and, if so, under what circumstances patients have a right to withdraw their data. Can patients request the deletion of data that has already been analyzed in aggregate form [53]?

12.4 Legal challenges

Many of the ethical issues discussed above have legal solutions or ramifications; while there is nothing sacrosanct in our division between the two, we now shift to challenges we associate more directly with the legal system.

12.4.1 Safety and effectiveness

As we discussed previously (Section 3.2), it is of uttermost importance that AIs are safe and effective. Stakeholders can contribute to a successful implementation of AI in clinical practice by making sure that the datasets are reliable and valid, perform software updates at regular intervals, and being transparent about their product, including shortcomings such as data biases. In addition, an adequate level of oversight is needed to ensure the safety and effectiveness of AI. How this plays out varies between the US and Europe. So, how is AI regulated in the US and Europe? How can AI makers bring their products to the US and European markets? The initial step of the analysis as to whether AI products need to undergo review is whether such products are medical devices.

12.4.1.1 United States
Let us start with the legal regulation in the US.

12.4.1.1.1 Medical devices
The FDA regulates medical devices in the US. A medical device is defined in Section 201(h) Sentence 1 of the US Federal Food, Drug, and Cosmetic Act (FDCA) as "an instrument, apparatus, implement, machine, contrivance, implant, in vitro reagent, or other similar or related article, including any component, part, or accessory, which is
1. recognized in the official National Formulary, or the United States Pharmacopeia, or any supplement to them;

2. intended for use in the diagnosis of disease or other conditions, or in the cure, mitigation, treatment, or prevention of disease, in man or other animals; or
3. intended to affect the structure or any function of the body of man or other animals; and
which does not achieve its primary intended purposes through chemical action within or on the body of man or other animals and which is not dependent upon being metabolized for the achievement of its primary intended purposes."

For example, medical devices include simple tongue depressors, pacemakers with microchip technology, and in vitro diagnostic products such as reagents and test kits [74].

12.4.1.1.2 Medical and certain decision support software

The 21st Century Cures Act (Pub. L. No. 114−255) was signed into law by the former President, Barack Obama, on December 13, 2016. Initially, it was hoped by some that the FDA would start to regulate medical advisory tools such as Watson for Oncology fully [75]. Ross and Swetlitz [75] reported, however, that IBM had a large team of lobbyists pushing for proposals to prevent regulatory hurdles facing health software. Indeed, on November 29, 2016—a day before the US House of Representatives passed the 21st Century Cures Act—the company expressed its strong support for the Act in a press release, emphasizing that it "will support health innovation and advance precision medicine in the United States" [75,76]. The 21st Century Cures Act (Sec. 3060) introduced an exemption in Section 520(o) of the FDCA for medical and certain decisions support software that does not fulfill the device definition. Section 201(h) of the FDCA was also amended by adding a second sentence which explicitly states that software functions under Section 520(o) FDCA do not fall under the term "device."

12.4.1.1.2.1 Software functions under Section 520(o)(1)(A)−(D) of the FDCA Section 520(o)(1)(A)−(D) of the FDCA contains the following four categories of software functions that shall generally *not* fall under the device definition in Section 201(h) of the FDCA:

1. The software function is intended "for administrative support of a healthcare facility" (including business analytics, appointment schedules, and laboratory workflows);

2. The software function is intended "for maintaining or encouraging a healthy lifestyle and is unrelated to the diagnosis, cure, mitigation, prevention, or treatment of a disease or condition";

3. The software function is intended "to serve as electronic patient records" (and "is not intended to interpret or analyze patient records"); or

4. The software function is intended "for transferring, storing, converting formats, or displaying clinical laboratory test or other device data and results."

The FDA has also published nonbinding *Guidance for Industry and Food and Drug Administration Staff on Changes to Existing Medical Software Policies Resulting from Section 3060 of the 21st Century Cures Act* [77] to provide clarification of the interpretation of Section 520(o)(1)(A)−(D) of the FDCA.

12.4.1.1.2.2 Software functions under Section 520(o)(1)(E) of the FDCA Section 520(o)(1)(E) of the FDCA exempts specific CDS software from the device definition in Section 201(h) of the FDCA. In order to be generally exempt from the device definition, a software function must meet the following four criteria simultaneously:

1. The software function is *not* "intended to acquire, process, or analyze a medical image or a signal from an in vitro diagnostic device or a pattern or signal from a signal acquisition system".

2. The software function is intended "for the purpose of (. . .) displaying, analyzing, or printing medical information about a patient or other medical information (such as peer-reviewed clinical studies and clinical practice guidelines)."

3. The software function is intended "for the purpose of (. . .) supporting or providing recommendations to a healthcare professional about prevention, diagnosis, or treatment of a disease or condition."

4. The software function is intended "for the purpose of (. . .) enabling such healthcare professional to independently review the basis for such recommendations that such software presents so that it is not the intent that such healthcare professionals rely primarily on any of such recommendations to make a clinical diagnosis or treatment decision regarding an individual patient"

[Sec. 520(o)(1)(E) of the FDCA; [78], pp. 6, 7].

In September 2019, the FDA [78] issued *Draft Guidance for Industry and Food and Drug Administration Staff on Clinical Decision Support Software* that contains nonbinding recommendations on the interpretation of the criteria in Section 520(o)(1)(E) of the FDCA.

In particular, the FDA clarifies that the term "clinical decision support" (CDS) is defined broadly and means software functions that meet the first two criteria and part of the third criterion [i.e., intended "for the purpose of (...) supporting or providing recommendations"] ([78], p. 8). A CDS function is only exempt from the device definition when the fourth criterion is additionally fulfilled ([78], pp. 8, 9). Thus it is decisive to determine whether the software function enables the "healthcare professional to independently review the basis for such recommendations that such software presents." The FDA clarifies in its draft guidance that "the software developer should describe the underlying data used to develop the algorithm and should include plain language descriptions of the logic or rationale used by an algorithm to render a recommendation. The sources supporting the recommendation or the sources underlying the basis for the recommendation should be identified and available to the intended user (e.g., clinical practice guidelines with the date or version, published literature, or information that has been communicated by the CDS developer to the intended user) and understandable by the intended user (e.g., data points whose meaning is well understood by the intended user)" ([78], p. 12). The FDA also states that healthcare professionals rely primarily on software recommendations—and thus are unable "to independently review the basis for such recommendations"—if they cannot be expected to independently understand the meaning of the information on which the recommendations are based ([78], p. 12). An example includes when inputs that are used to generate a recommendation are not described ([78], p. 12).

The FDA also makes clear that it does not intend at this time to enforce compliance with applicable regulatory requirements with respect to certain software functions that are intended for healthcare professionals, caregivers, or patients and may meet the device definition but are low risk ([78], pp. 9, 16−18). For example, even if the fourth criterion is *not* fulfilled and healthcare professionals rely primarily on software recommendations, the FDA does not intend at this time to enforce compliance with the relevant device requirements as long as the device CDS functions inform clinical management for nonserious healthcare situations or conditions ([78], p. 16). The agency thus focuses its oversight especially on device CDS software functions that inform clinical management for serious or critical healthcare conditions or situations ([78], p. 17). The FDA also clarifies in its draft guidance that it also intends to focus its regulatory oversight on software functions that are devices but are not classified as CDS ([78], pp. 24−27).

12.4.1.1.3 Other FDA initiatives

There are many other important FDA initiatives we cannot do justice here, including its *Guidance on Software as a Medical Device (SAMD): Clinical Evaluation* [79] and the launch of the so-called *Software Pre-Cert Pilot Program.* The latter will enable some digital health developers to become precertified based on excellence in identified criteria (e.g., patient safety, clinical responsibility, and product quality) and bring their lower-risk software-based medical devices with more streamlined FDA review to market or no review at all ([80], pp. 5−7; [81]). The FDA also published a Working Model that contains suggestions for the main components of the Pre-Cert Pilot Program [81,82]. Although there are still many open questions, the program is an innovative regulatory experiment that may hold lessons for peer countries and should be closely followed.

In particular, the FDA [83] has only recently, in April 2019, proposed a regulatory framework for public comment for *Modifications to Artificial Intelligence/Machine Learning (AI/ML)—Based Software as a Medical Device (SaMD).* SaMD is "software intended to be used for one or more medical purposes that perform these purposes without being part of a hardware medical device" (IMDRF, 2013, p. 6). The FDA's discussion paper proposes a new, total product lifecycle regulatory approach for AI/ML-based SaMD that are *medical devices* to allow those devices to adapt and optimize their performance in real time to continuously improve while ensuring their safety and effectiveness [83]. While we praise the FDA's efforts in the field, it will be essential for regulators to focus especially on the development of a process to continuously monitor, manage, and identify risks due to features that are closely tied to AI/ML systems' reliability (e.g., concept drift, instability, and covariate shift) [84]. Moreover, when there is substantial human involvement in decision-making, it becomes even more challenging for regulators to determine the effects of the update of such devices (Gerke, Babic, Evgeniou, and Cohen, 2020) [85].

12.4.1.2 Europe

Let us now shift to the legal particularities in Europe.

12.4.1.2.1 Medical devices and new legal developments

There are also new legal developments in the EU: Two new EU Regulations entered into force on May 25, 2017, namely, the *Medical Device Regulation* [2017/745—MDR; see Art. 123(1) of the MDR] and the *Regulation on in vitro diagnostic medical devices* [2017/746—IVDR; see

Art. 113(1) of the IVDR]. With some exceptions, the MDR was supposed to become effective on May 26, 2020 [Art. 123(2) and (3) of the MDR]. However, due to the need for medical devices to combat COVID-19, the European Parliament [83] postponed the MDR's application by one year (i.e., May 26, 2021). The MDR will repeal the *Medical Device Directive* (93/42/EEC − MDD) and the *Directive on active implantable medical devices* (90/385/EEC − AIMD) (Art. 122 of the MDR). The IVDR will become effective as planned on May 26, 2022 (Art. 113(2) and (3) of the IVDR), thereby especially repealing the *Directive on in vitro diagnostic medical devices* (98/79/EC − IVDD) (Art. 112 of the IVDR) [86].

12.4.1.2.2 MDR

The new MDR will bring some changes in the classification process of medical devices. Software that does not fall under the medical device definition of the MDD may soon be classified as a medical device under the MDR. In particular, the new medical device definition in Art. 2(1) of the MDR also considers software that is used for human beings for the *Medical purpose of prediction or prognosis* of disease as a medical device. However, the MDR also explicitly clarifies in Recital 19 that "software for general purposes, even when used in a healthcare setting, or software intended for lifestyle and well-being purposes is not a medical device."

Similar to the MDD, medical devices under the MDR will be classified into four categories, namely, classes I, IIa, IIb, and III, based on the intended purpose of the medical devices and their inherent risks [Art. 51 (1) of the MDR] (Fig. 12.1).

The MDR also introduces new implementing and classification rules for software in Chapters II and III of Annex VIII. In particular, the MDR contains a new classification rule that focuses explicitly on software. According to this rule, "software intended to provide information which is used to take decisions with diagnosis or therapeutic purposes is classified as class IIa, except if such decisions have an impact that may cause:

- death or an irreversible deterioration of a person's state of health, in which case it is in class III or

Figure 12.1 Classification of medical devices.

- a serious deterioration of a person's state of health or a surgical intervention, in which case it is classified as class IIb.

Software intended to monitor physiological processes is classified as class IIa, except if it is intended for monitoring vital physiological parameters, where the nature of variations of those parameters is such that it could result in immediate danger to the patient, in which case it is classified as class IIb.

All other software is classified as class I"

(Rule 11 in Chapter III of Annex VIII of the MDR).

This new rule will also lead to reclassifications, meaning software that was originally classified as a medical device under the MDD may be classified in another class category under the MDR. For example, CDS software such as Watson for Oncology will probably be at least classified as a class IIa medical device under the MDR since it "provide(s) information which is used to take decisions with diagnosis or therapeutic purposes." Depending on the decision's impact, the AI-based CDS software could even be classified as a class III (if it may cause "death or an irreversible deterioration of a person's state of health") or class IIb device (if it may cause "a serious deterioration of a person's state of health or a surgical intervention") (Rule 11 in Chapter III of Annex VIII of the MDR). In October 2019, the Medical Device Coordination Group also released nonbinding guidance on qualification and classification of software under the MDR and IVDR [87].

A CE marking (similar to the current MDD) will especially indicate the conformity with the applicable requirements set out in the MDR so that a medical device can move freely within the EU and be put into service in accordance with its intended purpose [Recital 40 and Art. 2(43) of the MDR]. In particular, manufacturers of medical devices shall undertake an assessment of the conformity of their devices prior to placing them on the market (Art. 52 and Annexes IX–XI of the MDR). The applicable conformity assessment procedure is based on the classification (class I, IIa, IIb, or III) and type (e.g., implantable) of the particular device (Art. 52 of the MDR). For example, class I devices have a low level of vulnerability and thus the conformity assessment procedure can generally be carried out under the sole responsibility of the manufacturers [Recital 60 and Art. 52 (7) of the MDR]. In contrast, class IIa, IIb, and III devices that have a higher risk than class I devices entail the involvement of a notified body, a conformity assessment body designated in accordance with the MDR [Recital 60 and Art. 2(42) of the MDR].

12.4.2 Liability

New AI-based technologies also raise challenges for current liability regimes. It will be crucial to creating an optimal liability design that figures out responsibilities.

12.4.2.1 United States

Imagine the following case: An AI-based CDS software (see Section 4.1.1.2) gives an incorrect treatment recommendation (in the sense that it is not one a non-AI clinician would have arrived at) that the clinician adopts resulting in harm to the patient. In this situation, the clinician would likely be liable for *medical malpractice*. Clinicians must treat patients with due expertise and care; they need to provide a standard of care that is expected of relevant members of the profession. At present, it appears that clinicians could thus be held liable even though they engaged in good faith reliance on a "black-box" ML algorithm because AI-based CDS software is considered a tool under the control of the health professional who makes the ultimate decision; she remains the captain of the ship and thus responsible for its course. But should that result obtain in a case where the software function does not enable the "healthcare professional to independently review the basis for such recommendations that such software presents" [see Sec. 520(o)(1)(E)(iii) of the FDCA]? In the other direction, could we imagine a future where the use of AI-based technology becomes the standard of care, and thus the choice *not* to use such technology would subject the clinician to liability [88]? At the moment, however, using advanced AI does not (yet) appear to be part of the standard of care. Thus, to avoid medical malpractice liability, physicians *can* use it as a confirmatory tool to assist with existing decision-making processes as opposed to *needing* to follow its recommendations out of fear of liability [89].

Setting the optimal liability regime depends heavily on what one thinks the "problem" is. If one is concerned that the deployment of AI-based technology in the clinical space is associated with a high risk for patients to get hurt, one might want to keep the current medical malpractice regime that attempts to meet both tort law's two functions: (1) deterrence and (2) compensation of the victims. By contrast, if one believes that over the run of cases, reliance on AI promotes patient health, then it may be a problem if physicians prove reluctant to rely on these algorithms, especially the more opaque ones, when they remain on the hook for resulting liability (see also [90], p. 12). This might drive the policy-maker to a different model.

Some have proposed *product liability* against the makers of AI, a tort that generally entails a strict liability of the manufacturer for defects. However, there are considerable challenges to win such a claim in practice. Courts have hesitated to apply or extend product liability theories to healthcare software developers since such software is currently primarily considered as a tool to support clinicians make the final decision ([90], pp. 11, 12).

A different approach would be to focus on *compensation, even without deterrence*. An example of such a system in the US is vaccine *compensation*. Vaccine manufacturers pay into a fund, and the system collectivizes the risk by paying out to those that are harmed by vaccines. AI manufacturers could do the same, which would compensate patients and spread the risks across the industry, but may give individual makers of AI less incentives to ensure the product's safety.

Beyond clinicians and AI makers, one must also consider the liability of the hospitals that purchase and implement the AI systems. Lawsuits might be brought against them under the theories of *corporate negligence and vicarious liability*. One interesting theory for hospital liability is "negligent credentialing"—just as hospitals may be liable if they do not adequately review the credential and practice of physicians and other staff they employ [91], they may have similar duties when they "hire" an AI.

Still another possibility would be to pair a liability shield with a more rigorous *pre-approval scheme* that would immunize healthcare professionals and manufacturers from some forms of liability because of the approval process. Whether this is desirable depends in part on one's view of litigation versus administrative law regimes: is ex ante approval by a regulator preferable to ex post liability at the hands of a judge or jury?

12.4.2.2 Europe

Europe is also not (yet) ready for the new liability challenges that AI-based technology will bring along with it. There is currently no fully harmonized EU regulatory framework for liability on AI and robotics such as care and medical robots in place. However, Europe has taken several steps to address the issue of liability.

One first step in the right direction was the publication of a resolution by the European Parliament called *Civil Law Rules on Robotics: European Parliament resolution of 16 February 2017 with recommendations to the*

Commission on Civil Law Rules on Robotics (2015/2103(INL)). This resolution, among other things, questions whether the current liability rules are sufficient and whether new rules are required "to provide clarity on the legal liability of various actors concerning responsibility for the acts and omissions of robots" (Sec. AB). It also points out that the current scope of *Council Directive concerning liability for defective products (85/374/EEC—Product Liability Directive)* may not adequately cover the new developments in robotics (Sec. AH). The resolution emphasizes "that the civil liability for damage caused by robots is a crucial issue which also needs to be analyzed and addressed at Union level in order to ensure the same degree of efficiency, transparency and consistency in the implementation of legal certainty throughout the European Union for the benefit of citizens, consumers and businesses alike" (Sec. 49). It thus asks the European Commission for "a proposal for a legislative instrument on legal questions related to the development and use of robotics and AI foreseeable in the next 10−15 years, combined with non-legislative instruments such as guidelines and codes of conduct" (Sec. 51). The resolution recommends that the European Commission should define in this legislative instrument which of the two approaches should be applied: either *strict liability* (i.e., which "requires only proof that damage has occurred and the establishment of a causal link between the harmful functioning of the robot and the damage suffered by the injured party") or the *risk management approach* (i.e., which "does not focus on the person 'who acted negligently' as individually liable but on the person who is able, under certain circumstances, to minimize risks and deal with negative impacts") (Secs. 53−55 and Annex to the resolution). It also recommends an obligatory insurance scheme and an additional compensation fund to ensure that damages will be paid out in situations where no insurance cover exists (Annex to the resolution).

As a second step, in April 2018, the European Commission adopted its AI strategy (see Section 2.2). A first mapping of liability challenges for emerging digital technologies, such as AI, advanced robotics, and the IoT, was provided in a *Commission Staff Working Document on Liability for Emerging Digital Technologies* also published in April 2018 together with the AI strategy [92].

Further, in November 2019, the independent Expert Group on Liability and New Technologies—New Technologies Formation (NTF) that was set up by the European Commission released a report on liability for AI and other emerging digital technologies such as IoT [93]. The

NTF's findings include that liability regimes are mainly regulated by the EU Member States except for strict liability of producers for defective products that is regulated by the Product Liability Directive at the EU level ([93], p. 3). The NTF's opinion is that the Member States' liability regimes are a good starting point for new technologies and provide at least basic protection of victims ([93], p. 3). However, the NTF also identifies several points in its report that need to be changed at national and EU levels ([93], p. 3). For example, the NTF emphasizes that "a person operating a permissible technology that nevertheless carries an increased risk of harm to others, for example AI-driven robots in public spaces, should be subject to strict liability for damage resulting from its operation" ([93], p. 3). It also states, for instance, that "a person using a technology which has a certain degree of autonomy should not be less accountable for ensuing harm than if said harm had been caused by a human auxiliary" ([93], p. 3).

Only recently, in February 2020, the European Commission also published a report on the safety and liability implications of AI, the IoT, and robotics [39]. The Commission understands the importance of these technologies and aims to make "Europe a world-leader in AI, IoT, and robotics" ([39], p. 1). To achieve this aim, the Commission states that "a clear and predictable legal framework addressing the technological challenges is required" ([39], p. 1). The Commission, in accordance with the NTF, argues that "in principle the existing Union and national liability laws are able to cope with emerging technologies" ([39], p. 17). However, it also identifies some challenges raised by new digital technologies such as AI that need to be addressed by adjustments in the current national and EU regulatory frameworks such as the Product Liability Directive ([39], pp. 16, 17).

We welcome the European Commission efforts to identify and address the liability issues raised by AI and other emerging digital technologies. As a next consequent step, changes need to be made at national and EU levels to implement the NTF's and European Commission's findings. Such updates of the liability frameworks should be carried out as soon as possible to have provisions in place that adequately deal with these new technological developments. Updated frameworks are needed to create clarity, transparency, and public trust.

12.4.3 Data protection and privacy

In the world of big data, it is of pivotal importance that there are data protection laws in place that adequately protects the privacy of individuals, especially patients. In the following, we will give an overview of relevant provisions and legal developments on data protection and privacy in the US and Europe.

12.4.3.1 United States

The Health Insurance Portability and Accountability Act (HIPAA) Privacy Rule (45 C.F.R. Part 160 as well as subparts A and E of Part 164) is the key federal law to protect health data privacy ([94], p. 38). However, HIPAA has significant gaps when it comes to today's healthcare environment since it only covers specific health information generated by "covered entities" or their "business associates." HIPAA does not apply to nonhealth information that supports inferences about health such as a purchase of a pregnancy test on Amazon ([95], p. 232; [94], p. 39). Moreover, the definition of "covered entities" also limits it scope; it generally includes insurance companies, insurance services, insurance organizations, healthcare clearinghouses, and healthcare providers (45 C.F.R. §§ 160.102, 160.103), but not much beyond that ([95], p. 231; [94], p. 39). In particular, much of the health information collected by technology giants such as Amazon, Google, IBM, Facebook, and Apple that are all investing heavily in the field of AI in healthcare, and are not "covered entities," will fall outside of HIPAA ([94], p. 39). HIPAA also does not apply in cases of user-generated health information ([95], p. 232; [94], p. 39). For example, a Facebook post about a disease falls outside of HIPAA's regime ([95], p. 232).

A different problem with HIPAA is its reliance on de-identification as a privacy strategy. Under HIPAA de-identified health information can be shared freely for research and commercial purposes [[95], p. 231; 45 C.F. R. § 164.502(d)(2)]. It provides two options for de-identification: (1) a determination by someone with appropriate knowledge of and experience with usually accepted scientific and statistical methods and principles; or (2) the removal of 18 identifies (e.g., names, social security numbers, and biometric identifiers) of the individual or of relatives, household members, or employers of the individual, and no actual knowledge of the covered entity that the information could be used to identify an individual [45 C. F.R. § 164.514(b)]. But this may not adequately protect patients because of the possibility of data triangulation—to re-identify data thought to be de-identified under the statute through the combination of multiple datasets ([94], pp. 39, 40; [96]). The problem of data triangulation has also recently been featured in a lawsuit, Dinerstein v. Google [70], in which the plaintiffs alleged that the defendants shared medical records with Google containing enough information that enabled Google to potentially re-identify patients given all of its other data at hand.

For all these reasons, HIPAA is not adequate to protect the health privacy of patients. It is time for federal law to take seriously the protection of health-relevant data that is not covered by HIPAA ([95], p. 232; [97], pp. 9, 16). Such a federal law should facilitate both innovations, including health AI applications, and adequate protection of health privacy of individuals.

While HIPAA preempts less protective state law, it does not preempt states whose laws are more protective. Inspired by the EU GDPR, California recently has taken action at the state level: The California Consumer Privacy Act of 2018 (CCPA) became effective on January 1, 2020 (Cal. Civ. Code § 1798.198). The CCPA grants various rights to California residents with regard to personal information that is held by businesses. The term *business* is defined in Section 1798.140(c) of the California Civil Code and applies to "a sole proprietorship, partnership, limited liability company, corporation, association, or other legal entity that is organized or operated for the profit or financial benefit of its shareholders or other owners that collects consumers' personal information or on the behalf of which that information is collected and that alone, or jointly with others, determines the purposes and means of the processing of consumers' personal information, that does business in the State of California, and that satisfies one or more of the following thresholds:

A. Has annual gross revenues in excess of twenty-five million dollars (...).
B. Alone or in combination, annually buys, receives for the business's commercial purposes, sells, or shares for commercial purposes, alone or in combination, the personal information of 50,000 or more consumers, households, or devices.
C. Derives 50 percent or more of its annual revenues from selling consumers' personal information."

The CCPA defines the term *personal information* broadly as "information that identifies, relates to, describes, is reasonably capable of being associated with, or could reasonably be linked, directly or indirectly, with a particular consumer or household," including a real name, alias, postal address, social security number, and biometric information [Cal. Civ. Code § 1798.140(o)(1)]. In particular, personal information is *not* "publicly available information"—"information that is lawfully made available from federal, state, or local government records" [Cal. Civ. Code § 1798.140(o)(2)].

The CCPA does not apply to protected health information that is collected by HIPAA covered entities or their business associates [Cal. Civ. Code § 1798.145(c)(1)]. However, it applies to a great deal of information in so-called "shadow health records"—health data that is collected outside of the health system ([98], p. 449). Thus the CCPA is a welcome attempt to at least partially fill in legal gaps and improve the data protection of individuals.

12.4.3.2 Europe

The General Data Protection Regulation (GDPR—2016/679) has been applied since May 25, 2018 [Art. 99(2) of the GDPR] in all EU Member States and introduced a new era of data protection law in the EU.

The GDPR particularly aims to protect the right of natural persons to the protection of personal data [Art. 1(2) of the GDPR]. It applies to the "processing of personal data in the context of the activities of an establishment of a controller or a processor" *in the EU*, notwithstanding of whether the processing takes place in an EU or non-EU country, such as in the US [Arts. 2, 3(1) of the GDPR]. In addition, the GDPR may also have implications for US companies. For example, the Regulation applies in cases where the processor or controller is *established in a non-EU country* and processes "personal data of data subjects who are in the Union" for "the offering of goods or services" (e.g., newspapers and affiliated websites for free or for a fee) to such data subject in the EU or for the "monitoring" of the data subjects' behavior [Art. 3(2) of the GDPR; [99]]. The GDPR also applies where a controller processes personal data and is *established in a non-EU country*, but "in a place where Member State law applies by virtue of public international law" [Art. 3(3) of the GDPR] (Table 12.2).

The term "personal data" is defined as "any information relating to an identified or identifiable natural person ('data subject')" [Art. 4(1) of the GDPR]. The GDPR defines "processing" as "any operation or set of operations which is performed on personal data or on sets of personal data, whether or not by automated means," including collection, structuring, storage, or use [Art. 4(2) of the GDPR]. Whereas a "controller" is "the natural or legal person, public authority, agency or other body which, alone or jointly with others, determines the purposes and means of the processing of personal data," a "processor" means "a natural or legal

Table 12.2 GDPR's territorial scope.

Art. 3(1)	Art. 3(2)	Art. 3(3)
Processing of personal data	Processing of personal data of **data subjects who are in the EU**	Processing of personal data
In the context of the activities of a **EU establishment** of a controller or a processor	**Non-EU establishment** of a controller or a processor	**Non-EU establishment** of a controller
Processing takes place **within or outside the EU**	The processing activities are related to: a. the **offering of goods or services** (paid or for free) to such data subjects in the EU; or b. the **monitoring of the data subjects' behavior** as far as their behavior takes place within the EU	But in a place where Member State law applies by virtue of **public international law**

person, public authority, agency or other body which processes personal data on behalf of the controller" [Arts. 4(7), (8) of the GDPR].

In the healthcare context, the definition of "data concerning health" under Article 4(15) of the GDPR is, in particular, relevant: "personal data related to the physical or mental health of a natural person, including the provision of healthcare services, which reveal information about his or her health status." The EU's GDPR is thus a lot broader in its scope compared to US' HIPAA, which only covers specific health information generated by "covered entities" or their "business associates" (discussed in Section 4.3.1).

According to Article 9(1) of the GDPR, the processing of special categories of personal data such as genetic data [Art. 4(13) of the GDPR], biometric data [Art. 4(14) of the GDPR], and data concerning health is prohibited. But Article 9(2) of the GDPR contains a list of exceptions to paragraph 1 [99]. For example, the prohibition in Article 9(1) of the GDPR shall usually not apply in cases where "the data subject has given explicit consent (...) for one or more specified purposes" or where the "processing is necessary for reasons of public interest in the area of public health" or "for archiving purposes in the public interest, scientific or historical research purposes or statistical purposes" [Art. 9(2)(a), (i), and (j) of the GDPR; [99]]. The EU Member States can also decide to introduce or maintain further requirements, including limitations, but only "with regard to the processing of genetic data, biometric data or data concerning health" [Art. 9(4) of the GDPR].

Noncompliance with these GDPR's conditions shall result in administrative fines up to 20 million EUR or—if higher—up to 4% of an undertaking's annual global turnover of the previous year [Art. 83(5) of the GDPR]. The first fines in the healthcare context have already been imposed under the GDPR. For example, a hospital in Portugal was charged 400 thousand EUR for two breaches of the GDPR: First, 300 thousand EUR for the permit of "indiscriminate access to a set of data by professionals, who should only be able to access them in specific cases"; and second, 100 thousand EUR for the incapacity to "ensure the confidentiality, integrity, availability and permanent resilience of treatment systems and services" [100].

The GDPR also contains provisions that are especially relevant to AI-infused medicine. For example, where personal data are collected, the controllers must generally provide data subjects with *information* about "the existence of automated decision-making, including profiling, referred to in Article 22(1) and (4) and, at least in those cases, meaningful information about the logic involved, as well as the significance and the envisaged consequences of such processing for the data subject" [Arts. 13(2)(f), 14(2)(g) of the GDPR]. In addition, data subjects have the *right of access* to the personal data concerning them that are being processed and the information about "the existence of automated decision-making, including profiling, (...) and (...) meaningful information about the logic involved, as well as the significance and the envisaged consequences of such processing for the data subject" [Art. 15(1)(h) of the GDPR].

"Automated decision-making" means a decision that is made—without any human involvement—solely by automated means ([101], p. 20). The term "profiling" is defined in Article 4(4) of the GDPR as "any form of automated processing of personal data consisting of the use of personal data to evaluate certain personal aspects relating to a natural person, in particular to analyze or predict aspects concerning that natural person's performance at work, economic situation, health, personal preferences, interests, reliability, behavior, location or movements." Thus the term "profiling" is a subset of the term "processing" with two additional requirements, namely, the processing must be (1) automated and (2) for evaluation purposes ([102], p. 52).

Under Article 22(1) of the GDPR, data subjects shall also "have the right not to be subject to a decision based solely on automated processing, including profiling, which produces legal effects concerning him or her or similarly significantly affects him or her." Article 22(2) of the GDPR lists some exceptions to Article 22(1) of the GDPR, but these exceptions do generally not apply where decisions are based on genetic and biometric data as well data concerning health [Art. 22(4) of the GDPR].

It is highly controversial, however, whether the GDPR actually grants a "right to explanation" and what such a right means [102—105]. Recital 71 of the GDPR explicitly mentions "the right (...) to obtain an explanation of the decision reached after such assessment." Some scholars doubt the legal existence and the feasibility of such a right to explanation of *specific automated decisions*, inter alia, because Recital 71 of the GDPR is not legally binding, and a right to explanation is not mandated by the legally binding requirements set out in Article 22(3) of the GDPR [103]. Thus, according to this view, there is from the outset *no* legally binding right of the data subject to receive insight into the internal decision-making process of algorithms [106], and thus to open the "black boxes" of health AI applications. However, if a legally binding right to explanation of specific automated decisions does *not* exist, Articles 13(2)(f), 14(2)(g), and 15(1)(h) of the GDPR at least entitle data subjects to obtain "meaningful information about the logic involved, as well as the significance and the envisaged consequences" of automated decision-making systems [103]. This information includes the purpose of an automated decision-making system, how the system works in general, the predicted impact as well as *other system functionality* such as decision trees and classification structures [103].

It is also likely that companies that are controllers under the GDPR must carry out a data protection impact assessment for new AI-based technologies that shall be deployed in the clinical space. In general, Article 35 (1) of the GDPR requires such an assessment, prior to the processing, for "new technologies" where the processing "is likely to result in a high risk to the rights and freedoms of natural persons." Article 35(3) of the GDPR explicitly states when a data protection impact assessment shall especially be required such as in cases of "a systematic and extensive evaluation of personal aspects relating to natural persons which is based on automated processing, including profiling, and on which decisions are based that produce legal effects concerning the natural person or similarly significantly affect the natural person" or "processing on a large scale of special categories of data" (e.g., genetic data and data concerning health). Recital 91 of the GDPR clarifies that personal data should *not* be considered "on a large scale if the processing concerns personal data from patients (...) by an individual physician." Article 35(7) of the GDPR contains a list of what the assessment shall at least include, such as a description of the envisaged processing operations, an assessment of the risks to the freedoms and rights of data subjects, and the measures envisaged to address the risks.

As complementation to the GDPR, the Regulation (EU) 2018/1807 entered into force in December 2018 and has been directly applicable since May 28, 2019 (Art. 9 of Regulation 2018/1807). This Regulation contains a framework for the free flow of *nonpersonal data* in the EU by laying down rules to the availability of data to competent authorities, data localization requirements, and the porting of data for professional users (Art. 1 of Regulation 2018/1807). It applies to the *processing of electronic data* [other than personal data as defined in Art. 4(1) of the GDPR] in the EU, which is either "provided as a service to users residing or having an establishment in the Union," irrespective of whether the service provider is established in an EU or non-EU country or "carried out by a natural or legal person residing or having an establishment in the Union for its own needs" [Arts. 2(1), 3(1) and (2) of Regulation 2018/1807]. In cases of datasets composed of personal and nonpersonal data, the Regulation (EU) 2018/1807 does also apply to the nonpersonal data part of such datasets [Art. 2(2) of Regulation 2018/1807]. However, the GDPR applies in cases where the personal and nonpersonal data in datasets are inextricably linked [Art. 3(2) of Regulation 2018/1807].

12.4.4 Cybersecurity

Cybersecurity is another important issue we need to consider when addressing legal challenges to the use of AI in healthcare. In the future, much of the healthcare–related services, processes, and products will operate within the IoT. Unfortunately, much of the underlying infrastructure is vulnerable to both cyber and physical threats and hazards [107]. For example, sophisticated cyber actors, criminals, and nation-states can exploit vulnerabilities to steal or influence the flow of money or essential (healthcare) information [107]. Such actors are increasingly developing skills to threaten, harm, or disrupt the delivery of vital (medical) services [107]. Targets in the health sector may include hospital servers, diagnostic tools, wearables, wireless smart pills, and medical devices [108]. All can be infected with software viruses, Trojan horses, or worms that risk patients' privacy and health [53]. Moreover, corrupted data or infected algorithms can lead to incorrect and unsafe treatment recommendations [53]. Hostile actors could get access to sensitive data such as health information on patients or could threaten patients' safety by misrepresenting their health. AIs are, in particular, vulnerable to manipulation [109]. For example, Finlayson et al. [110] have shown in a recent publication that the system's output can completely be changed so that it classifies a mole as malignant with 100% confidence by making a small change in how inputs are presented to the system [109].

The need for increased cybersecurity was shown in the "WannaCry" ransomware attack, a global cyberattack using sophisticated hacking tools that crippled the National Health Service (NHS) in the UK, hit the international courier delivery services company FedEx and infected more than 300,000 computers in 150 countries [111]. Events like these not only resulted in reactions at the national level such as in the UK [112] but also prompted a *new Cybersecurity Act* [Regulation (EU) 2019/881] that came into force on June 28, 2019.

The new Cybersecurity Act's goals are to achieve a high level of cyber resilience, cybersecurity, and trust in the EU while ensuring the internal market's proper functioning [Art. 1(1)]. In particular, it lays down a *European cybersecurity certification framework* to ensure that certified information and communications technology (ICT) products, ICT services, and ICT processes in the EU fulfill an adequate level of cybersecurity [Art. 1 (1)(b)]. The Act also lays down the tasks, objectives, and organizational matter relating to the European Union Agency for Cybersecurity (ENISA) [Art. (1)(a)].

There is also new progress in the US: The Cybersecurity and Infrastructure Security Act of 2018 (H.R.3359) was signed into law by President Donald Trump on November 16, 2018 [107]. This Act (Sec. 2) amended the Homeland Security Act of 2002, and, in particular, redesignated the National Protection and Programs Directorate of the Department of Homeland Security as the Cybersecurity and Infrastructure Security Agency (CISA) (Sec. 2202; 6 U.S.C. 652; [113]). CISA augments the US national capacity to defend against cyberattacks and will help the federal government provide cybersecurity tools, assessment skills, and incident response services to safeguard sensitive networks [107].

While the latest legal developments in the US and Europe will hopefully promote the safety of AI-driven products, services, and processes in the healthcare sector, cyberattacks are often a global issue; data sharing and breaches frequently do not stop at the US or European borders but occur around the world [53]. Thus there is the need for an internationally enforceable, large-scale regulatory framework on cybersecurity that ensures a high level of cybersecurity and resilience across borders [53]. It will not be easy to set up such a framework since it will require to properly balance the different interests of all stakeholders involved [53].

12.4.5 Intellectual property law

Translating AI and big data into safe and effective "real-world" products, services, and processes is an expensive and risky venture. As a result, the commercial protection of AI and data-driven healthcare/life science technologies have become an exceedingly important topic [114—117]. At the same time, there are continuing discussions about open science and innovation and the primary objective of more data sharing as well as increasing debates over access to such technologies and the pertinent data [115,117].

AI—and the data that fuels it—can be protected by various intellectual property rights (IPRs), typically involving a combination of long contracts, copyright, trade secrets/the law of confidence, and/or—in Europe—database rights, as well as may also comprise competition law and personal data integrity rights ([118,119], p. 123). The result is that data are frequently the subject of litigation ([118,119], p. 123). Thus it has been suggested that more regulations for data-generating internet giants are necessary as well as that the new data economy requires a better approach to competition and antitrust rules ([118,119], p. 123).

The combination of big data and IPRs creates challenges that need to be addressed such as access to data and ownership rights ([120], p. 311). In particular, in cases of data mining and data analytics, various forms of IPRs might protect the references to or copying of databases and information ([120], p. 311). However, users will need to rely on an exemption to IPR infringement where data is *not* licensed or owned ([120], p. 311). This circumstance has led to vigorous disputes between stakeholders, especially data scientists and data "owners" ([120], p. 311). Moreover, in the context of big data applications, there is a lot of misunderstanding about the nature, the availability, and legal effects of overlapping rights and remedies.

For example, copyrights might protect the software that helps to collect and process big datasets. However, due to the somewhat unstructured nature of the nonrelational databases—a typical characteristic of big datasets and the material they contain—the traditional role and purpose of copyrights and the EU's sui generis right in databases have been called into question [121].

With regard to patents, recent case law in Europe (e.g., the German Federal Supreme court case on receptor tyrosine kinase and the UK *Illumina* case) and the US (e.g., the landmark cases *Mayo*, *Myriad*, and *Alice*) might have an impact on precision medicine with its aim to better tailoring treatment to the need of patients in three areas, namely, (1) biomarkers and nature-based products, (2) diagnostics, and (3) algorithms, big data, and AI [122]. In the US, recent patent law decisions made it harder—but not impossible—to obtain patent protection for precision medicine inventions, whereas in Europe, a less stringent standard of patent eligibility is applied such as for nature-based biomarkers [122].

Drug companies will most likely use AI systems to expand their traditional drug patent portfolio [121,123]. However, AI systems could also be used by competitors or patent examiners to predict incremental innovation or to reveal that a patent was ineligible for patent protection due to, for example, the lack of novelty or inventive step [121,123]. Furthermore, trade secret law, in combination with technological protection measures and contracts, can protect complex algorithms, as well as datasets and sets of insights and correlations generated by AI systems [121].

Some rights, such as copyrights and trade secrets, are becoming more and more crucial for the commercial protection of big data ([120], p. 323). Other rights, such as patents, may not always be applicable, or they may be tactically used in novel ways ([120], p. 323). While more flexible data exclusivity regimes could perhaps address some of the issues posed by traditional IP protections for chemical and pharmaceutical products, it is clear that these developments raise considerable doctrinal and normative challenges to the IPR system and the incentives it creates in a variety of areas [120,121]. Moreover, the full effect and purpose of some IPRs (e.g., as data aggregators) are unclear in the context of big data innovation and need additional study ([120], p. 323; [124]). The un/availability of such rights could not only lead to underinvestment in some areas due to a lack of incentives but also block effects for anticommons scenarios and open innovation in other areas ([120], p. 323). Furthermore, the interaction between IPRs and data transparency initiatives and their possible impact on public—private partnerships or open innovation scenarios should be clarified ([120], p. 323). For different technological applications, differentiated approaches and IPR user modalities will need to be taken into account and discussed ([120], p. 323).

It becomes apparent that more data sharing is necessary in order to achieve the successful deployment of AIs in healthcare on a large scale. Stakeholders such as companies, agencies, and healthcare providers need to increasingly consider with whom they are going to collaborate and what datasets under what conditions they are going to share. Some stakeholders are reluctant and refuse to share their data due to, for example, a lack of trust, previous spending on data quality or the protection of commercial and sensitive personal data ([119], p. 123). To resolve these tensions, legal frameworks would be desirable that promote and incentivize data sharing through, for example, data sharing intermediaries [125] and public—private partnerships, while ensuring adequate protection of data privacy. In cases where stakeholders such as companies act unfairly and collude to entirely control a market where competition and access are essential for healthcare, the hope is that more refined competition and antitrust law tools can intervene. To serve this role, competition and antitrust law will need to become more future-oriented to better understand and predict the dynamics and developments of big data and AI in the healthcare sector. The value of data differs and often depends on multiple factors, including its usage and uniqueness ([126], p. 2). For instance, diverse data that provides a multitude of signals appears to be more useful and thus valuable since ML is a dynamic experimentation process ([126], p. 2). It could also be the case that particular combinations where patient data or other medical data is a crucial asset may result in market power if the data is unique and *not* replicable ([126], p. 2).

12.5 Conclusion

In this chapter, we have given an overview of *what AI is* and have discussed the *trends and strategies* in the US and Europe, thereby focusing on the ethical and legal debate of AI in healthcare and research. We have seen that the US has taken a more free market approach than Europe and that several AI products such as IDx-DR—the first FDA-authorized autonomous AI diagnostic system—have already entered the US market. According to one forecast, AI has the potential to contribute up to 13.33 trillion EUR to the worldwide economy in 2030, and the regions that are estimated to gain the most from AI are likely to be China and North America, followed by Southern Europe ([127], pp. 2, 3). In contrast, Europe emerges as a global player in AI ethics. In particular, the European Commission's High-Level Expert Group on AI published *Ethics Guidelines for Trustworthy AI* in April 2019.

We have also discussed four primary *ethical challenges* that need to be addressed to realize the full potential of AI in healthcare: (1) informed consent to use, (2) safety and transparency, (3) algorithmic fairness and biases, and (4) data privacy. This has been followed by an analysis of five *legal challenges* in the US and Europe, namely, (1) safety and effectiveness, (2) liability, (3) data protection and privacy, (4) cybersecurity, and (5) intellectual property law. In particular, it is crucial that all stakeholders, including AI makers, patients, healthcare professionals, and regulatory authorities, work together on tackling the identified challenges to ensure that AI will be successfully implemented in a way that is ethically and legally. We need to create a system that is built on *public trust* to achieve a desirable societal goal that AI benefits everyone.

Informed consent, high levels of data protection and privacy, cyber resilience and cybersecurity, algorithmic fairness, an adequate level of transparency and regulatory oversight, high standards of safety and effectiveness, and an optimal liability regime for AIs are all key factors that need to be taken into account and addressed to successfully create an AI-driven healthcare system based on the motto *Health AIs for All of Us*. In this regard, we not only need to rethink current regulatory frameworks and update them to the new technological developments. But it is also important to have public and political discussions centered on the ethics of AI-driven healthcare such as its implications on the human workforce and the society as a whole. AI has tremendous potential for improving our healthcare system, but we can only unlock its potential by already starting now to address the ethical and legal challenges facing us.

Acknowledgements

This research was supported by a Novo Nordisk Foundation-grant for a scientifically independent Collaborative Research Programme in Biomedical Innovation Law (grant agreement number NNF17SA0027784).

References

[1] Accenture. Artificial intelligence (AI): Healthcare's new nervous system, https://www.accenture.com/_acnmedia/pdf-49/accenture-health-artificial-intelligence.pdf; 2017 [accessed 08.08.19].

[2] Mehta N, Devarakonda MV. Machine learning, natural language programming, and electronic health records: The next step in the artificial intelligence journey? J Allergy Clin Immunol 2018;141:2019—21. Available from: https://doi.org/10.1016/j.jaci.2018.02.025 e1.

[3] Yu KH, Beam AL, Kohane IS. Artificial intelligence in healthcare. Nat Biomed Eng 2018;2:719—31. Available from: https://doi.org/10.1038/s41551-018-0305-z.

[4] US Government. Preparing for the future of artificial intelligence, https://obama-whitehouse.archives.gov/sites/default/files/whitehouse_files/microsites/ostp/NSTC/preparing_for_the_future_of_ai.pdf; 2016 [accessed 08.08.19].

[5] US Government. The National Artificial Intelligence Research and Development Strategic Plan, https://obamawhitehouse.archives.gov/sites/default/files/whitehouse_files/microsites/ostp/NSTC/national_ai_rd_strategic_plan.pdf; 2016 [accessed 08.08.19].

[6] US Government. Artificial intelligence, automation, and the economy, https://obamawhitehouse.archives.gov/sites/whitehouse.gov/files/documents/Artificial-Intelligence-Automation-Economy.PDF; 2016 [accessed 08.08.19].

[7] Dutton T. An overview of national AI strategies, https://medium.com/politics-ai/an-overview-of-national-ai-strategies-2a70ec6edfd; 2018 [accessed 08.08.19].

[8] White House. Summary of the 2018 White House summit on artificial intelligence for American Industry, https://www.whitehouse.gov/wp-content/uploads/2018/05/Summary-Report-of-White-House-AI-Summit.pdf; 2018 [accessed 30.04.19].

[9] Executive Office of the President. Memorandum for the Heads of Executive Departments and Agencies. M-18-22, https://www.whitehouse.gov/wp-content/uploads/2018/07/M-18-22.pdf; 2018 [accessed 08.08.19].

[10] White House. Executive order on maintaining American leadership in artificial intelligence, https://www.whitehouse.gov/presidential-actions/executive-order-maintaining-american-leadership-artificial-intelligence; 2019 [accessed 08.08.19].

[11] Knight W. Trump has a plan to keep America first in artificial intelligence. MIT Technol Rev, https://www.technologyreview.com/s/612926/trump-will-sign-an-executive-order-to-put-america-first-in-artificial-intelligence; 2019 [accessed 08.08.19].

[12] US Government. Accelerating America's leadership in artificial intelligence, https://www.whitehouse.gov/articles/accelerating-americas-leadership-in-artificial-intelligence; 2019 [accessed 30.04.19].

[13] White House. Draft memorandum for the Heads of Executive Departments and Agencies. Guidance for regulation of artificial intelligence applications, https://www.whitehouse.gov/wp-content/uploads/2020/01/Draft-OMB-Memo-on-Regulation-of-AI-1-7-19.pdf; 2020 [accessed 16.03.20].

[14] White House. American artificial intelligence initiative: year one annual report, https://www.whitehouse.gov/wp-content/uploads/2020/02/American-AI-Initiative-One-Year-Annual-Report.pdf; 2020 [accessed 16.03.20].

[15] OECD. OECD principles on AI, https://www.oecd.org/going-digital/ai/principles; 2019 [accessed 16.03.19].

[16] G20. G20 Ministerial statement on trade and digital economy, https://www.mofa.go.jp/files/000486596.pdf#targetText = a)%20AI%20actors%20should%20respect,and%20internationally%20recognized%20labor%20rights; 2019 [accessed 17.03.20].

[17] FLI Team. AI policy—United States, https://futureoflife.org/ai-policy-united-states; 2019 [accessed 08.08.19].

[18] FLI Team. State of California endorses Asilomar AI principles, https://futureoflife. org/2018/08/31/state-of-california-endorses-asilomar-ai-principles; 2018 [accessed 08.08.19].

[19] The Medical Futurist. FDA approvals for smart algorithms in medicine in one giant infographic, https://medicalfuturist.com/fda-approvals-for-algorithms-in-medicine; 2019 [accessed 16.03.19].

[20] Topol EJ. High-performance medicine: the convergence of human and artificial intelligence. Nat Med 2019;25:44–56.

[21] FDA. Summary K163253, https://www.accessdata.fda.gov/cdrh_docs/pdf16/ K163253.pdf; 2017 [accessed 30.04.19].

[22] Marr B. First FDA approval for clinical cloud-based deep learning in healthcare. Forbes. https://www.forbes.com/sites/bernardmarr/2017/01/20/first-fda-approval-for-clinical-cloud-based-deep-learning-in-healthcare/#6af107d161c8; 2017 [accessed 08.08.19].

[23] Arterys. https://www.arterys.com; 2019 [accessed 08.08.19].

[24] FDA. FDA permits marketing of artificial intelligence-based device to detect certain diabetes-related eye problems, https://www.fda.gov/newsevents/newsroom/pressannouncements/ucm604357.htm; 2018 [accessed 08.08.19].

[25] IDx Technologies Inc. https://www.eyediagnosis.net/idx-dr; 2018 [accessed 08.08.19].

[26] FDA. DeNovo summary DEN180001, https://www.accessdata.fda.gov/cdrh_docs/ reviews/DEN180001.pdf; 2018 [accessed 08.08.19].

[27] FDA. DeNovo summary DEN180005, https://www.accessdata.fda.gov/cdrh_docs/ reviews/DEN180005.pdf; 2018 [accessed 08.08.19].

[28] FDA. FDA permits marketing of artificial intelligence algorithm for aiding providers in detecting wrist fractures, https://www.fda.gov/NewsEvents/Newsroom/ PressAnnouncements/ucm608833.htm; 2018 [accessed 08.08.19].

[29] European Commission. Communication from the Commission to the European Parliament, the European Council, the Council, the European Economic and Social Committee and the Committee of the Regions. Artificial intelligence for Europe. COM(2018) 237 final, https://eur-lex.europa.eu/legal-content/EN/TXT/PDF/? uri = CELEX:52018DC0237&from = EN; 2018 [accessed 08.08.19].

[30] AI HLEG. Ethics guidelines for trustworthy AI, https://ec.europa.eu/digital-single-market/en/news/ethics-guidelines-trustworthy-ai; 2019 [accessed 08.08.19].

[31] AI HLEG. A definition of AI. Main capabilities and disciplines, https://ec.europa.eu/digital-single-market/en/news/ethics-guidelines-trustworthy-ai; 2019 [accessed 16.03.20].

[32] AI HLEG. Policy and investment recommendations for trustworthy AI, https://ec. europa.eu/digital-single-market/en/news/policy-and-investment-recommendations-trustworthy-artificial-intelligence; 2019 [accessed 08.08.19].

[33] UK Department of Health and Social Care. New code of conduct for artificial intelligence (AI) systems used by the NHS, https://www.gov.uk/government/news/ new-code-of-conduct-for-artificial-intelligence-ai-systems-used-by-the-nhs; 2019 [accessed 08.08.19].

[34] UK Government. Industrial strategy. Artificial intelligence sector deal, https://assets. publishing.service.gov.uk/government/uploads/system/uploads/attachment_data/ file/702810/180425_BEIS_AI_Sector_Deal__4_.pdf; 2018 [accessed 08.08.19].

[35] House of Lords. AI in the UK: ready, willing and able?, https://publications.parliament.uk/pa/ld201719/ldselect/ldai/100/100.pdf; 2018 [accessed 08.08.19].

[36] German Federal Government. Strategie Künstliche Intelligenz der Bundesregierung, https://www.bmwi.de/Redaktion/DE/Publikationen/Technologie/strategie-kuenstliche-intelligenz-der-bundesregierung.pdf?__blob = publicationFile&v = 8; 2018 [accessed 08.08.19].

[37] European Commission. Member States and Commission to work together to boost artificial intelligence "made in Europe", http://europa.eu/rapid/press-release_IP-18-6689_en.htm; 2018 [accessed 08.08.19].

[38] European Commission. Communication from the Commission to the European Parliament, the Council, the European Economic and Social Committee and the Committee of the Regions. A European strategy for data, https://ec.europa.eu/info/sites/info/files/communication-european-strategy-data-19feb2020_en.pdf;   2020 [accessed 16.03.20].

[39] European Commission. Report from the Commission to the European Parliament, the Council and the European Economic and Social Committee. Report on the safety and liability implications of Artificial Intelligence, the Internet of Things and robotics,   https://ec.europa.eu/info/sites/info/files/report-safety-liability-artificial-intelligence-feb2020_en_1.pdf; 2020 [accessed 16.03.20].

[40] European Commission. White Paper on artificial intelligence—a European approach to excellence and trust, https://ec.europa.eu/info/sites/info/files/commission-white-paper-artificial-intelligence-feb2020_en.pdf; 2020 [accessed 16.03.20].

[41] Ada. Your personal health guide, https://ada.com; 2020 [accessed 16.03.20].

[42] De Fauw J, Ledsam JR, Romera-Paredes B, Nikolov S, Tomasev N, Blackwell S, et al. Clinically applicable deep learning for diagnosis and referral in retinal disease. Nat Med 2018;24:1342−50. Available from: https://doi.org/10.1038/s41591-018-0107-6.

[43] Moorfields Eye Hospital. Next phase of Moorfields work with Google Health, https://www.moorfields.nhs.uk/news/next-phase-moorfields-work-google-health; 2019 [accessed 16.03.20].

[44] Ultromics. http://www.ultromics.com; 2019 [accessed 08.08.19].

[45] Corti A. Co-pilot for medical interviews, https://corti.ai; 2019 [accessed 16.03.20].

[46] Vincent J. AI that detects cardiac arrests during emergency calls will be tested across Europe this summer. Verge, https://www.theverge.com/2018/4/25/17278994/ai-cardiac-arrest-corti-emergency-call-response; 2018 [accessed 08.08.19].

[47] Maack MM. Europe launches a heart arrack-detecting AI for emergency calls, https://thenextweb.com/artificial-intelligence/2018/04/25/europe-launches-heart-attack-detecting-ai-emergency-calls; 2018 [accessed 08.08.19].

[48] Cohen IG, Amarasingham R, Shah A, Xie B, Lo B. The legal and ethical concerns that arise from using complex predictive analytics in health care. Health Aff 2014;7:1139−47. Available from: https://doi.org/10.1377/hlthaff.2014.0048.

[49] Cohen IG. Petrie-Flom Center launches project on Precision Medicine, Artificial Intelligence, and the Law (PMAIL). Harv Law Today, https://today.law.harvard.edu/petrie-flom-center-launches-project-precision-medicine-artificial-intelligence-law-pmail; 2018 [accessed 08.08.19].

[50] UK Nuffield Council on Bioethics. Artificial intelligence (AI) in healthcare and research, http://nuffieldbioethics.org/wp-content/uploads/Artificial-Intelligence-AI-in-healthcare-and-research.pdf; 2018 [accessed 08.08.19].

[51] Klugman CM, Dunn LB, Schwartz J, Cohen IG. The ethics of smart pills and self-acting devices: autonomy, truth-telling, and trust at the dawn of digital medicine. AJOB 2018;18:38−47. Available from: https://doi.org/10.1080/15265161.2018.1498933.

[52] Cohen IG, Pearlman A. Smart pills can transmit data to your doctors, but what about privacy? N Scientist, https://www.newscientist.com/article/2180158-smart-pills-can-transmit-data-to-your-doctors-but-what-about-privacy; 2018 [accessed 08.08.19].

[53] Gerke S, Minssen T, Yu H, Cohen IG. Ethical and legal issues of ingestible electronic sensors. Nat Electron 2019;2:329−34. Available from: https://doi.org/10.1038/s41928-019-0290-6.

[54] IBM. IBM Watson for oncology, https://www.ibm.com; 2020 [accessed 17.03.20].

[55] Brown J. IBM Watson reportedly recommended cancer treatments that were 'unsafe and incorrect'. Gizmodo, https://gizmodo.com/ibm-watson-reportedly-recom-mended-cancer-treatments-tha-1827868882; 2018 [accessed 08.08.19].

[56] Ross C, Swetlitz I. IBM's Watson supercomputer recommended 'unsafe and incor-rect' cancer treatments, internal documents show. STAT, https://www.statnews.com/2018/07/25/ibm-watson-recommended-unsafe-incorrect-treatments; 2018 [accessed 08.08.19].

[57] Figure Eight. What is training data?, https://www.figure-eight.com/resources/what-is-training-data; 2020 [accessed 17.03.20].

[58] Wahl B, Cossy-Gantner A, Germann S, Schwalbe NR. Artificial intelligence (AI) and global health: how can AI contribute to health in resource-poor settings? BMJ Glob Health 2018;3:e000798. Available from: https://doi.org/10.1136/bmjgh-2018-000798.

[59] Short E. It turns out Amazon's AI hiring tool discriminated against women. Silicon Repub, https://www.siliconrepublic.com/careers/amazon-ai-hiring-tool-women-discrimination; 2018 [accessed 08.08.19].

[60] Cossins D. Discriminating algorithms: 5 times AI showed prejudice. N Scientist, https://www.newscientist.com/article/2166207-discriminating-algorithms-5-times-ai-showed-prejudice; 2018 [accessed 08.08.19].

[61] Fefegha A. Racial bias and gender bias examples in AI systems, https://medium.com/thoughts-and-reflections/racial-bias-and-gender-bias-examples-in-ai-systems-7211e4c166a1; 2018 [accessed 08.08.19].

[62] Obermeyer Z, Powers B, Vogeli C, Mullainathan S. Dissecting racial bias in an algo-rithm used to manage the health of populations. Science 2019;366:447−53. Available from: https://doi.org/10.1126/science.aax2342.

[63] Sharkey N. The impact of gender and race bias in AI. Humanitarian Law Policy, https://blogs.icrc.org/law-and-policy/2018/08/28/impact-gender-race-bias-ai; 2018 [accessed 08.08.19].

[64] Price II WN. Medical AI and contextual bias, Harv J Law Technol, available at SSRN: https://papers.ssrn.com/sol3/papers.cfm?abstract_id = 3347890; forthcoming 2019 [accessed 08.08.19].

[65] Wexler R. Life, liberty, and trade secrets: intellectual property in the criminal justice system. Stanf Law Rev 2018;70:1343−429.

[66] London AJ. Artificial intelligence and black-box medical decisions: accuracy versus explainability. Hastings Cent Rep 2019;49:15−21. Available from: https://doi.org/10.1002/hast.973.

[67] Minssen T, Gerke S, Aboy M, Price N, Cohen IG. Regulatory responses to medical machine learning. Journal of Law and the Biosciences 2020;1−18. Available from: https://doi.org/10.1093/jlb/lsaa002.

[68] ICO. RFA0627721—provision of patient data to DeepMind, https://ico.org.uk/media/2014353/undertaking-cover-letter-revised-04072017-to-first-person.pdf; 2017 [accessed 08.08.19].

[69] ICO. Royal Free—Google DeepMind trial failed to comply with data protection law, https://ico.org.uk/about-the-ico/news-and-events/news-and-blogs/2017/07/royal-free-google-deepmind-trial-failed-to-comply-with-data-protection-law; 2017 [accessed 08.08.19].

[70] Dinerstein v. Google. No. 1:19-cv-04311; 2019.

[71] Copeland R. Google's 'Project Nightingale' gathers personal health data on millions of Americans, https://www.wsj.com/articles/google-s-secret-project-nightingale-gathers-personal-health-data-on-millions-of-americans-11573496790; 2019 [accessed 17.03.20].

[72] Cohen IG. Is there a duty to share healthcare data? In: Cohen IG, Lynch HF, Vayena E, Gassner U, editors. Big data, health law, and bioethics. Cambridge: Cambridge University Press; 2018. p. 209–22.

[73] Roberts JL, Cohen IG, Deubert CR, Lynch HF. Evaluating NFL player health and performance: legal and ethical issues. Univ Pa Law Rev 2017;165:227–314.

[74] FDA. Is the product a medical device?, https://www.fda.gov/medicaldevices/deviceregulationandguidance/overview/classifyyourdevice/ucm051512.htm; 2018 [accessed 08.08.19].

[75] Ross C, Swetlitz I. IBM to Congress: Watson will transform health care, so keep your hands off our supercomputer. STAT, https://www.statnews.com/2017/10/04/ibm-watson-regulation-fda-congress; 2017 [accessed 08.08.19].

[76] IBM. IBM letter of support for the 21st Century Cures Act, https://www.ibm.com/blogs/policy/ibm-letter-support-21st-century-cures-act; 2016 [accessed 08.08.19].

[77] FDA. Changes to existing medical software policies resulting from Section 3060 of the 21st Century Cures Act, https://www.fda.gov/media/109622/download; 2019 [accessed 17.03.19].

[78] FDA. Clinical Decision Support Software. Draft guidance for Industry and Food and Drug Administration Staff, https://www.fda.gov/media/109618/download; 2019 [accessed 17.03.20].

[79] FDA. Software as a Medical Device (SAMD): clinical evaluation, https://www.fda.gov/media/100714/download; 2017 [accessed 08.08.19].

[80] FDA. Digital Health Innovation Action Plan, https://www.fda.gov/downloads/MedicalDevices/DigitalHealth/UCM568735.pdf; 2017 [accessed 08.08.19].

[81] FDA. Digital Health Software Precertification (Pre-Cert) Program, https://www.fda.gov/medicaldevices/digitalhealth/digitalhealthprecertprogram/default.htm; 2019 [accessed 08.08.19].

[82] FDA. Developing a Software Precertification Program: a working model (v.1.0), https://www.fda.gov/media/119722/download; 2019 [accessed 18.03.20].

[83] FDA. Proposed regulatory framework for modifications to artificial intelligence/machine learning (AI/ML)-based Software as a Medical Device (SaMD), https://www.fda.gov/media/122535/download; 2019 [accessed 08.08.19].

[84] Babic B, Gerke S, Evgeniou T, Cohen IG. Algorithms on regulatory lockdown in medicine. Science 2019;366:1202–4. Available from: https://doi.org/10.1126/science.aay9547.

[85] Gerke S, Babic B, Evgeniou T, Cohen IG. The need for a system view to regulate artificial intelligence/machine learning-based software as medical device. npj Digital Medicine 2020;3. Article number: 53. Available from: https://doi.org/10.1038/s41746-020.

[86] European Parliament. Parliament decides to postpone new requirements for medical devices. https://www.europarl.europa.eu/news/en/press-room/20200415IPR77113/parliament-decides-to-postpone-new-requirements-for-medical-devices. 2020 [accessed 23.04.20].

[87] MDCG. Guidance on qualification and classification of software in regulation (EU) 2017/745—MDR and Regulation (EU) 2017/746—IVDR, https://ec.europa.eu/docsroom/documents/37581; 2019 [accessed 18.03.20].

[88] Froomkin AM, Kerr I, Pineau J. When AIs outperform doctors: confronting the challenges of a tort-induced over-reliance on machine learning. Ariz Law Rev 2019;61:34–99.

[89] Price II WN, Gerke S, Cohen IG. Potential liability for physicians using artificial intelligence. JAMA 2019;322:1765–11766. Available from: https://doi.org/10.1001/jama.2019.15064.

[90] Price II WN. Artificial intelligence in health care: applications and legal implications. SciTech Lawyer 2017;14:10–13.

[91] Bezaire J, Felton KW, Greve P, Allen D. Hospital negligent credentialing liability, https://www.willistowerswatson.com/en-US/Insights/2017/08/hospital-negligent-credentialing-liability; 2017 [accessed 08.08.19].

[92] European Commission. Commission Staff Working Document. Liability for emerging digital technologies. *Accompanying the document* Communication from the Commission to the European Parliament, the European Council, the Council, the European Economic and Social Committee and the Committee of the Regions. Artificial Intelligence for Europe. SWD(2018) 137 final, https://eur-lex.europa.eu/legal-content/EN/TXT/PDF/?uri = CELEX:52018SC0137&from = en; 2018 [accessed 08.08.19].

[93] NTF. Liability for artificial intelligence and other emerging digital technologies, https://ec.europa.eu/transparency/regexpert/index.cfm?do = groupDetail.groupMeetingDoc&docid = 36608; 2019 [accessed 18.03.20].

[94] Price II WN, Cohen IG. Privacy in the age of medical big data. Nat Med 2019;25:37−43. Available from: https://doi.org/10.1038/s41591-018-0272-7.

[95] Cohen IG, Mello MM. HIPAA and protecting health information in the 21st century. JAMA 2018;320:231−2. Available from: https://doi.org/10.1001/jama.2018.5630.

[96] Gerke S, Yeung S, Cohen IG. Ethical and legal aspects of ambient intelligence in hospitals. JAMA 2020;323:601−2. Available from: https://doi.org/10.1001/jama.2019.21699.

[97] Arney D, Senges M, Gerke S, Canca C, Haaber Ihle L, Kaiser N, et al. A user-focused transdisciplinary research agenda for AI-enabled health tech governance, https://papers.ssrn.com/sol3/papers.cfm?abstract_id = 3385398; 2019 [accessed 08.08.19].

[98] Price II WN, Kaminski MG, Minssen T, Spector-Bagdady K. Shadow health records meet new data privacy laws. How will research respond to a changing regulatory space? Science 2019;363:448−50. Available from: https://doi.org/10.1126/science.aav5133.

[99] Gerke S. The EU's GDPR in the health care context. Bill Health, http://blog.petrieflom.law.harvard.edu/2018/05/30/the-eus-gdpr-in-the-health-care-context; 2018 [accessed 08.08.19].

[100] Lusa. Barreiro hospital disputes judicially fine of 400 thousand euros Data Commission, https://translate.google.com/translate?depth = 1&hl = en&ie = UTF8&prev = _ t&rurl = translate.google.com&sl = auto&sp = nmt4&tl = en&u = https://www.publico.pt/2018/10/22/sociedade/noticia/hospital-barreiro-contesta-judicialmente-coima-400-mil-euros-comissao-dados-1848479&xid = 17259,15700022,15700124,15700149,15700186, 15700190, 15700201; 2018 [accessed 08.08.19].

[101] Working Party on the Protection of Individuals with Regard to the Processing of Personal Data. Guidelines on automated individual decision-making and profiling for the purposes of Regulation 2016/679. WP251rev.01; 2018.

[102] Goodman B, Flaxman S. European Union regulations on algorithmic decision making and a "Right to Explanation". AI Mag 2017;38:50−7. Available from: https://doi.org/10.1609/aimag.v38i3.2741.

[103] Wachter S, Mittelstadt B, Floridi L. Why a right to explanation of automated decision-making does not exist in the general data protection regulation. Int Data Priv Law 2017;7:76−99. Available from: https://doi.org/10.1093/idpl/ipx005.

[104] Burt A. Is there a 'right to explanation' for machine learning in the GDPR?, https://iapp.org/news/a/is-there-a-right-to-explanation-for-machine-learning-in-the-gdpr; 2017 [accessed 08.08.19].

[105] Kaminski MA. The Right to Explanation, Explained. U of Colorado Law Legal Studies Research Paper No. 18-24, 1–25; 2018.

[106] Wachter S, Mittelstadt B, Russell C. Counterfactual explanations without opening the Black Box: automated decisions and the GDPR. Harv J Law Technol 2018;31:842–87.

[107] US Department of Homeland Security. Cybersecurity, https://www.dhs.gov/topic/cybersecurity; 2019 [accessed 08.08.19].

[108] Pinsent Masons. New 'digital' pills pose data protection and cybersecurity challenges for drugs manufacturers and health bodies, says expert, https://www.out-law.com/en/articles/2017/november/new-digital-pills-pose-data-protection-and-cybersecurity-challenges-for-drugs-manufacturers-and-health-bodies-says-expert;   2017 [accessed 08.08.19].

[109] Gerke S, Kramer DB, Cohen IG. Ethical and legal challenges of artificial intelligence in cardiology. AIMed Mag 2019;2:12–17.

[110] Finlayson SG, Bowers JD, Ito J, Zittrain JL, Beam AL, Kohane IS. Adversarial attacks on medical machine learning. Science 2019;363:1287–9. Available from: https://doi.org/10.1126/science.aaw4399.

[111] Graham C. NHS cyber attack: Everything you need to know about 'biggest ransomware' offensive in history, https://www.telegraph.co.uk/news/2017/05/13/nhs-cyber-attack-everything-need-know-biggest-ransomware-offensive;   2017 [accessed 08.08.19].

[112] Smart W. Lessons learned review of the WannaCry Ransomware Cyber Attack, https://www.england.nhs.uk/wp-content/uploads/2018/02/lessons-learned-review-wannacry-ransomware-cyber-attack-cio-review.pdf; 2018 [accessed 08.08.19].

[113] Congress. Summary: H.R.3359—115th Congress (2017-2018), https://www.congress.gov/bill/115th-congress/house-bill/3359?q = %7B%22search%22%3A%5B%22cybersecurity + and + infrastructure%22%5D%7D&r = 1; 2018 [accessed 08.08.19].

[114] The Economist. The world's most valuable resource is no longer oil, but data, https://www.economist.com/leaders/2017/05/06/the-worlds-most-valuable-resource-is-no-longer-oil-but-data; 2017 [accessed 08.08.19].

[115] Carrier MA. Innovation for the 21st century: harnessing the power of intellectual property and antitrust Law. New York, NY: Oxford University Press; 2009.

[116] Katz ML, Shapiro C. Product introduction with network externalities. J Ind Econ 1992;40:55–83.

[117] Lemley MA, Shafir Z. Who chooses open-source software? Univ Chic Law Rev 2011;78:139–64.

[118] Burbidge R. Medical data in a twist—Technomed v Bluecrest, http://ipkitten.blogspot.com/2017/09/medical-data-in-a-twist-technomed-v.html; 2017 [accessed 08.08.19].

[119] Minssen T, Schovsbo J. Big data in the health and life sciences: what are the challenges for European Competition Law and where can they be found? In: Seuba X, Geiger C, Pénin J, editors. Intellectual property and digital trade in the age of artificial intelligence and big data. CEIPI-ICTSD; 2018. p. 121–30.

[120] Minssen T, Pierce J. Big data and intellectual property rights in the health and life sciences. In: Cohen IG, Lynch HF, Vayena E, Gasser U, editors. Big data, health law, and bioethics. Cambridge: Cambridge University Press; 2018. p. 311–23.

[121] Gervais D. Big data and intellectual property law, http://www.ceipi.edu/en/news/piece-of-news/?tx_ttnews%5Btt_news%5D = 10942&cHash = 2f3a1ea32cb69dd6b-7ed38659ef6e440; 2019 [accessed 08.08.19].

[122] Aboy M, Liddell K, Crespo C, Cohen G, Liddicoat J, Gerke S, et al. How does emerging patent case law in the US and Europe affect precision medicine? Nat Biotechnol 2019;37:1118−25. Available from: https://doi.org/10.1038/s41587-019-0265-1.

[123] Maloney D. AI patent trolls now on the job for drug companies, https://hackaday.com/2019/01/30/ai-patent-trolls-now-on-the-job-for-drug-companies; 2019 [accessed 08.08.19].

[124] Burk DL. Patents as data aggregators in personalized medicine. Boston Univ J Sci Technol Law 2015;21 https://papers.ssrn.com/sol3/papers.cfm?abstract_id = 2597525## [accessed 08.08.19].

[125] Richter H, Slowinski PR. The data sharing economy: on the emergence of new intermediaries. Int Rev Intellect Prop Compet Law 2019;50:4−29. Available from: https://doi.org/10.1007/s40319-018-00777-7.

[126] Kamenir E. Themes and takeaways from the FTC hearings on the intersection of big data, privacy, and competition, https://www.competitionpolicyinternational.com/wp-content/uploads/2018/11/North-America-Column-November-2018-7-Full.pdf; 2018 [accessed 08.08.19].

[127] European Commission. USA-China-EU plans for AI: where do we stand?, https://ec.europa.eu/growth/tools-databases/dem/monitor/sites/default/files/DTM_AI%20USA-China-EU%20plans%20for%20AI%20v5.pdf; 2018 [accessed 08.08.19].

Concluding remarks

It is important to conclude by stating that this book is the direct result of years of research into various biological, medical, engineering, social, and computer science fields. The most important take-home message is that artificial intelligence serves as a tool that can be utilized by each of the above disciplines to improve and enhance the function of multiple applications. AI then, at its core, is truly cross-disciplinary.

Further, increasing evidence suggests that nearly all aspects of healthcare can benefit from AI-based solutions. While healthcare has been lagging behind other technology sectors including the financial sector, retail, and media, when it comes to execution of AI, it can have a significant impact on the healthcare sector and this will lead to better outcomes for healthcare workers, patients, and society. As described throughout this book, healthcare technologies are producing vast amount of data in different formats that are fed to AI-based technologies to streamline procedures. In addition to providing new opportunities, these tools make existing processes more cost and time effective while releasing the immense pressure on healthcare workers.

Additionally, with an overall aging population, the threat of the current and future pandemics, the growing pressure on healthcare systems, and the shortage of skilled medical professionals, there is a dire need for change. Although narrow AI applications can replace human skills for certain tasks, these tools are more likely suited for enhancing human skills by augmenting the capabilities of an already skilled physician or by providing less skilled healthcare workers with guidance on how to provide sufficient care. Therefore, professionals can provide expert care and patients in critical and high-risk situations can receive high-quality care while the machines help conduct the routine work. Outside of the clinic, AI can aid in early detection of illnesses and further contribute toward the increased monitoring that can lead to prevention of disease development and thus a reduction in the need for direct interaction and a decreased chance of clinical intervention.

Finally, it is worth noticing that if a single sector were to receive the benefits of technological advances, it should be the healthcare sector. As humans, we go about our daily lives often oblivious to the extraordinary advances that we have made throughout the years. In the previous century, our life expectancies have risen significantly, and our technological advances are beyond imagination even for those who observed the moon landing. However, to sustain success, good health, and technological innovation, the society demands a system that protects and mends our biological bodies so that a fertile ground for creative and imaginative endeavors is prepared for the present and future generations. Artificial intelligence in healthcare can support humans with this worthy goal.

Adam Bohr
Kaveh Memarzadeh

Index

Note: Page numbers followed by "*f*" and "*t*" refer to figures and tables, respectively.

Printed in the United States
By Bookmasters